AF522147

ENCYCLOPEDIA OF BIOPROCESS TECHNOLOGY

ENCYCLOPEDIA OF BIOPROCESS TECHNOLOGY

Vol. 3

ELEMENTS OF BIOPROCESS TECHNOLOGY

By

Dr. Arvind N. Shukla

School of Studies of Zoology & Biotechnology

Vikram University

Ujjain

(India)

DISCOVERY PUBLISHING HOUSE PVT. LTD.

NEW DELHI-110 002

Published by:
Tilak Wasan
DISCOVERY PUBLISHING HOUSE PVT. LTD.
4383/4A, Ansari Road, Darya Ganj
New Delhi-110 002 (India)
Phone: +91-11-23279245, 43596064-65
Fax : +91-11-23253475
E-mail: parul.wasan@gmail.com
discoverypublishinghouse@gmail.com
web : www.discoverypublishinggroup.com

***First Edition:* 2012**

ISBN: 978-93-5056-026-6 (Set)
978-93-5056-027-3 (Vol. 1)
978-93-5056-028-0 (Vol. 2)
978-93-5056-029-7 (Vol. 3)
978-93-5056-030-3 (Vol. 4)
978-93-5056-031-0 (Vol. 5)

Encyclopedia of Bioprocess Technology

Printed at:
Shree Balaji Art Press
Delhi

Preface

The present title **"Encyclopedia of Bioprocess Technology"** has been written for undergraduate, post-graduate students and those engaged in pharmaceutical, environmental biotechnology, industrial and clinical research. Actually the explosion of new technologies with their vast potential has brought with it the need for emerging scientists to equip themselves and their laboratories with a whole array of new expertise. With the high discriminating power of the DNA systems has come high potential in evidentiary terms, high profile status for many investigations, and not least, a high degree of professional scruting of evidence produced by such technology.

This book is addressed to all who are curious and who want to understand fundamental biological processes that ensure our survival. The person who might be interested is probably someone just starting research who wants to learn about aspects of regulation, outside his chosen field. It could become a supplementary text for graduate students and post-graduate in biochemistry and molecular biology.

To make the work more comprehensive and informative, the author has consulted many authoritative books, research journals, abstracts, monographs etc., so there can be no claim to originality except in the manner of treatment.

The author express his thanks to his friends and colleagues whose continue inspirations have initiated him to bring out this book.

The author expresses her gratitude to Mr. Wasan and staff of M/s Discovery Publishing House Pvt. Ltd. for their whole hearted co-operation in the publication of this book.

The author will remain sincerely responsible for any shortcomings of the book and be grateful to the readers for their suggestions and constructive criticism for the continuous betterment of the book. He takes this opportunity to appeal to the readers to send their suggestions straightaway to this Publisher.

Arvind N. Shukla

CONTENTS

CELLULAR DISINTEGRATION

Almost all cells excrete biological products of commercial interest into the surrounding medium, but the larger proportion of cellular products, many of which have a real or potential value, remain inside the cell. These intracellular products must be released into the medium before they can be recovered and purified. There are several strategies for accomplishing this: the cells can be genetically engineered to excrete the products, the cell wall can be made more permeable, cell lysis may be induced, or the cell wall can be disrupted. In preparative-scale applications the disruption of the cell wall is the most common method. The cell wall is a rigid and complex structure that protects the cell from the environment, maintaining cell shape and providing osmotic stability, while performing biochemical functions of regulation and selective control of the exchange of nutrients and metabolites with the medium. However, microbial cell wall characteristics vary widely in composition and structure according to species; it is important to recognize the effects of these differences on the strength of the wall and their implications which disruption techniques should be adopted.

Mammalian tissues, halophilic bacteria (which do not possess a true rigid cell wall), and the Mycoplasma bacteria (which lack a wall) have the lowest mechanical strength and can be easily lysed under osmotic stress in hypotonic media. The more structured and stronger cell walls in some bacteria and fungi are more difficult to disrupt, and more energy is required to overcome the toughness of the wall. Moreover, the cell age and physiological state are also important differences that contribute to the strength of a given species' cell wall.

CELL BOUNDARY

Biochemistry

The diversity of cell wall composition and structure is well documented. Purified cell walls can be hydrolyzed, and the different components can be identified.

Bacterial cell walls

Bacteria differ with respect to their ability to retain a crystal violet–iodine stain when treated with organic solvents (alcohol or acetone). Accordingly, they can be classified into two main groups: Gram positive if they retain the stain, and Gram negative if the stain is not retained. This difference reflects the diversity in cell wall composition and structure as well as differences in biochemical, physiological, or genetic features. Common to all cell walls is the element of rigidity, which is provided by a supramolecular structure known as a peptidoglycan, or murein. This is a regular heteropolymer of alternating *N*-acetylglucosamine and *N*-acetylmuramic acid (the lactyl ether of *N*-acetylglucosamine) linked by β-(1–4) glycosidic bonds in linear chains of up to 65 disaccharide residues. A tetrapeptide side chain, consisting of at least three amino acids (some of the unusual D-amino acids), is attached to the carboxyl group of the *N*-acetylmuramic acid.

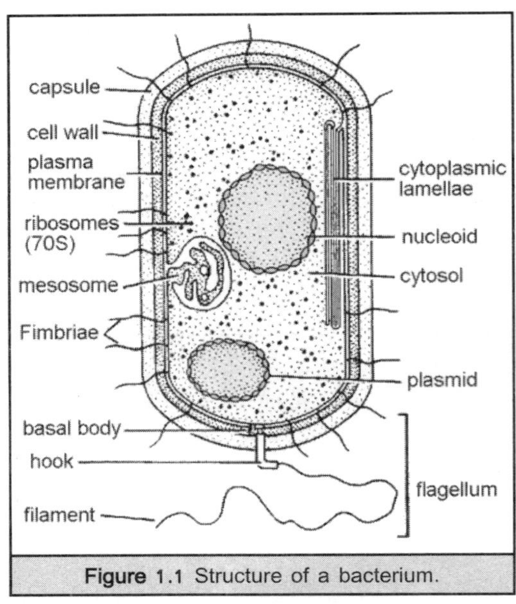

Figure 1.1 Structure of a bacterium.

The tetrapeptide chain of one polysaccharide chain is linked to a tetrapeptide chain of an adjacent polysaccharide chain, either directly (the terminal D-alanine of one to the *meso*-diaminopimelic acids in position 3 of the other, as in *Escherichia coli*), or indirectly through a pentaglycine peptide (the terminal D-alanine of one to the L-lysine in position 3 of the other, as in *Staphylococcus aureus*). Not all pairs of these tetrapeptide chains are involved in the cross-linking of the glycan chains; the degree of cross-linking, as well as the nature of the amino acids in positions 2 and 3 are dependent on the species. Murein is the skeletal component of the cell wall in Gram-positive bacteria; it has a multilayered architecture that is 20–80 nm thick. In addition, teichoic acids (polymers of glycerol or ribitol phosphate) are associated with murein on the external side of the wall and are unique to Gram-positive cells.

Teichoic acids do not contribute to wall strength but provide antigenic activity or serve as an attachment anchor. Gram-positive cell walls are devoid of significant amounts of lipids and proteins. In Gram-negative bacteria the cell wall is thinner and more complex and stratified. Mechanical resistance is provided by an inner layer of murein, which is much thinner than in Gram-positive bacteria. This murein is covered externally by a bilayered lipid membrane, the internal leaflet being lipoprotein, containing phospholipids, and the outer leaflet being lipopolysaccharide (LPS), which has toxic properties and is able to resist the chemical aggressions of the environment.

Fungi and yeasts

Under the electronic microscope, the walls of fungi and yeasts are seen to have a fibrillar structure that results from the presence of certain components such as cellulose or chitin. The main constituents are homo- or heteropolysaccharides, accounting for more than 80% of the wall, built up from different sugars (mainly D-glucose, D-mannose and *N*-acetylglucosamine, but also galactose, galactosamine, fucose, xylose, etc.). Usually, pairs of carbohydrates dominate the composition of the fungal walls, and a vast majority of fungi have a chitin–glucan combination, whereas the majority of yeasts have a glucan–mannan combination. Cellulose is also dominant in the walls of many lower fungi.

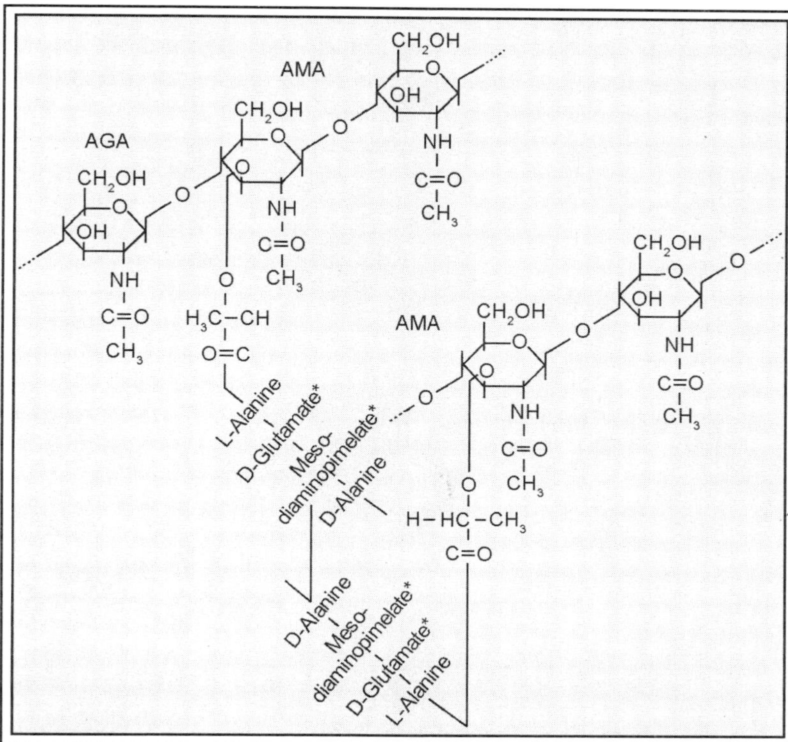

Figure 1.2 Chemical structure of peptidoglycan in *E. coli*. The nature of the amino acids at positions 2 and 3, as well as the degree of cross-linking differ among species. AGA, *N*-acetylglucosamine; AMN, *N*-acetylmuramic acid.

Glucans are usually considered to be the main skeletal component in yeasts and consist of linear chains of β-(1–6)-linked D-glucose and side chains of D-glucose linked to the first ones through β-(1–3) bonds. Chitin is a polymer of *N*-acetylglucosamine and is always found in yeast cells, although in small amounts. Lipids and proteins are also present in fungi and yeasts cell walls in different amounts. Proteins are now regarded as true structural wall components. In yeasts the wall appears stratified into two layers: the protein is associated with mannans in the form of mannoproteins that form an external layer covering an internal layer of glucans.

Algae

Cellulose is also the main skeletal component of the algal cell wall, but xylans and mannans (depending on which sugars are the most abundant) are also found. Algae differ considerably from the fungi with respect to cellulose content and crystallinity. The physical aspect is also fibrillar, as with other eukaryotes.

Effects of Upstream Conditions

The conditions that affect the synthesis of the cell wall constituents during the growth process will be reflected in the strength of the cell wall. For instance, cells of *E. coli* grown in a complex medium are less easily disrupted than cells grown in a simple medium. The ease of disintegration also depends on the batch growth of *E. coli;* the cells are weaker during the logarithmic phase and become stronger in the transition to the stationary phase. This corresponds to an increase in the murein thickness and cross-linking.

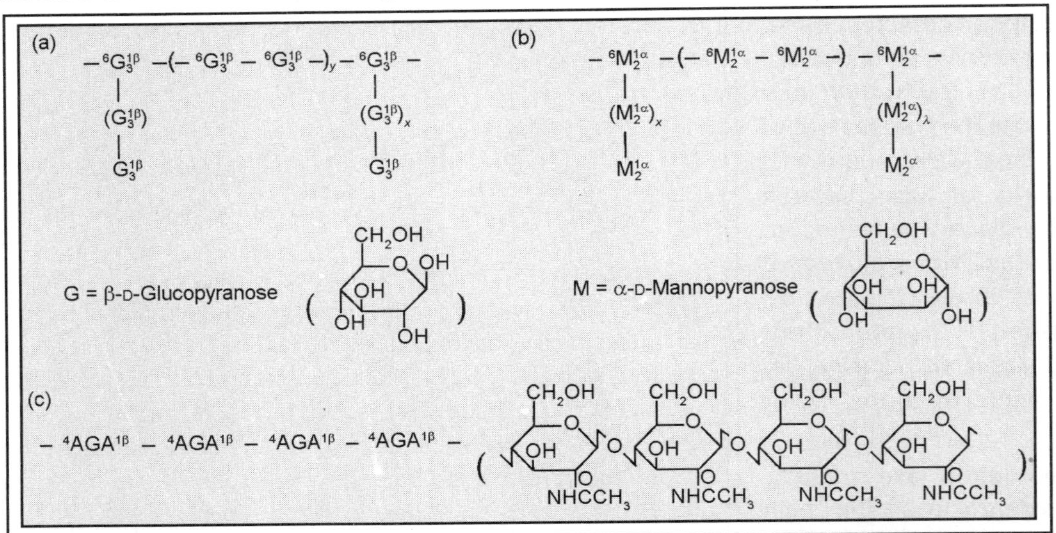

(a) $-{}^{6}G_{3}^{1\beta}-(-{}^{6}G_{3}^{1\beta}-{}^{6}G_{3}^{1\beta}-)_{y}-{}^{6}G_{3}^{1\beta}-$

G = β-D-Glucopyranose

(b) $-{}^{6}M_{2}^{1\alpha}-(-{}^{6}M_{2}^{1\alpha}-{}^{6}M_{2}^{1\alpha}-)-{}^{6}M_{2}^{1\alpha}-$

M = α-D-Mannopyranose

(c) $-{}^{4}AGA^{1\beta}-{}^{4}AGA^{1\beta}-{}^{4}AGA^{1\beta}-{}^{4}AGA^{1\beta}-$

Figure 1.3 Structure of the building blocks of (a) glucan, (b) mannan, and (c) chitin in yeasts.

A quantitative relationship has been established between mean cell strength and cell length and degree of cross-linkage. A higher specific rate of growth, as attained in continuous culture of *Candida utilis,* leads to a weaker cell wall, whereas a lower growth rate, as in a batch operation, allows the cells to build a stronger cell wall. Wall thickness and resistance to disruption in bakers' yeast increase with age, possibly due to exhaustion of the carbon source. Bud scars, much more frequent in a younger yeast population, may introduce local areas of lower wall strength. The compressive force required for the disruption of brewers' yeast cells was measured by using a micromanipulation technique and ranged from 40 μN for cells in the exponential phase to 90 μN for cells in the stationary phase.

DEGREE OF DISINTEGRATION

Suitable analytical techniques are required to assess the extent of cell disruption in order to develop models for process selection, scale-up, and optimization. Different methods have been used to quantify the degree of disintegration at specified conditions; each one has its advantages and drawbacks. The number of disrupted cells must be estimated, which can be done by direct counting of nondisrupted cells or, indirectly, by determination of any intracellular metabolite whose quantitative release into the medium is proportional to the number of disrupted cells. Direct counting of suitably diluted samples can be achieved by plating out techniques, microscopic counting, electronic counting, or using an analytical disk centrifuge. Plating out techniques are time consuming, and only viable cells can be counted. However, nonviable intact cells can be present in a significant proportion and are uncountable. The situation is worsened if the cells have a tendency to agglomerate.

Microscopic counting is a faster and simpler alternative. Whole cells in suitably diluted samples of original and treated suspensions are counted in microscope hemocytometers. However, the counting process, though easier for yeasts, is tedious and difficult with very small bacterial cells, and a distinction between intact cells and cells that have been slightly damaged is generally impossible to make. Proportional counts of stained smears may be advantageous as, for instance, in the case of yeasts. By adopting the staining technique of Bianchi, it is possible to distinguish among whole intact cells, broken cells contaminated with cytoplasm, and empty debris, at 1,000x magnification. The main difficulty lies in finding a suitable staining technique for the cells concerned. Electronic particle counters can be used to evaluate the number of whole yeast cells. The Coulter counter, for instance, was used to assess the concentration of yeast cells, but it is not a sensitive method for bacteria, and the machine tends to be fouled by cellular debris.

The analytical disk centrifuge was recently introduced for monitoring the disruption of *E. coli* containing recombinant inclusion bodies. This technique provides a size distribution in the homogenate and requires that the size distribution of the inclusion bodies be subtracted from the homogenate sample distribution by numerical or graphical deconvolution of both distributions. Indirect counting is based on the determination of certain compounds (e.g., soluble protein, enzyme activities) that are released into the solution when the cell wall is broken: the yield of disintegration is the percentage ratio of the released material, R, to the maximum amount, R_{max}, that can theoretically be released into the solution.The most common material that has been used for this purpose is total soluble protein assayed in the sample supernatant. The "dilution method," using the Biuret method, was based on the comparison of the protein concentrations in the supernatants obtained by centrifugation of diluted and undiluted samples of homogenized suspensions. However, dilution interferes with protein solubility, because of partial denaturation. The method also fails when the protein is produced as an inclusion body. Overheating, particularly during a prolonged operation to determine R_{max}, can cause thermal denaturation or enzyme deactivation.

Other bioactive products and, in particular, enzyme activities released into the medium can be useful. However, the assessed value of the disruption yield under fixed conditions depends largely on the enzyme selected for analysis, its location within the cell, whether it is bound to membranous materials, and its heat or shearing lability, which can affect the values of both R and R_{max}. The most adequate material would be the product to be extracted and isolated, if it can be easily assayed. Continuous monitoring of the disruption process and the performance of the equipment, particularly if wearing problems are severe, is advantageous. For instance, a relationship between the degree of cell disruption and the temperature increase of a baker's yeast suspension when passing through a mechanical homogenizer can be established. Therefore, by inserting a thermocouple in the homogenizer, continuous monitoring of its performance can be achieved.

DISINTEGRATION TECHNIQUE

There are both mechanical and nonmechanical methods. Mechanical methods are based on shear forces that deform and rupture the cell wall. Nonmechanical methods induce the lysis of the cell by chemical, physical, or enzymatic means. Many of these methods were developed for laboratory-scale use, with the objective of purifying biomolecules or organelles while maintaining as much integrity as possible for fundamental studies. For preparative or commercial applications, the strategy is to maximize product release, keeping the biological activity and minimizing the formation of fine cell debris, which can interfere with further downstream operations. The mechanical methods, based on solid or liquid shear, are best suited for this purpose.

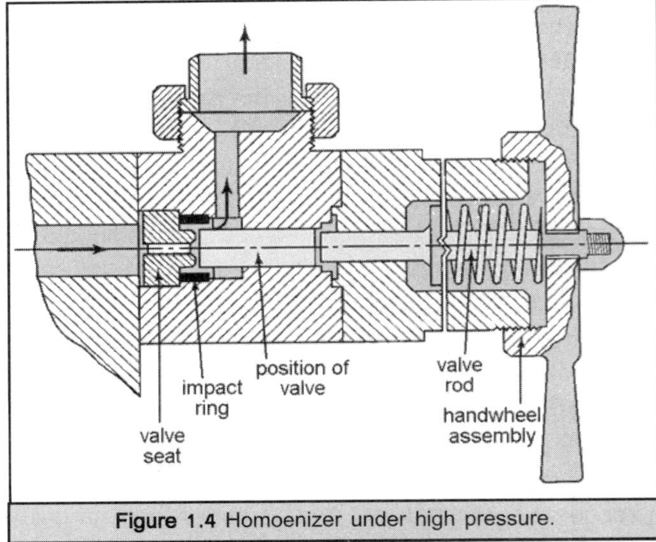

Figure 1.4 Homoenizer under high pressure.

Physico-chemical Homogenization

High-pressure mechanical homogenizers, typified by the APV Manton–Gaulin homogenizer, have been one of the most widely used devices for the disruption of microbial cell walls in large-scale recovery of intracellular bioproducts. These devices consist of a positive displacement pump capable of generating high pressures to force the cell suspension through an adjustable discharge needle valve under high shear stresses, which disrupts the cells. The body of the valve is positioned against a valve seat by a spring-loaded rod, and the disrupting pressure is adjusted by the load that is applied, hydraulically or mechanically, to the spring. The treated suspension can be either collected or recycled for more passes through the valve. In an earlier report such a device was used to disrupt baker's yeast, and the degree of disruption was proportional to the number of passes, *N*, through the valve according to the model (first-order kinetics relative to *N*),

$$\ln \frac{1}{1-X} = kNP^{a} \qquad \ldots(1)$$

where $X = R/R_{max}$ is the fraction of disrupted cells, *k* is a dimensional, machine-dependent (it depends on the geometry of the needle valve) specific rate constant $[(k) = (P)^{-a} = \text{MPa}^{-a}]$; the constant *a* is a measure of the cell resistance to disruption (the value $a = 2.9$ was reported for baker's yeast, and 2.2 for *E. coli*) and is dependent on the type of organism and the physiologic conditions and rate of growth.

Besides the pressure, the type of organism and the cell growth conditions, other variables such as the temperature and the cell concentration affect the specific rate of disintegration. The process has been found to be independent of cell concentration in the range 84-168 g dry matter/L and for pressures extending from 19.6 to 53.9 MPa, although lower protein recoveries for higher cell concentrations at higher pressures were obtained. Further studies extended the analysis of the influence of the cell concentration up to 240 g dry matter/L, and no significant effect was detected. However, a theoretical analysis based on the turbulence theory of Kolmogoroff showed that the amount of soluble protein release in a Manton–Gaulin homogenizer was dependent on cell concentration according to

$$R = 1 - \exp\{-[(P - P_0)/z]^{\beta}\} \qquad \text{...(2)}$$

where z is a constant, P is the disrupting pressure, P_0 is a threshold pressure, and β is a parameter that depends on cell concentration. More recently, the disruption of *E. coli* in another type of homogenizer, the Microfluidizer, was correlated by

$$\ln\frac{1}{1-X} = kN^{b}P^{a} \qquad \text{...(3)}$$

where the exponent b was found to be a function of the cell concentration and the specific growth rate.

The rate of cell wall disruption has been shown to be highly temperature dependent, but a temperature increase when the cells are disrupted can lead to severe thermal denaturation of heat-labile products. An increase of about 2–3°C (depending on cell concentration) can be expected for every 10 MPa increase in the back pressure of the needle valve. In fact, the suspension has to be pre-cooled in order to keep the temperature under control, and cooled immediately after leaving the homogenizer.

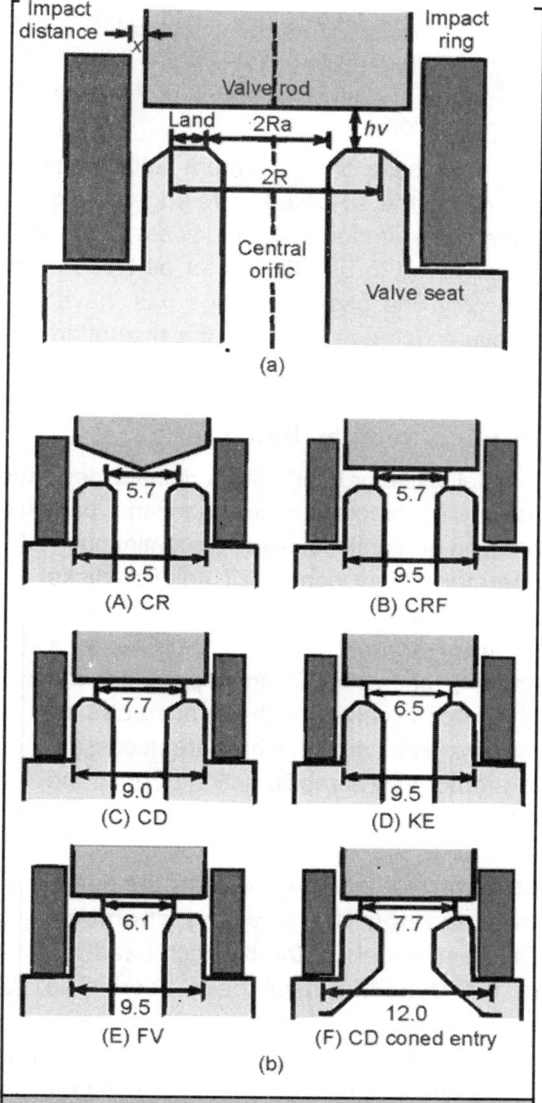

Figure 1.5 (a) Diagrammatic representation of a typical valve unit. (b) Schematic diagram of valve units used: (A) CR, cell rupture with coned valve rod; (B) CRF, cell rupture with fla valve rod; (C) CD, cell disruption; (D) KE, knife edge; (E) FV, flat valve unit; (F) CD-type unit with coned entry-dimensions in millimeters.

In most cases no losses of enzyme activity or protein solubility were detected if the temperature was not allowed to rise above 30°C. Increased productivity can be attained with higher pressures, allowing the recovery of more product with fewer passes. However, a higher disrupting pressure demands more energy and can cause severe wearing problems in the valve seat. Erosion of the valve seat is, in fact, of great concern in using this type of equipment. Optimization of the equipment design and operation requires a deep understanding of the mechanisms of cell disruption. Breakage of the cell wall occurs as a result of normal stresses or shear forces developed during the passage of the suspension through the needle valve. One mechanism considers the magnitude and the rate of pressure drop to be responsible for cell breakage. A hydrodynamic mechanism in which turbulent eddies of smaller dimensions than the cells cause the intracellular liquid to oscillate with a kinetic energy large enough to rupture the cell. Impingement of a high-velocity jet against a flat surface, as occurs in the needle valve when the direction of the jet is suddenly and abruptly changed, has also been considered to be a cause of cell disruption.

In the case of the Manton–Gaulin–type of homogenizers, this led to the redesign of the valve geometry; the so-called knife-edge configuration showed the best performance, whereas the flat valve configuration had the lowest yields in the disruption of baker's yeast. These homogenizers have proved to be efficient for all types of microorganisms, with high degrees of disruption even with only one pass. Blockage has, nevertheless, remained a major concern in the use of high-pressure homogenizers for the disruption of filamentous microorganisms. The use of a Manton–Gaulin homogenizer resulted in a severe degradation of plasmid recovered from *E. coli*.

Disintegration by Ball Mills

Ball mills, or bead mills, originally designed for nonbiotechnological applications, have also been used to disintegrate microorganisms, particularly yeasts. They are the most suitable for scaling up and can be continuously or discontinuously operated. They consist of agitator disks mounted concentrically or excentrically on a motor-driven central shaft inside a grinding chamber. During operation, the chamber is filled with small glass or stainless steel beads, which are necessary to promote an abrasive action to grind the cells. In continuous operation, a bead retainer (a sieve plate, a vibrating slot, or a special rotating disk) prevents the beads from leaving the grinding chamber. Different models of the Netzsch bead mill or the Dyno-Mill have been used. The mechanism of disintegration is very complex and results from both a liquid shearing and a solid shearing, resulting from the impacts of the beads in a field of intense turbulence. The disruption process can be properly described by a first-order kinetics law and, for a batch operation or continuous plug flow system, the integrated rate equation is

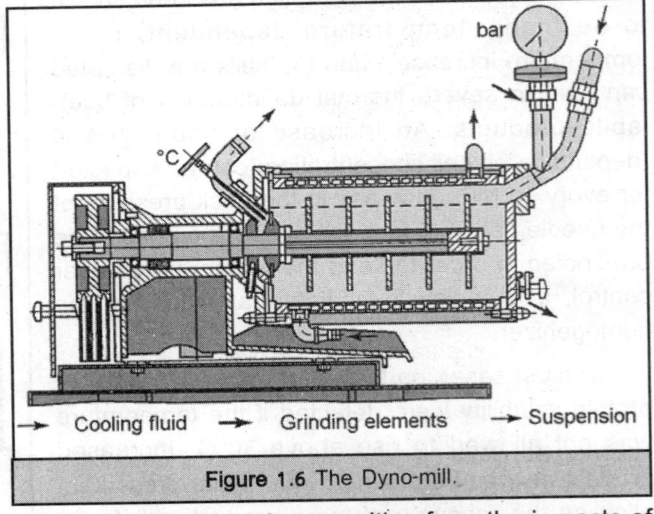

Figure 1.6 The Dyno-mill.

$$\ln \frac{R_{max}}{R_{max} - R} = kt \qquad \ldots(4)$$

where R is the amount of protein released in time t, R_{max} is the maximum amount that could be released if 100% disrupted had occurred, and k is the specific velocity of disintegration.

For a continuous operation there is a certain degree of back mixing, and n tanks (CSTR) in a series model is a better representation of the hydrodynamics in the chamber. In this case a mass balance to the released protein gives

$$\frac{R_{max}}{R_{max} - R} = [1 + (k\theta/n)^n] \quad \ldots(5)$$

where $\theta = V/Q$ is the mean residence time, V is the total free volume of the grinding chamber, and Q is the flow rate of the microbial suspension.

Several process parameters affect the yield and the rate of cell wall disruption. Besides the design characteristics of the mill (volume and geometry of the grinding chamber, the number of disks per unit volume of the mill, and geometry and construction materials of the disks), the rate constant is influenced by a number of process parameters such as the agitator speed, cell concentration, bead size, bead load, temperature, and flow rate of the suspension. The speed of the agitator disks determines the frequency of bead collisions and the intensity of shear, and, within certain limits, the specific rate of disintegration is proportional to the disk tip speed, U:

$$k = KU \quad \ldots(6)$$

Practical limitations arising from high energy consumption, high heat generation, erosion of the beads, and inactivation of shear-labile products require that the disk tip speeds remain between 5 and 15 ms^{-1}.

The influence of cell concentration requires a more detailed investigation because contradictory results have been reported. The disruption of baker's yeast was reported to be independent of the cell concentration in the range of 4–20% w/v on a dry matter basis, as expected for a first-order process. However, others have reported a maximum protein release for cell concentrations of 35–40% (wet wt), and decreasing thereafter, depending on the disk materials, such effect being more pronounced when the agitator speed was low. The viscosity of the suspension increases with the cell concentration and is further increased with the progress of the disruption process, therefore increasing the rate at which the turbulence is damped out. However, insufficient data are generally generated or reported regarding the rheology of the suspension.

The optimal cell concentration varies from case to case, and one must also consider that less heat is generated but the power consumption per unit weight of treated cells is higher for lower cell concentrations, and that for low cell concentrations (<5% solids) increased wear is observed on the mill and beads. Hence, an optimum must be found by experiment and is generally in the range of 30–60% or, more usually, in the range of 40–50% (cell wet wt). The bead diameters that have been reported range from 0.1 to 1.5 mm. Usually the disruption is faster with smaller beads, but the tendency of very small beads to float and difficulties in retaining them inside the grinding chamber impose practical limitations on the lower limits of size. In practice, the lower limit for batch operation of laboratory mills is 0.2 mm, whereas for production equipment working continuously the lower sizes are ≥0.4 mm. For the same bead load, larger particles correspond to fewer beads and reduced abrasive action or bead collisions. For the same bead load, there are also reports of an optimum bead size, which also has to be found by experiment and depends on the location of the product (enzyme) inside the cells; larger beads are preferred for the recovery of enzymes located in the periplasmic region, and smaller ones for cytoplasmic material.

The rate of protein release is also enhanced by increasing the bead load (volume percentage of bead material relative to the total volume of the grinding chamber). Bead loads ranging from 30 to 90% are reported in the literature, but the packing density, depending on the bead diameter inside the chamber, is generally in the range of 80–90%. The upper limit is generally imposed by the high power consumption and difficulties in removing the heat released in the operation. The rate constant is, obviously, dependent on the temperature that results from the balance between the heat generated during the milling process and the heat removed by a cooling liquid, brine, or a cooling liquid circulating through a cooling jacket of the grinding chamber. The heat generated in the process, if accumulated, would raise the temperature to unacceptable levels; in fact, the suspension has to be precooled and cooled immediately after leaving the chamber to avoid inactivation of heat-labile products.

The effects of the flow rate should mirror the effects of the residence time of the cells inside the grinding chamber, and assuming a first-order process, the yield after one passage should decline as the feed rate increases. However, degree of caution should be exercised because the degree of longitudinal dispersion or back mixing increases as the flow rate decreases, and the yield of disruption is not a simple function of the flow rate. On the other hand, the power consumption per unit mass of processed cells strongly decreases as the feed rate increases, and for this reason, a high flow rate should be adopted and the decline in the yield of disruption could be alleviated by recycling part of the suspension. It should be emphasized that by increasing the residence time the viscosity may be decreased because prolonged shear may cause degradation of the shear-sensitive DNA polymer. In one study, the disintegration rate was correlated with the specific energy as determined by the following combination of parameters: small grinding beads, moderate to high cell concentrations, and/or low to moderate agitator speeds. In this case the utilization of energy input reaches its maximum. In other words, high cell concentrations are energetically favorable. This result is considered to be the basis for scale-up. For cases of low cell concentrations in the suspension, high agitator speeds, and larger beads, the disintegration rate was no longer correlated with specific energy input but was a function of the stress frequency (which is proportional to the product of the agitator speed and disintegration time), which was experimentally proved.

Sonicating Method of Disintegration

Ultrasound has been used on the laboratory scale for analytical and preparative extraction and purification of cell constituents of bacteria or yeasts. The mechanism of cell disruption derives from the intensive shear induced by sonicating the suspension at sound frequencies above 20 kHz. A magnetorestrictive or a piezoelectric transducer converts the alternating current of an electric oscillator into mechanical waves that are transmitted to the suspension through a metallic probe (usually made from titanium) vibrating with the same frequency as the oscillator. The sound waves create many microbubbles at various nucleation sites in the suspension, which collapse implosively during the rarefaction period of the sound waves. This phenomenon of cavitation (formation, growth, and collapse of the vapor-filled bubbles) produces intense local shock waves, and intense local shear gradients are generated that cause the cells to deform beyond the limit of elasticity and rupture. The rate of microbial cell disintegration is usually low for preparative purposes, and the inclusion of small stainless steel or glass (Ballotini) beads will help in the nucleation for cavitation as well as with a grinding action (solid shear) that will increase the efficiency of disintegration.

The kinetics of disruption (protein release) has been reported to follow a first-order law, and for a well mixed pattern inside the disruption chamber the integrated equation is

$$1 - X = \exp(-kt) \qquad \text{...(7)}$$

where X is the fraction of released protein and k is the rate constant.

There are a number of variables that can markedly affect either the sample or the efficiency of the sonication process. One of these is the temperature rise. Most of the energy that is released when the bubbles collapse is absorbed by the liquid as heat. Although the value of the specific rate k increases with the temperature, the thermal energy has to be extracted to avoid temperatures that could destroy the activity of the useful bioproducts. To control the temperature within acceptable limits, the suspension should be precooled to 0-5°C, and jacketed disruption vessels should be used with a continuous circulation of a cooling fluid. Alternatively or concomitantly, short periods of sonication can alternate with short periods of cooling. The ratio of the time of sonication to the time of cooling is known as the duty factor.

The integrity of the bioproduct can also be adversely affected if free radicals are allowed to accumulate during the intense cavitation. The addition of radical scavengers (e.g., cysteine or glutathione) or pregassing the suspension with hydrogen can alleviate this problem, particularly if a prolonged sonication is expected. Other operating parameters affect the efficiency of the sonication process. One is the amplitude of vibrations of the sound waves (directly related to the acoustic power supplied to the suspension), which affects the rate constant according to the following relation:

$$k \alpha (W - W_0)^{0.9} \qquad \text{...(8)}$$

where W is the acoustic power and W_0 is the cavitation threshold power. However, a linear relationship between the rate constant and the acoustic power was recently reported.

The viscosity of the suspension affects the hydrodynamics and the rate of energy dissipation and inhibits the phenomenon of cavitation. When the viscosity is high, the operation should be started at a lower power, and the power should be gradually increased. The initial cell concentration that affects the viscosity of the suspension was reported to have no effects on the rate constant. Problems of foaming can result from the presence of surface active agents (e.g., proteins released from the disrupted cells) that lower the surface tension of the liquid medium. An intense foaming, particularly when high power is applied, contributes to cavitation "unloading" (therefore decreasing the efficiency of the process) and protein denaturation at the gas–liquid interface.

High working volumes require high acoustic power, which results in violent turbulence and in the formation of large bubbles that cause cavitation unloading. Rate constants were found to decrease linearly with increasing working volumes. The difficulties associated with higher working volumes (100–200 mL) can be alleviated by good design of the disruption vessel. Commonly a jacketed glass beaker with a continuous flow of cooling fluid is used; inlet and outlet tubes can be inserted for continuous operation. In continuous operation, the flow rate determines the residence time of the cells in the vessel and affects the overall yield of the process. For tougher cells the inclusion of Ballotini beads is necessary, and for further increasing the grinding action, the vessel should have internal tubing loops to circulate the suspension under the pumping energy of the sonication. The load and size of beads must in any case be optimized, but the value of the rate constant increases to a maximum and then decreases with increases in the bead diameter for the same bead load.

The shape of the probe is of particular concern. For a constant power level the amplitude of the sound waves at the probe tip is inversely proportional to the tip area. Therefore large tips dissipate the acoustic energy over a larger area, and the amplitude is smaller. However, if the tip diameter is too small the acoustic energy is dissipated over a smaller field and is less efficiently used, this is particularly important if the working volume is small because intense turbulence can lessen the power absorbed near the tip. Manufacturing materials dictate the severity of the wearing problems that are also common with the probe tips.

Titanium is a good material because of its good acoustic and mechanical properties and low toxicity to bioactive materials. However, titanium probes are expensive and difficult to machine. Pitting erosion produces deep cavities on the tip surface, particularly when Ballotini beads are included, which lessens the efficiency of sonication. The tip can usually be detached for servicing to produce a new flat surface, which is difficult with titanium, stainless or hard steels are better alternatives despite their poorer acoustic and mechanical properties. Sonicating equipment can be found in many laboratories. In preparative applications the performance is generally not as good as with the high-pressure homogenizers or bead mills. In the case of plasmid recovery from *E. coli,* even when the disruption yield was good, sonication failed to produce intact plasmids. The application of this technique to large-scale purification of bioproducts is rather limited: it is difficult to transmit high acoustic power to large volumes, it is difficult to extract the large amounts of heat generated during the cavitation process, and the equipment has to be shielded to protect humans.

Non-mechanical Methods

The mechanical methods are, generally, effective in disrupting tough microbial cell walls and much more efficient than the nonmechanical methods. Concomitantly, there are also several disadvantages: they are energy intensive, generate high temperatures and shear forces that can damage labile products, are nonspecific with respect to the organisms and the product to be released, and cause intense cell fragmentation. Although several attempts have been made to optimize operations and minimize these effects, process modifications that could result in a significant reduction in the overall energy requirements are still being pursued. Nonmechanical methods are not efficient on their own, and very often they only increase the permeability of the cell envelope without breaking the wall. However, by weakening the cell wall, lytic methods can be combined with the more efficient mechanical methods, and therefore require less harsh conditions. Among the physical methods of lysing cells, two have been reported more frequently, although they are not production oriented: osmotic shock and freeze–thawing. Osmotic stresses are generated when a cell is placed in either a hypertonic or a hypotonic medium; in the first case the cell tends to shrink (plasmolyze), and in the second case to expand.

In both cases the cells can suffer potentially serious injuries. Usually the cells have osmoregulating mechanisms in addition to tough cell walls that protect them from lysis. However, if suddenly placed in a hypotonic medium (a medium of very low salt concentration), the cells suffer an osmotic shock, absorb water, and swell; eventually they reach a point of unsustainable internal pressure and burst. This bursting process is effective and up-scaleable with weak animal cells. However, osmotic shocks are rather inefficient with tougher bacterial, fungal, or yeast cells, though some periplasmic enzymes and even some proteins/enzymes located near the inner side of the cytoplasmic membrane can be selectively released. Osmotic shock is frequently used to prepare membrane vesicles for studies of membrane proteins. The important operational parameter is the time to fully resuspend the cells in the hypotonic medium, which should be as short as possible to avoid the osmoadaptation of the organisms. The release of intracellular components can also be induced by freeze–thawing the cells. The ice crystals that form and grow during the freezing step will mechanically disrupt the integrity of the inner cell membranes, which become permeable and allow several enzymes to be extracted into appropriate buffers during the thawing step.

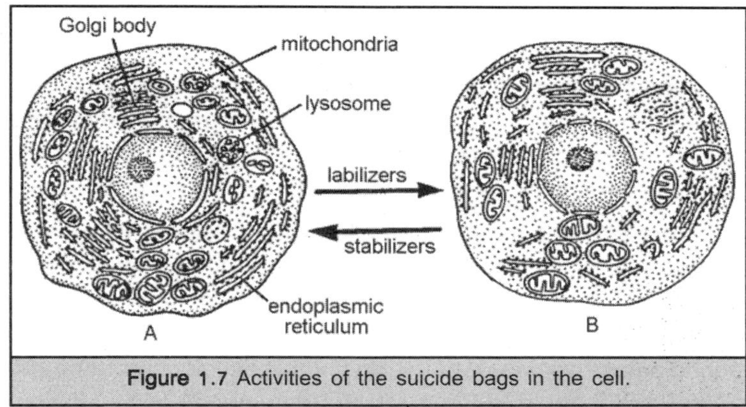

Figure 1.7 Activities of the suicide bags in the cell.

The precise mechanism for the release of proteins has not yet been elucidated; however, the formation of small transient pores in the membranes is a logical explanation consistent with the facts that protein solubility is important in governing its release to the medium, the inclusion bodies remain unreleased, and large molecular weight proteins are excreted with very low yields. The efficiency of this method depends largely on the rates of thawing. With a large mass of frozen suspension, the rate of thawing is higher in the outside region, while the interior of the block remains frozen, allowing chemical or enzymatic (eventually, proteolytic) reactions to proceed to different degrees in different parts of the block. On the other hand, the lifetime of the transient pores (regions of discontinuity) depends on the temperature of thawing that affects the rate of membrane resealing and is longer at low temperatures. This may explain the lack of reproducibility of the results in the ball mill disintegration of brewers' yeast samples that were kept frozen and separately thawed. The freeze–thawing process is not easily scaled up and find application mainly in laboratory-scale recovery of recombinant proteins. It can also be useful in disrupting pathogenic cells in a sealable package. Chemical methods on lysis are milder and often anticipate an autolytic enzymatic activity.

Organic solvents (e.g., toluene, ethanol, *n*-butanol, isobutanol, acetone, methylethyl ketone, etc.) are believed to extract, nonselectively, lipids from the cell wall or cytoplasmic membrane, therefore increasing the permeability to intracellular components. However, organic solvents might cause protein denaturation, and by diffusing into the cell, they also destroy the membranes of intracellular organelles, thus allowing destructive enzymes, such as proteases, to be liberated. These harmful effects of the organic solvents were minimized in the extraction of β-D-galactosidase by mixing yeast cells (*Kluyveromyces fragilis*) with concentrated selected solvents and, following filtration, resuspending and agitating the cells in an adequate buffer for extraction; 95% of the enzyme was released at the end of a 20-h extraction period. The process was claimed to be suited for scale-up, and the organic solvents were beneficial in killing the viable yeast cells, therefore avoiding the need to separate them from the lysate and the addition of antimicrobial agents. The use of concentrated organic solvents, however, imposes a fireproof environment. Other chemicals, such as detergents, chaotropic agents (e.g., guanidine HCl or urea), EDTA, phenethyl alcohol (PEA), or alkaline solutions have been used to increase the permeability of the cell walls.

All chemical methods suffer from long holding times and, thus, have long been considered unattractive for larger- scale applications. Addition of glycine in low concentrations to the medium at an appropriate pH could also increase the permeability of recombinant *E. coli* immobilized in κ-carrageenan gel beads to α-amylase. However, the recent progress in recombinant DNA technologies has renewed interest in the development of methods to increase the permeability of the cell envelope in order to avoid a detrimental accumulation of plasmid-encoded proteins in the cytoplasm. If these proteins are allowed to accumulate, further synthesis is reduced while proteolytic degradation may cause loss of protein activity, or the plasmids may loose stability. The benefits of whole viable cell immobilization may also be lost if the foreign proteins are not excreted and the cells would require a disruption step for the product to be released and further purified. Chemical methods to recover these active recombinant products in situ, with increased selectivity concerning the intracellular location, have been investigated, with great promise. *E. coli,* a Gram-negative bacterium, has typically been used as the host organism, and many high- value recombinant proteins are secreted into its periplasmic space. Treatment of *E. coli* with guanidine HCl (a chaotropic agent) increases the permeability of the outer membrane, and periplasmically expressed proteins such as β-lactamase can thus be leaked into the surrounding medium with high yield (better than 80% in 5 h), free of many intracellular contaminants.

After treatment of the same cells with Triton X-100, which attacks the inner membrane, the cytoplasmic proteins could be released but with low yield and only when the cells were treated with both the guanidine HCl and Triton X-100 was the yield of intracellular protein higher than 50%. The cost of Triton X-100 is high but, nevertheless, this process was claimed to be less expensive than enzymatic methods. The release of β-lactamase was also enhanced by treating recombinant *E. coli* with EDTA and PEA at appropriate concentrations; cell viability and synthesis of the plasmid encoded β-lactamase were retained. Pretreatment with guanidine HCl and Triton X-100 was also used in conjunction with a high-pressure homogenization step to disrupt recombinant *E. coli* expressing human growth hormone inclusion bodies. A reduced number of passes through the homogenizer at lower disrupting pressure were required to recover the protein from the inclusion bodies; due to their synergistic effects, Triton X-100 was used to reduce the concentration of guanidine HCl required to weaken the cell wall and to wash the inclusion bodies from contaminant proteins.

Pretreatment of *Candida lipolytica* with an alkaline solution resulted in a weakened cell wall and increased protein solubility, and further mechanical homogenization was thus facilitated at lower pressures than with non-treated cells. Enzymic methods of cell wall hydrolysis are also becoming increasingly more attractive. The advantages associated with enzymic lysis include an increased selectivity of product release (which is desirable for further product isolation and purification), an increased rate and yield of product release, the minimization of product damage, requirements for mild conditions of pH and temperature, and no debris being produced. More than any other methodology, enzymatic lysis requires a good understanding of the details of the chemical and physical structure of the cell walls. No single enzyme is capable of fully degrading the cell walls.

As already described, peptidoglycan is the structural element in bacterial cell walls. In Gram-positive bacteria a single enzyme could lyse the cell, but a mixture of several enzymes (glycosidases [endoacetylmuramidase, *endo-* and *exo-*acetylglucosaminidase], *N*-acetylmuramyl L-alanine amidase or peptidases) acting synergistically would assist in the lysis of the bacterial peptidoglycan; in Gram-negative species a pretreatment with a surface-active agent is usually required to remove the outer membrane. For yeast cells a wall lytic protease and β-(1-3)-glucanase are essential, but the inclusion of β-(1-6)-glucanase, mannase, and chitinase assist the lytic process. Most lytic enzyme systems are active against pretreated nonliving cells (lyophilized or heat-killed cells) or against cell walls.

Table 1.1 Enzymatic disintegration of cell wall

Organisms	*Enzymes*	*Type of hydrolyzed linkage*
Bacteria	Glycosidases	β-(1,4)-linkages between AGA and AMA residues in a peptidoglycan
	Acetylmuramoyl-L- alanine amidases	Link between *N*-acetylmuramoyl residues and L-amino acid residues in certain glycopeptides
	Peptidases	Gly-gly, ala-gly, etc., peptide bonds
Fungi, yeasts	β-(1,3)-Glucanases	Random β(1,3)-linkages in glycans
	β-(1,6)-Glucanases	Random β(1,6)-linkages in glycans
	Mannanases	(1,2)- (1,3)-, (1,6)-β-D-Mannosidic linkages
	Chitinase	*N*-Acetyl-*b*-D-glucosaminide β(1,4)-B-linkages in chitin and chitodextrins
	Proteases	
Algae	Cellulases	α(1,4)-Linkages in cellulose

The main restrictions in using multienzyme systems are the availability and high costs associated with their preparation. Different sources of lytic enzymes have been investigated. Various species of *Arthrobacter* can be induced, by whole yeast cells or cell walls, to produce yeast lytic activity (β-(1-3)-glucanase, mannase, and protease) subject to catabolite repression by glucose. Yeast lytic activity was also identified in *Cytophaga* sp. and *Oerskovia xanthineolytica*. Different strains of *Streptomyces* sp., *Pseudomonas* sp., *Staphilococcus* sp., and *Cytophaga* sp. have a wide spectrum of bacteriolytic activities. Lysozyme from hen egg white is bacteriolytic against only a few bacteria. For the extraction of a given product from an organism, the selection of the most appropriate lytic system depends on the nature of the wall to be lysed and the nature and location of the product inside the cell.

The composition of a multienzyme system determines the efficiency and the selectivity of the lytic process. If the product is proteinaceous, a highly proteolytic system is undesirable. Thus, if the organism to be lysed is a yeast, a high proteolytic activity should be provided in only the early part of the lytic process, when the substrate is the mannan–protein complex of the wall; the protease should then be inactivated, or a protease inhibitor should be added. However, the composition of lytic systems can be manipulated in part in the production phase, using different inducers and growth conditions. For instance, in the case of the lytic enzyme system from *Cytophaga* sp., high β-(1-3)-glucanase activities and high glucanase: protease ratios were obtained when produced in continuous culture at low dilution rates under carbon-limited conditions.

The quantitative data concerning the effects of pH, temperature, concentrations of lytic enzyme and of cells on the lysis rate should also be obtained to optimize a lytic process. As with all the nonmechanical methods, extensive enzymatic lysis is a very slow process, but in conjunction with others, the rate and the yield of product extraction can be much increased. Mutants that are defective in β-glucan can be produced by mutagenesis and are, therefore, more susceptible to β-(1-3)-glucanase (commercially known as zymolyase) attack. Enhanced disruption of *C. utilis* was obtained by high-pressure homogenization (using a microfluidizer) when the cells were pretreated with zymolyase. *Bacillus cereus* pretreated with 0.5 mg cellosyl (a lytic enzyme produced by *Streptomyces coelicolor*) per gram of cells (wet wt), at pH 5.5 and 30°C, was disrupted with 98% yield in one single passage in a Manton–Gaulin homogenizer at 7×10^7 Pa, as compared with less than 40% yield when the cells were untreated. The same enzymatic treatment facilitated disruption in a bead mill. Autolysis is the process by which the cells use their own lytic capabilities and produce lytic enzymes. Autolysis, as a self-digestion process, is very slow and often requires several days, which is very inconvenient. Because of that, it is not a common method of cell disruption; however, yeasts are autolyzed on an industrial scale for the production of yeast autolysates. The process can be speeded up by drying or freeze-thawing the cells.

As already mentioned, autolysis can also be induced by plasmolysis with surfactants, thiol reagents, and organic solvents (e.g., toluene or butanol) or by aging or temperature. β-Lactam antibiotics are known to cause the activation of autolytic enzymes (autolysins), and this has also been suggested to be the mechanism by which surfactants induce the autolytic process. Autolysis enzymes are important for cell growth and reproduction, but once they are deregulated, they attack bacterial peptidoglycan. A key factor is the pH, which is generally optimum around 8.0 when yeast is to be plasmolyzed by organic solvents or detergents. Microbial cells differ in their potential for autolysis; often they lack the genetic information to produce lytic enzymes, in which case the genetic coding must be introduced from outside. Lytic enzyme systems that can be cloned include not only the autolysins (several peptidoglycan hydrolases) but also colicin enzymes (a class of antibiotic proteins) and phages.

EFFECTS OF CELLULAR FRACTIONATION

One of the most challenging problems in solid–liquid separation is the removal of cell debris, which includes cell wall fragments, cell membranes, and protein precipitates. Disruption and lytic treatments cause a reduction in the sedimentation properties of the cells and fine debris (<0.1 μm), leading to inefficient centrifugation and/or filtration due to changes in the surface properties, high compressibility of the solids, high viscosity of the disrupted suspension, and a small difference between the densities of the solids and the suspending medium. *Centrifugation* has been used, but it is unsatisfactory when the particles are smaller than 0.5 μm, particularly in viscous homogenates, therefore requiring very low flow rates (on the order of 10 times less than for whole cells). A previous flocculation can improve the centrifugation efficiency. A highly selective flocculation of cell debris by borax (sodium tetraborate) was improved by further addition of poly(ethylene imine) to the clarified homogenate supernatant. *Rotary vacuum precoat filtration* has been used for the clarification of disrupted yeasts, but *cross-flow microfiltration* has the advantage of preventing cake formation and avoiding the need for expensive filter aids.

The separation of cell debris is, however, the most complex problem that crossflow microfiltration has been tested against. Membrane fouling and concentration polarization are some of the difficulties that contribute to a rapid flux decline and limit the concentration of debris that can be attained. Apart from the influence of the operational parameters (pressure, feed velocity, fouling factors) on the flux and transmission, which are well documented in the literature, we have to consider that strong adsorption of proteins to cell debris may occur in the lysate, depending on the concentration, temperature, pH, or ionic strength. *Aqueous two-phase extraction* is attractive for the recovery of whole cells and debris from protein solutions but involves high costs and difficult recovery of reagents. Phase components include water-soluble polymers and electrolytes, the best-known systems being PEG-dextran and PEG-phosphate. Separation techniques depend on many specific aspects, but the average particle size and the size distribution are crucial to all of them: the maximum amount of released protein increases with increasing micronization of the cell debris, but these become increasingly more difficult to separate.

Mechanical methods are generally, effective in disrupting tough microbial cell walls. The major difference among them lies in the different sizes of cell fragments that are generated; some methods disrupt the cell envelope with less overall destruction than others. This is significant because the size of the fragments has an important impact on downstream processing when they are removed from the lysate or separated from other subcellular species: the submicron particles are easily resuspended in the supernatant after centrifugation, hinder the fractional precipitation, and tend to foul chromatographic columns or membranes that follow. The behavior of *E. coli* cell debris from a bead mill and high-pressure homogenizers was investigated. No significant differences in the yield and rate of release of protein were reported.

However, the characteristics of the debris (particle size distribution, separability by centrifugation, and dead-end filtration) were quite dependent on the disruption method. The mean residence time in the bead mill did not significantly affect the particle size distribution, but the number of passes in the mechanical homogenizer broadened the debris size distribution. The effect of operating conditions on the cell debris size distribution was investigated for bakers' yeast disrupted by high-pressure homogenization, and a two-parameter model was developed, incorporating both the operating pressure and the number of passes, to simulate the particle size distribution in the homogenate. A threshold pressure of 11.5 MPa was determined, below which no disruption of yeast cells will occur.

Yeast cell debris particle size distribution was also reported to depend on the growth conditions: in batch cultivation the median diameter, d_{50}, was less in the exponential phase and greater in the stationary phase; in continuous culture the median diameter was less for higher dilution rates (younger cells). High-pressure homogenization of of *E. coli* cells treated with a combination of guanidine HCl and Triton-X-100 produced cell debris of larger sizes than when the pretreatment was made with only guanidine HCl, which is highly desirable when the debris are to be separated by cross-flow microfiltration. The cell suspension should also be conditioned to avoid unwanted degradation of labile products liberated from the disrupted cells. Protease inhibitors are often required and should be added to the cell slurry.

Antioxidants are useful to avoid oxidation of S—S bonds. An appropriate buffer should be included to maintain the pH within the desired range. All the chemicals that are added to the system in all phases of the process (starting from the medium preparation and fermentation steps) may have effects on the product and subsequent operations; this should be taken into account. Evolving from the recent developments in recombinant DNA technology, a large number of proteins of therapeutical importance, now in use or undergoing clinical evaluation, are produced as inclusion bodies (or "refractile" bodies) in the cytoplasm, mainly in *E. coli,* where they accumulate due to the absence of a secretory mechanism. These inclusion bodies are dense and amorphous aggregates of protein, consisting of misfolded (frequently denatured) polypeptides, often as large (0.8–1.5 μm) as a normal bacterium cell, with a relatively higher density ($\sim 1.2 - 1.35$ kg/L) than the rest of the cell material. They require the use of chaotropic agents or detergents to be solubilized in an aqueous environment.

The standard procedure for the purification of a protein expressed as an inclusion body begins with the isolation of the inclusion bodies, which itself begins with the disruption of the cell. The best cell disruption technique is high-pressure-homogenization. The aggregates are then separated from the soluble components and from the debris of the lysed cells, usually by centrifugation, typically using disc-stack centrifuges, at 5,000-10,000 *g* during 5-10 min; the lighter cell debris is removed along with the supernatant. Cross-flow membrane filtration has also been investigated. The pellets are then washed with either dilute detergents or chaotropic reagents (urea or guanidine HCl) to remove contaminants, and often reducing agents (e.g., dithiothreitol or β-mercaptoethanol) are added to disrupt incorrect disulfide bonds. The solubilized and reduced protein is then ready for subsequent purification and refolding.

2

MECHANICAL SEPARATION

The first application of centrifugal separation other than in milk processing was in the harvesting of cells in baker's yeast production. Centrifugation is still the only choice for cell separation and protein recovery in many fermentation processes. Centrifugal separation is a sedimentation operation accelerated by centrifugal force. Thus, a prerequisite for the separation is a difference in density between the phases. This applies to both solid–liquid and liquid–liquid separation. The settling velocity (V_g) of a solids particle (or droplet) under the influence of gravity alone is given by Stokes' law:

$$V_g = \frac{\rho_p - \rho_f}{18 \cdot \eta} \cdot d_p^2 \cdot g \qquad \ldots(1)$$

In the centrifuge field, the settling velocity becomes:

$$V_c = V_g \cdot Z \qquad \ldots(2)$$

where

$$Z = \frac{\omega^2 \cdot r}{g} \qquad \ldots(3)$$

is called relative centrifugal force (RCF) or G-number.

It is, however, not the RCF only that determines the performance of a given centrifugal separator. The residence time of the particle or droplet in the centrifugal force field is important, so a larger volume of the rotor will increase the possible flow rate. In the most common centrifugal separators, the disk bowl machines, a given bowl material can cope with a certain peripheral speed—the stronger the bowl material, the higher the speed. This leads to the fact that the maximal RCF is inversely proportional to the bowl diameter, but not necessarily that the lower RCF machine has a lower performance.

TYPES OF CENTRIFUGAL SEPARATORS

What determines the design of a centrifuge is the method by which the solids phase in the process material is handled. The first two types of centrifuges developed were applied to duties where the resulting products were in liquid form, such as cream and skim milk and yeast cream and cell-free wort. In the beginning of this century, centrifuges were applied to processes with solids, amorphous or crystalline.

In the biotechnological industry, the solids phase is often recovered in a diluted suspension or is, even at high solids concentrations, a slurry with good flow properties. Types of machines are described in the work by Rushton et al. The most comprehensive description of centrifuges can be found in the book by Sokolov. Brochures and technical publications from sedimenting centrifuge manufacturers such as Alfa-Laval Separation AB also provide information on machine types and sizes. In this article, centrifugal filtration will not be dealt with because of the limited use in this industry, with the exception of crystal recovery in antibiotic production.

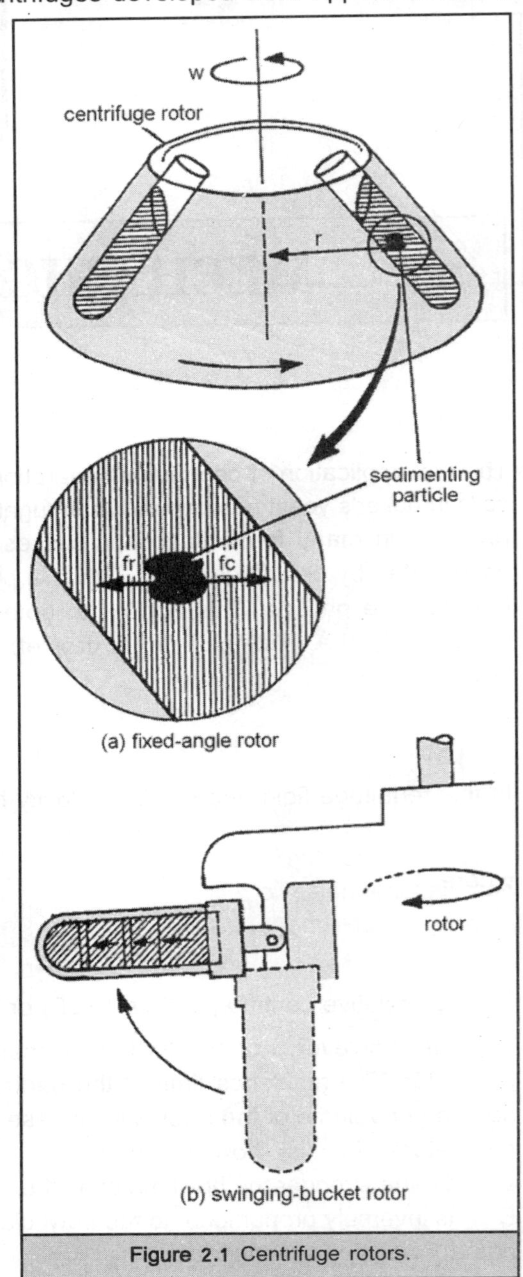

Figure 2.1 Centrifuge rotors.

Disk Bowl Machines

Solid bowl

A solid bowl machine has two liquid outlets, each with a paring disk. Other versions have none or one paring disk. The solids accumulate at the periphery and must be removed manually. In the separation of two nonmiscible liquids, a coalescence of the dispersed droplets takes place at an internal interface in the bowl. The positioning of this interface is very important. The nearer it is placed to the light liquid outlet, the more contaminated the light liquid will be with droplets of the heavy phase. The position is determined by a gravity disk, over which the heavy liquid flows. The gravity disk diameter is given by the relation:

$$D_h = \sqrt{D_1^2\left(1-\frac{\rho_{fl}}{\rho_{fh}}\right)+D_1^2\,\frac{\rho_{fl}}{\rho_{fh}}} \quad \ldots(4)$$

If properly designed and if the feed flow rate is high enough to prevent sedimentation in the bowl, this type of machine can be used to recover a dilute cell concentrate in the heavy liquid outlet. This is also valid for three-phase solids- ejecting machines. Solid-bowl machines are made in a large number of sizes, with bowl diameters ranging from 140 to 750 mm.

Solids-ejecting separators, radial discharge

Of all types of disk bowl machines, the solids-ejecting separators with radial discharge are the most common. With an automated, periodic partial discharge of the sediment, it is possible to obtain a considerably higher concentration of the solids than in the peripheral nozzle machine. The machine is discharged in about 0.1 s by pressing a sliding bowl bottom in the bottom part of the bowl downward, thereby opening large discharge ports in the outer bowl wall. To make the discharge at the right moment, the discharge can be initiated by a self-triggering mechanism that hydraulically senses the solids level in the solids space. Three-phase separation units are available. The solids-ejecting machines can be equipped with hermetic seals and a hermetic feed from below in a hollow spindle. In these true hermetic machines, the split between the exiting light and heavy liquids can be made by controlling the back pressure in the outlet pipes. The heavy liquid can contain large amounts of solids (cells or proteins) if the machine is properly designed. The solids-ejecting separators are available in about 10 sizes, with bowl diameters between 180 and 900 mm. The smaller units are more often available in versions equipped for BL2-LS requirements than are larger units.

Solids-ejecting separators, axial discharge

When comparing axial and radial discharging separators, one finds that the axial machine has a different geometry. The lock ring can be placed at a smaller diameter, because no sliding bowl bottom needs to be fitted. This reduces the stresses in the winding of the lock ring. Also, the solids outlet takes place through a few small axial channels that do not induce such large stresses in the bowl shell. These two factors make it possible to increase the bowl speed considerably above that of the radially discharging machines, so that the RCF is at least doubled. The discharge system is operated by pressurized air, which forces a ring slide, equipped with valve seats, downward, opening the axial channels. To facilitate the transport of solids toward the channels, the bowl is machined to have a star-shaped inner contour. This type of machine exists at present in three sizes, with bowl diameters ranging from 500 to 900 mm. BL2-LS versions are available.

Nozzle machines with pressurized discharge of concentrate

The apparent viscosity of bacteria and yeast suspensions as well as some protein precipitates decreases with increasing velocity gradient. Therefore, they are able to flow even at high concentrations in channels or tubes from the periphery of the bowl in nozzle machines with pressurized discharge of concentrate. The tubes empty into a central chamber, where a centripetal pump, a paring tube, picks up the concentrate and pumps it out of the bowl. The nozzles are placed at the end of the concentrate tubes, just in front of the chamber. This machine type is suitable only for applications where the solids have the specific flow properties mentioned earlier and in the absence of other type of solids. In some machines, of this type of vortex chamber is placed in front of the nozzle. Liquid enters the chamber tangentially, creating a viscosity-dependent whirl, and leaves through the nozzle in the centre of the chamber. The viscosity effect is giving a self-regulating function so that varying solids flow rates (within limits) to the machine will result in a constant concentration of the concentrate, thereby reducing the risk of clogging the machine. One of the units of this type is available in a sterilizable, contained, and aseptic version, conforming to BL2-LS. To make this possible, the cleaning-in-place (CIP) has to be efficient. Therefore, the sterilizable machine uses the discharge system of the machine during CIP. This system is also used in noncontained versions. Bowl diameters between 500 and 900 mm are available.

Nozzle machines with peripheral nozzles

In nozzle machines with peripheral nozzles, nozzles with a diameter of 0.5 to 3 mm are situated at the periphery of the bowl, which has sloping walls toward the nozzles. The number of nozzles varies between 4 in small bowls and 20 in the largest bowls. It can be equipped with a recirculation device for the sediment, a separate central tube that is divided into tubes along the bowl bottom ending just in front of the nozzle. This makes it possible to increase the concentration of the solids phase without the need to decrease the nozzle size, leading to increased risk of nozzle clogging. The nozzle flow rate is proportional to number of nozzles, distance from the center of rotation, bowl speed, and square of the nozzle diameter. The peripheral nozzle machines are used for the largest flow rates; the bowl diameters are up to 1,000 mm.

Decanter Centrifuges

In the 1950s, the solid-bowl separator with scroll discharge (decanter centrifuge) was developed for large quantities of solids. It is equipped with a screw conveyor that rotates at a speed slightly higher or lower than the bowl speed. The differential speed can be varied. Careful control of this parameter, torque control, and some newer design features (e.g., the baffle disk), have recently made the decanter interesting to use also for biological sludges. However, the conventional mechanical design of these machines does not allow more than 5,000 G for a medium-diameter

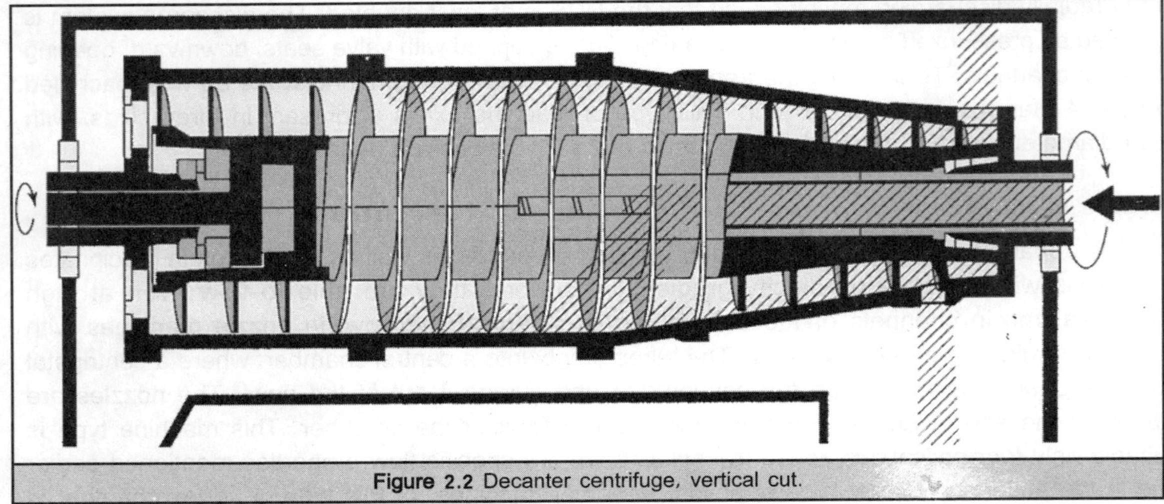

Figure 2.2 Decanter centrifuge, vertical cut.

machine. To reach higher *g*-forces new design features are needed. With suspended main bearings, a separate gear box, and a swimming conveyor, held in place by the liquid, 10,000 G is feasible.

For three-phase decanters, the gravity disk dimensioning follows in principle equation 4. It is, however, very important to take the weir head into account, because a variation of the light-to-heavy phase ratio gives a very unstable interface position. The range of bowl diameters is very wide, from 150 to 1,200 mm; the bowl length is often 4 to 5 times the diameter. The newest addition to the decanter family is a vertical decanter equipped with a disk stack. The thought is to combine the best characteristics of a decanter (dry solids) and a disk bowl machine (good clarification).

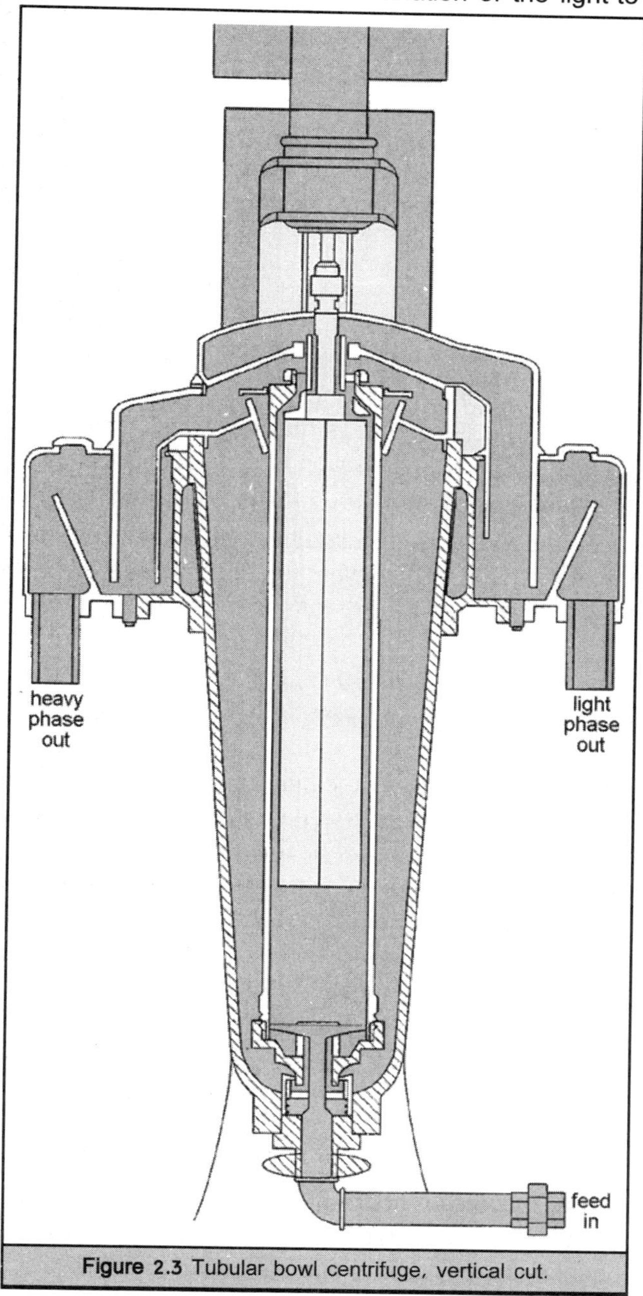

Figure 2.3 Tubular bowl centrifuge, vertical cut.

Other Classical Bowl Designs

Tubular bowl

In the tubular bowl the rotor, driven from above, consists of a long cylinder, into which liquid is pumped from below. Most tubular centrifuges are used for separating very valuable solids from a liquid. The liquid is pumped out by a paring disk, avoiding aerosol and foam formation. To remove the solids accumulated at the periphery, the machine must be stopped and dismantled. The bowl diameters are up to 130 mm with a solids volume of up to 6 L. In the industrial models, 20,000 G are generated. Sterilizable and contained units are available.

Multichamber bowl

In the multichamber bowl machine, the process liquid flows in the annuli between concentric cylinders. Having a relatively low separation capability, this type of machine is used for recovery of valuable solids, when the solids volume of the tubular bowl is too small. Some designs include cooling channels in the bowl wall, a cooling jacket around the bowl, and a cooled centripetal pump. The maximum bowl diameter is about 500 mm.

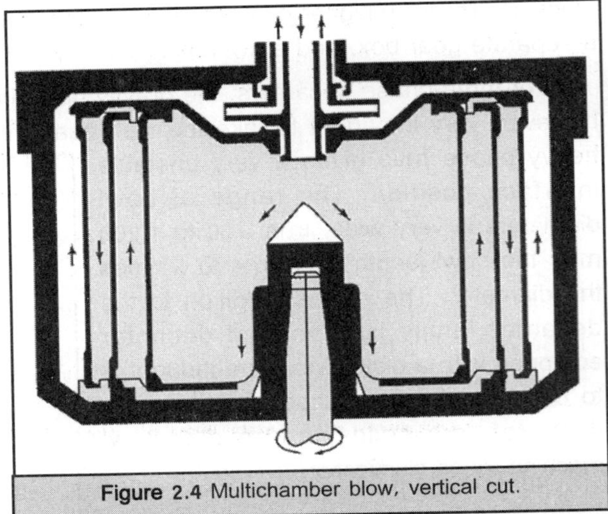

Figure 2.4 Multichamber blow, vertical cut.

New Bowl Designs

Centritech machines

In the Centritech machine, a disposable separation insert in plastic is placed in a rotor. The feed liquid enters at the top of the insert, and the main liquid outflow is at the opposite end of the top of the insert. The heavy liquid phase (cell concentrate) is withdrawn intermittently through the bottom outlet. Through the principle of the inverted comma no seals between rotating and nonrotating parts are needed. The G-number is maximum 100. The mechanics are further described in Apelman et al. Its only described application is separation of mammalian cells from culture fluid. Two sizes of the machine are available. The diameter of the insert is about 300 mm in the larger unit.

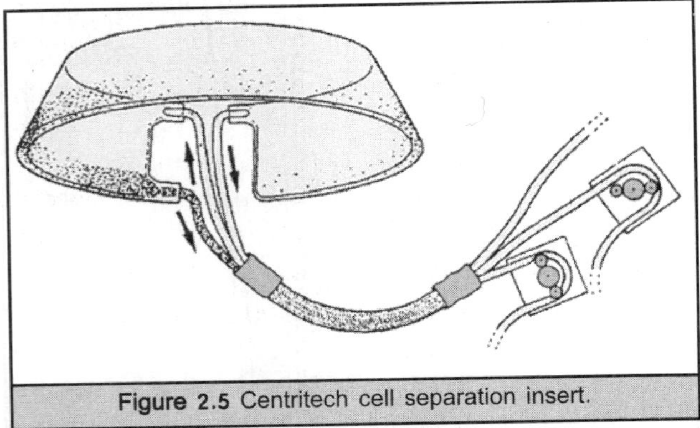

Figure 2.5 Centritech cell separation insert.

Inverted chamber bowl with intermittently operated scraper

The inverted chamber bowl is a relatively new type of machine, sold under the name Powerfuge. Its bowl has its opening downward, where the centrate leaves under gravity and is collected in a chute. The feed is from the top. When the bowl is full of solids, it is decelerated. At low speed, the asymmetrically positioned scraper is engaged and removes the solids. Four sizes are available, with bowl diameters from 150 to 600 mm. They are designed for CIP and sterilization-in-place (SIP). The smallest unit can develop 20,000 G. It has to be remembered that the hydraulic capacities given by centrifuge manufacturers are the maximum possible feed flow rates that often are far above the flow rates that will give satisfactory separation results for difficult separations, such as bacteria and protein separations.

FLUID AND PARTICLE DYNAMICS

A successful scale-up method should be based on a mathematical model that accurately describes the separation process, the hydrodynamics and particle dynamics inside the bowl. The phenomena going on inside the centrifuge bowl are, however, so complicated that no mathematical model is able to predict in absolute terms the separation efficiency as a function of machine and process media parameters. Axelsson reviewed of some of the flow phenomena, which are briefly discussed in this section, along with examples of more recent findings.

Centrifuges with Conical Disks

Fluid and particle dynamics in the disk stack

The disks in the disk bowl split the flow in thin layers. The Reynold number for the flow between the disks is usually low, and the flow is laminar, resulting in an easy particle settling. The critical Reynolds number for transition to turbulent flow increases with decreasing disk spacing and increasing bowl speed, which is beneficial for the separation efficiency because the settling distance at a small disk spacing will be short, and the high bowl speed gives a large settling velocity. Also, a small disk spacing keeps the liquid rotating at or near the bowl speed, thus reducing the pressure drop in the disc stack. In Figure 2.7 is a vertical cut of two disks in the disk stack, where the flow is driven toward the center by a radial pressure gradient. The character of the velocity profile between the disks depends on a dimensionless number, λ, which is defined as

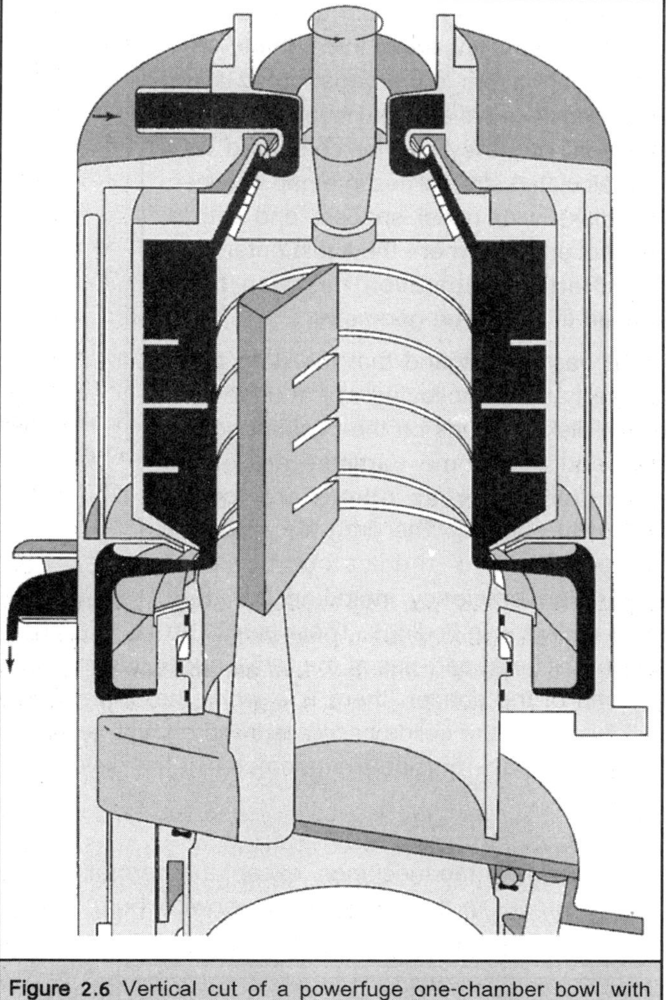

Figure 2.6 Vertical cut of a powerfuge one-chamber bowl with scraper.

$$\lambda = t \cdot \sqrt{\frac{\omega \cdot \sin\alpha}{\nu}} \qquad \ldots(5)$$

When λ is small, the velocity profile is parabolic, but for λ values above 5, normally encountered in industrial centrifuges, the radial velocity component splits up in two thin layers, Ekman layers. In these layers, all the net transport of liquid toward the center takes place. Because separated particles (heavy phase) are collected at the underside of the disk, they are subjected to a strong shear force from the Ekman layer that drags the particle toward the liquid outlet. The velocity in these layers increases with increasing λ, such as with increasing bowl speed. Therefore, a higher bowl speed cannot always be fully used because of the counteracting shear force. In between the Ekman layers is the geostrophic layer. Its streamlines have been visualized by dye injection, and they can be compared with the calculated streamlines in a gap between two disks with radial spacers and with a distribution hole where the liquid enters the gap. Particle separation has also been studied in the same geometry.

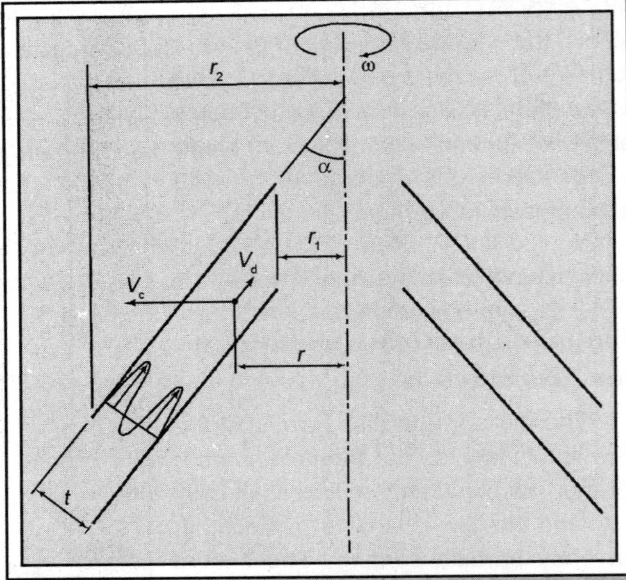

Figure 2.7 Flow profile between conical disks at high l value, partical separation.

It has been found that the drag force on the particles collected at the underside of the disk depends on the position on the disk and that some particles are easily separated, whereas others are forced toward the center. The drag force can be computed, and the calculations of separation efficiency including the drag force verify these findings. There is also an accumulation of particles at the radial spacer, along which the solids slide toward the periphery. At the end of the spacer, there is a wake that forces the particles back toward the center until the bulk weight of the solids becomes large enough to cause an intermittent discharge of particles into the solids space outside the perimeter of the disk.

Inlet

All bowls in the machines, except the hermetic machine and the Centritech machine, have an open chamber in the center of the bowl, where the nonrotating feed liquid is received. In most inlets, the liquid forms a cylindrical interface with the gas (usually air) in the receiving chamber. The nonrotating incoming liquid hitting the rotating interface results in a turbulence that entrains into the feed and splits particles that are susceptible to shear; the further out from the center the interface, the larger the turbulence. If it is not practical to use a hermetic inlet, the interface in the inlet must have as little an inner radius as possible. Also, the unavoidable energy dissipation must be distributed over a large volume, thereby diminishing the shear force. One design that allows smoother inlet conditions and a reduced air entrainment is the disk inlet. A mathematical model for feed zone breakage of shear-sensitive particles has been verified for a conventional pilot-

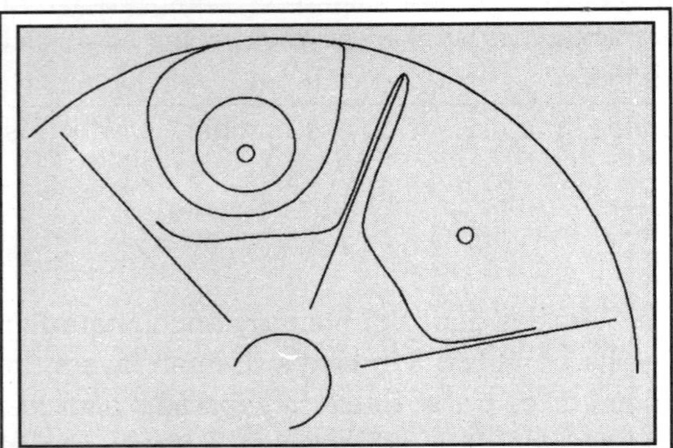

Figure 2.8 Comparison of calculated actual patterns in a disk bowl.

scale solids-ejecting separator. The extent of breakage was found to be independent of flow rate. This agrees with the finding that temperature increase in the separated liquid after the separator is also independent of flow rate.

Outlets

The liquid streams leaving the separator have picked up energy in the form of kinetic energy and heat caused by inner friction. When the liquid is decelerated outside the separator, the kinetic energy is converted to heat. The temperature increase is proportional to the square of the bowl speed and to the square of the radius at which the liquid leaves the separator. It is therefore not surprising that the discharge in a peripheral nozzle or a solids-ejecting machine causes breakage of cells because of the high kinetic energy of the fluid.

A typical degree of bacterial cell breakage is 30%. The consequence of this is that if the particles to be separated are susceptible to shear and cell integrity is important, the cell concentrate should leave the centrifuge as near to the center as possible. It has been reported that in such machines the breakage of *Escherichia coli* cells is low. The outlet from the centrifuge is often of the paring disk type. For a good function, this outlet must be designed to avoid air entrainment and large shear forces. This is especially important for chemically sensitive substances such as proteins. A hermetic design can be used in some applications.

Decanter Centrifuges

The objective of a decanter is usually to concentrate the solids present in concentrated suspensions. The hydrodynamics of a decanter are complicated. The conveyor adds turbulence and forces the liquid into a helical path. Below the top layer, vortices are formed. The transport of solids with low inner friction will influence the sedimentation process. The turbulence at the inlet zone, with possible re- suspension of separated solids, and the lower solids concentration toward the liquid outlet contribute to the fact that a slender decanter will separate better than a short one with the same theoretical size.

THEORETICAL SIZE OF A CENTRIFUGAL SEPARATOR

From the descriptions in the previous section it is evident that mathematical models describing the flow phenomena in centrifuges are very difficult to build. However, the attempts made are successful when machines of different sizes but with identical geometry are compared, for example, for scale-up purposes.

Σ Theory

Generalized Σ formula

The most-used quantity to characterize centrifuges, the Σ concept, was presented by Ambler. In its derivation he considered particles with a critical diameter (d_c) that were separated to 50%. Today, however, the most-used definition of the critical particle is the particle that is separated to

100%. This does not influence the formulas for the Σ value, but the value for the feed flow rate (Q) will be halved. Σ is defined as

$$Q_{theor} = V_g \cdot \Sigma \qquad \text{...(6)}$$

V_g is the Stokes settling velocity (equation 1). Σ has the general expression

$$\Sigma = \frac{V \cdot \omega^2 \cdot r_e}{g \cdot s_e} \qquad \text{...(7)}$$

The equation for the critical diameter becomes

$$d_c = \left(\frac{18 \cdot \eta \cdot Q_{theor}}{\rho_p - \rho_f} \cdot \frac{s_e}{V \cdot \omega^2 \cdot r_e} \right)^{1/2} = \left(\frac{18 \cdot \eta \cdot Q_{theor}}{\Sigma \cdot (\rho_p - \rho_f) \cdot g} \right)^{1/2} \qquad \text{...(8)}$$

The derivation in equation 8 is based on the following assumptions:

1. Viscous drag is determining the particle movement.
2. The flow in disk bowls between the disks is laminar and symmetrical.
3. The liquid rotates at the same speed as the bowl.
4. The particle concentration is low (no hindered settling).
5. The particle always moves at its final settling velocity.
6. This settling velocity (V_c) is proportional to the *g*-force.

Ambler uses the critical particle in his formula; therefore, his analysis is a special case of Svarovsky's grade efficiency function, where for each particle size the ratio of sedimented mass to the mass in the feed material is determined and integrated over all particle size to get a more realistic measure of the separation efficiency.

Σ for disk bowl centrifuges

The particle has a radial velocity component because of the centrifugal force (V_c), which increases with increasing *r*. It has also a drag- dependent velocity component (V_d), which can have any size or direction depending on the flow situation round the particle. In the Σ theory, it is assumed that V_d depends on uniform radial plug flow parallel to the disks. The derivation of the equation for Σ is found in the paper by Ambler.

Separation Efficiency

The assumptions in the Σ theory are not fulfilled in reality, as mentioned previously. Therefore, equation 6 should include an efficiency factor, so that in practice the flow rate is reduced at the requested separation performance. The efficiency factors have been estimated to have values according to Table 2.1.

This table was compiled in 1984 from data in literature. By careful measurement of the density difference and particle dimensions, it has been shown that in one case an efficiency factor of maximum 16% was found to fit the experimental data in a pilot-size disk bowl machine. However, this probably also included effects of hindered settling.

Table 2.1 Efficiency factors μ for various separator types (%)

Disk bowl machines	45–73
Decanter centrifuges	54–67
Tubular bowls	90–98

In the absence of hindered settling, the equation describing the relationship between the feed flow rate, the Σ value, and the settling rate, will thus be

$$Q_{act} = \mu \cdot V_g \cdot \Sigma \qquad \text{...(9)}$$

KQ Formula for Disk Bowl Centrifuges

In 1961 a semi-empirical equation for the separation area was published. It is based on the finding that separation data fitted the relation

$$\frac{V_c}{V_g} = \left(\frac{\omega^2 \cdot r}{g}\right)^k \qquad \text{...(10)}$$

with $k = 0.75$ better than the classical analysis (equations 2, 3, and 6), where $k = 1$. This fact is a consequence of the increasing shear forces in the Ekman layers. For practical purposes, the following relation is used:

$$KQ = 280 \cdot \left(\frac{n}{100}\right)^{1.5} \cdot N \cdot \cot\alpha \left(r_2^{2.75} - r_1^{2.75}\right) \qquad \text{...(11)}$$

where n = bowl speed in rpm and r_2 and r_1 are the radii of the disk in centimeters. This form of the KQ formula is dimensionally incorrect, but it is mathematically possible to make it correct, expressing the KQ value as an entity with the dimension area. In practice, the difference between the KQ and the Σ for scale-up purposes is negligible, compared to other inaccuracies.

Hindered Settling

One of the few experimental investigations into hindered settling of cells shows that complex phenomena occur at high concentrations. A model with several experimental constants was developed. It showed that the settling rates drop very fast with increasing solids content, by a factor of about 2 at 20% by volume, and 15 to 20 at 50% by volume. The classical model used in the work by Clarkson et al. gave similar results. In the tests by Aiba et al., it was also found that cells may settle flocculated at higher concentrations. Hindered settling is one further example of the difficulty to calculate theoretically the performance of centrifuges.

SELECTION OF TYPE AND SIZE OF CENTRIFUGE

The choice of type and size of centrifuges is a combined theoretical and practical exercise. In most cases the final choice for a new centrifuge application cannot be made without a test of the type of centrifuge preliminary selected, at between 25 and 100% of the final plant flow rate. There are at least two reasons for this. One is the elimination of the technical risk. The other is most common in the pharmaceutical industry. During the pre-approval inspection, those in the inspecting and license-giving institutions are concerned with the comparability between batches of product from the development phase and from the production phase. Because the validation procedures are costly and time consuming, it is very important to select at least the right type of centrifuge already in pilot scale. It is important that alternative process and plant designs are discussed before the centrifuge type and size are selected.

Main Process Parameters

The *feed flow rate* is often determined by the available or necessary time to process one batch. The *separability* can be estimated on a bench scale by test tube centrifugation (spin test). The *flow of solids* is the feed flow rate multiplied by the percent by volume of suspended solids. This solids flow rather than the suspended solids content alone influences the choice of machine type. Different machine types differ in their ability to produce high *concentration of the separated solids*. The *required degree of separation* has a strong influence on the possible flow rate in a given machine. To increase separation efficiency of *E. coli* from 90 to 99%, the flow rate must be decreased by 30 to 35%. To reach 99.9%, a further reduction of about 30% is necessary. Many cell suspensions, including yeast, have the same narrow particle size distribution and the numbers are the same. With the wider distribution of cell debris, for example, the flow rate reduction percentages are much higher. *Solids rheology,* which is the reaction of the solids to pressure and shear forces, is very important for the choice of machine type.

Pseudoplastic (shear thinning) solids (such as unflocculated yeast, rod-shaped bacteria, and cocci as well as some protein precipitates) will flow out of the machines at 80% by volume of suspended solids, which often has positive consequences (reduced cell breakage). Shear thinning solids may have an angle of repose of close to 0 degrees, but this is not always the case, such as with flocculated or mycelium-forming bacteria. On the other hand, dilatant solids (starch or calcium carbonate) will not be discharged in concentrated form without manual handling from any machines. The rheology of inclusion body concentrates is not much studied, but they are preferably discharged from the machine. Because of the content of a valuable dissolved substance in the entrained liquid, *washing of solids* may be necessary. Another reason for washing solids is, of course, to get rid of impurities. This is usually done by resuspension and further separation steps.

Test Tube Centrifugation (Spin Test)

Determination of separability

The most important task for a spin test is the determination of separability, that is, the settling velocity. In practice, it is difficult to calculate, because often the information needed in Stokes' law is not complete. The spin test is done by spinning a sample of the feed stream for different periods of time or different G-numbers or both. For each of these spins, the supernatant is analyzed with regard to the important process parameter, for example, residual solids, turbidity and cell count, or water content in the case of light solvent. Thereafter, the time of the spin with the acceptable supernatant quality is recorded, and the necessary Σ is calculated according to equation 16.

Spin test as an analytical tool

The spin test is also a valuable tool for measuring the phase-to-volume ratios. The solids content in percent by volume from the spin test will be a very important parameter for the next step, which is to make a crude mass balance around the centrifuge. The spin test performed under different time and *g*-force conditions gives also an indication of solids compressibility and rheology. It is also valuable to prepare a sample of the solids phase after spinning to get a good indication of its dry solids content after the centrifugation. In cases where the fermentation substrate contains suspended solids, residues of these may be present after fermentation. A spin test will make it possible to identify the nature and amount of these, which is very important because they may obstruct the solids phase flow in nozzle and solids-ejecting machines.

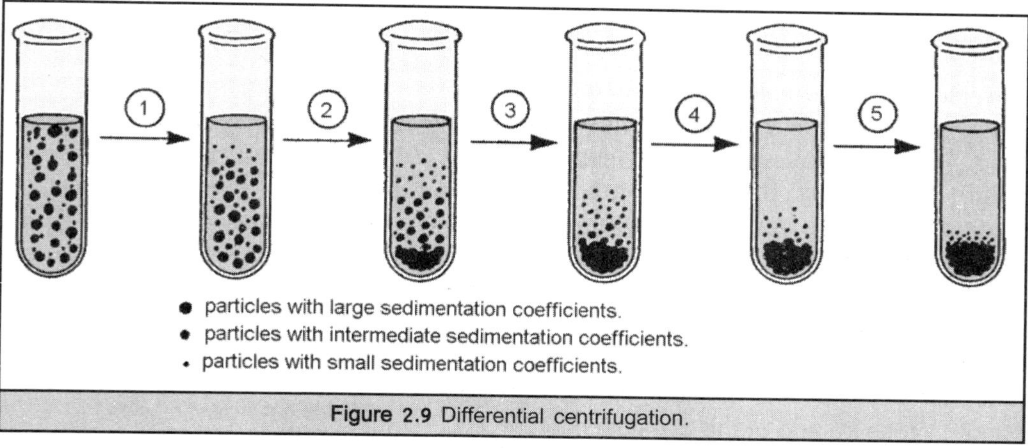

Figure 2.9 Differential centrifugation.

Theory of the test tube centrifuge

The swing-out test tube (bottle) centrifuge is shown schematically in Figure 2.10. In order for a particle to be separated out with 100% chance, it has to move from radius r_1 to radius r_2 in the spin time (T). The shaded volume with height (h) is the volume taken out for sampling of the result. The following equations describe the particle movement.

$$V_c = \frac{dr}{dt} \qquad ...(12)$$

$$V_c = V_g \cdot \frac{r \cdot \omega^2}{g} \qquad ...(13)$$

Insertion of equation 10 in equation 9 and rearranging and integration gives

$$\int_0^T dt = \frac{g}{V_g \cdot \omega^2} \int_{r_1}^{r_2} \frac{dt}{r} \qquad ...(14)$$

$$T = \frac{g}{V_g \cdot \omega^2} \cdot {}^e\log \frac{r_2}{r_1} \qquad ...(15)$$

In the test tube centrifuge, the efficiency factor μ in equation 9 is 100% because the settling process is considered ideal. Therefore, equation 6 is valid and

$$\Sigma = \frac{T \cdot Q_{theor} \cdot \omega^2}{g \cdot {}^e\log \frac{r_2}{r_1}} \qquad \ldots(16)$$

and

$$\frac{Q_{theor}}{\Sigma} = \frac{g \cdot {}^e\log \frac{r_2}{r_1}}{T \cdot \omega^2} = V_g \qquad \ldots(17)$$

For a test tube centrifuge with angle head, ${}^e\log(r_2/r_1)$ is exchanged for ${}^e\log(1 + D/r_1 \cdot \cos\beta)$.

Important Properties of the Process Streams

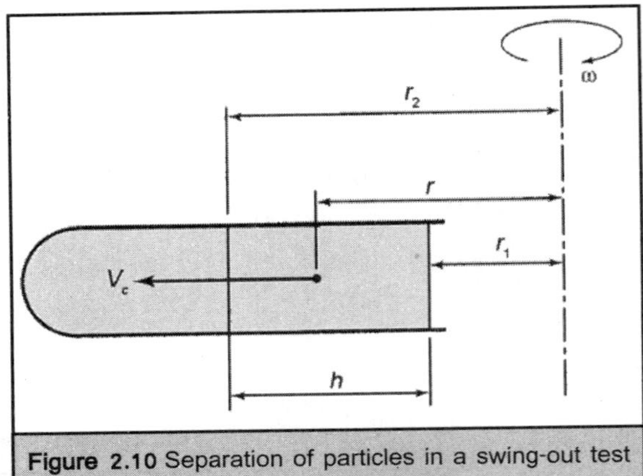

Figure 2.10 Separation of particles in a swing-out test tube centrifuge.

In Table 2.2, some solids phase data, such as *particle size* and *density difference,* have been collected. These data show that many important biological processes belong to the most demanding centrifuge applications regarding separability. The data are very approximate. Cell size will vary with fermentation conditions. Cell debris size distribution is very wide and depends on the method of cell breakage. The density difference between cell and medium depends on the osmotic pressure and of the possible content of inclusion bodies. Apparent density and the density difference of inclusion bodies vary with size, so that larger ones have lower density. As many as possible of the items in Table 2.3 should be defined before the selection of a centrifuge is made. Some of the items need further comments.

Apart from Stokes' law, the *density* is important for the stress in the bowl material. *Low pH* and *high temperature* may, together with even a moderate chloride concentration, lead to difficulties with *corrosion*. In the biological process industries, steels 316 and 316L are often preferred. These steels have, however, such a low tensile strength that they are less suitable for high-speed centrifuges. Instead, duplex steel has been developed, it has high tensile strength coupled with a good corrosion resistance. The corrosive properties of process materials can be very serious, but it is in many cases possible to reduce the corrosion rate by intermittent washing with caustic solution every 4 to 5 hours of running to reestablish a passivating layer on the metal surfaces.

Erosion may be a problem with silt-containing process water or other raw materials such as molasses. This may necessitate the use of hydrocyclones to remove these particles. The *angle of repose* is the angle a cone solid will form if allowed to settle, for example, between the nozzles in a nozzle bowl. If it is 0, or close to 0, degrees, the flow properties are extremely good. The *shear sensitivity* was mentioned in "Centrifuges with Conical Disks". The most shear-sensitive cells are mammalian cells, for which the shear forces must be minimized when the cells are used in perfusion systems.

Table 2.2 Size, density difference, and relative feed flow rate for some solid–liquid systems

Type of solids	*Particle size (μm)*	*Density difference (kg/m^3)*	*Relative throughput (relative to baker's yeast)*
Cell debris	0.2–0.5 (mean, bacteria)	—	0.2
	2 (median, yeast)	—	1
Streptococii	1 x 1	—	1–2
	1 (chain)	—	3–6
Clostridia	1 to 3	—	5
E. coli	1 x 2	70–85	2–6
Actinomyces spp.	10 x 10–143 x 143	1 (for large particles)	1–7
Inclusion bodies	0.7 x 0.7–1.3 x 1.3	29–154	5–8
Corynebacterium	—	—	6
Bacillus spp.	1 x 3	—	6–10
*Aspergillus*spp.	—	—	15
Mammalian cells	20 to 40	70	20
Flocculated material (e.g., debris)	—	—	20
Chlorella	4 x 4	—	30
Saccharomyces (alcohol yeast)	4 x 6	90	50–75
Saccharomyces (baker's yeast)	6 x 8	90	100

Table 2.3 Important process liquid properties

Chemical composition of liquid and solid phases
Densities of liquid and solid phases
Viscosities of liquid and solid phases
Particle size distribution
pH
Temperature
Vapor pressure
Solubility of one phase in the other
Corrosivity
Erosivity
Shear sensitivity
Angle of repose
Toxicity
Explosion hazard

In the Centritech machines, the RCF is limited to 300, more than one order of magnitude lower than for other centrifuges. The cells may be so vulnerable that the residence time in the centrifugal force field must be minimized. Some protein precipitates are also sensitive to shear, including those formed in the maturation tanks in breweries. Many bacteria are sensitive, especially starter culture bacteria, where the activity of the cultures depends on intact cells. Yeast cells are the most robust cells, but they are also sheared to some extent at discharge in radial solids-ejecting centrifuges.

Improvement of Separability

According to Stokes' law, the settling velocity improves when the density difference and the particle density increase and when the viscosity decreases. Bowden lists physical, chemical, and biological techniques for broth conditioning. The most-used techniques are coagulation and flocculation, but other methods that induce cell lysis can improve aggregation of cells, such as aging, pH, and change and addition of enzymes or other chemicals. In some cases it is possible to flocculate submicron suspensions, so that a decanter centrifuge can recover the precipitate (cell debris) with good efficiency and a dry cake.

Pilot Machine Testing

After maximizing the separability of the feedstream, the preliminary Q_{theor}/Σ is determined by spin tests. The machine type and size are tentatively chosen based on the preliminary mass balance and on the requested Q_{act}. With the separation efficiency μ, Q_{act}/Σ can be calculated from the equation

$$\frac{Q_{act}}{\Sigma} = \mu \cdot V_g \qquad \ldots(18)$$

It is often advisable to make a test with the chosen type of machine. The purpose of such a test is to obtain a more reliable Q_{act}/Σ value, but also to assess solids flow properties, emulsification, and particle breakup tendencies and clogging risks.

Scale-Up

For all types of centrifugal separators, the Q_{act}/Σ value can be used for scale-up of the sedimentation performance if some general considerations are taken into account:

1. The type of centrifuge is similar.
2. The geometry is similar; for decanters, feed inlet positions, ratio of length to diameter, or beach angle, and for disk machines, cone angle, disk spacing, and number of feed holes.
3. The bowl speed gives similar linear velocity of, for example, wings when treating shear-sensitive process feeds.
4. The λ value (equation 5) is almost the same in the machines.
5. The larger machine has no hydraulic limitations in relation to the calculated flows. The limitations could be inlet and outlet pressure drops, paring disk pumping capacities (upper and lower limit), and the upper and lower limit of solids flow.

6. There are no mechanical limitations, such as power transmission and gear-box torque in decanters.

The equation to use in scale-up is

$$\left(\frac{Q_{act}}{\Sigma}\right)_1 = \left(\frac{Q_{act}}{\Sigma}\right)_2 (= \mu \cdot V_g) \qquad \ldots(19)$$

where indexes 1 and 2 denote the two machines. Direct calculations of V_g and μ are thus avoided.

DESCRIPTION OF SOME APPLICATIONS

The most important centrifuge application in biotechnology is cell harvesting, where either the liquid phase or the solid phase contains (or is) the product. Downstream of the harvesting stage the number of applications for centrifuges is growing. The operations performed are many: classification, dewatering, liquid–liquid extraction, and precipitate recovery. The applications discussed in this section are examples only; there is an infinite variation in the downstream process design.

Cell Harvesting

The through-put capacity of a given centrifuge (i.e., Q_{act}/Σ) may vary by a factor of 100 depending on the microorganism used.

Yeasts

The second application of disk bowl centrifuges developed after milk skimming is concentration of *baker's yeast*. Today, nozzle machines with pressurized discharge are very common in the industry. The flow rates are the highest in the biotechnology industry, up to 150 m^3/h of fermented broth. Cell concentrations in the fermenters have increased over the years and have now reached above 25% by volume. The baker's yeast is washed, often twice and countercurrently, to increase shelf life. The pressurized discharge of the solids concentrate makes it easy to use an in-line mixer to mix the wash water with the cells. The Q_{act}/Σ is about 2×10^{-7} m/s for a 99.9% reduction of the cell count. In *distilleries*, nozzle bowl centrifuges have been used for a long time. The centrifuge has been the prerequisite for the development of new alcohol fermentation systems. In the continuous fermentation from spent sulfite liquor, centrifuges are used for recycling of the yeast. This is also the case in the batchwise Melle-Boinot process, where the recovered yeast inoculates the next batch. In the continuous Biostil process, in which the distillery slop is returned to the fermenter, the centrifuge recovers the yeast from the feed to the distillation column and returns it to the fermenter.

In *breweries*, centrifuges are used in several process steps. In the separation step after fermentation, green *beer separation*, the cell count can be as high as 80×10^6 cells per mL (2% by volume), but during the emptying of the fermentation tank the solids decrease. Radial solids-ejecting machines are used, equipped with self-triggering and special outlet devices to prevent air contact. At a cell reduction of 99% and 20×10^6 cells/mL, the Q_{act}/Σ is about 2×10^{-7} m/s, increasing somewhat with higher cell count. The flow rates are up to 75 m^3/h. In this step, the same machine type as for baker's yeast could be used if the solids flow to the machine could be controlled. New instruments measuring the dielectric constant or the inductive permittivity may be used. For the separation step after lagering and before final filtration, *preclarification,* the cell count is much smaller, and less solids handling capacity is needed. The hermetic inlet is an advantage because the separation of the shear-sensitive protein particles is helped by the gentle acceleration. The bowl has only one liquid outlet. The solids are ejected at the periphery. Because the emphasis is on cell reduction, the Q_{act}/Σ is about 5×10^{-8} m/s. Flow rates are up to 45 m^3/h. The corresponding separation step is also present in wine-making. Because the same machine unit is used also in must separation before fermentation, radial solids-ejecting machines are used. In rDNA yeast processes, the same type of machine as for baker's yeast can be used. Such machines are available for BL2-LS containment.

Bacteria

In modern biotechnology, solids-ejecting machines, nozzle machines with pressurized discharge, and decanters are used to harvest bacteria. For microorganisms with sizes about 1 μ, axial discharging machines are used. If the solids concentration is so high that the solids discharge frequency has to be more than 1 per 60 s, a radial discharging machine has to be used, but this is less cost effective. The vortex nozzle machine has the disadvantage of accepting only clear substrate media. Residues of substrate may clog the concentrate tubes. If a clear substrate is used, this type of machine is as cost effective as the axial discharging machine but without its solids-handling limitation.

A further possibility for clear media is the hermetic machine, where the cell concentrate leaves the machine continuously and under pressure. The intermittent discharge mechanism is used for CIP. A typical Q_{act}/Σ value for 99% recovery of *E. coli* K12 in a disk stack machine is 10^{-8} m/s. But the variation between various types of *E. coli* can be very large indeed; between 30 and 300% of the above value has been encountered by the author. *Corynebacterium,* somewhat larger cells used in amino acid production, are harvested in nozzle machines with pressurized discharge, often with washing of the bacteria (repulping) and a second separation to improve the yield. A typical Q_{act}/Σ is 5 times that of*E. coli. Actinomyces* mycelia can be harvested by radial discharging machines. As in nozzle machines, the cells are washed, often in two steps, to improve the yield.

There are cases where repulping and reseparation are not possible because the cells break up and may damage the rest of the process. To avoid losing yield, the cells must be recovered with maximum dryness. The obvious choice of machine is then the decanter. The low Σ value of this machine type makes it impossible to use it without flocculation of cells. Polymer addition gives a separation efficiency of 98 to 99% at a Q_{act}/Σ of 3×10^{-7} m/s for a spore-free *Bacillus subtilis* broth with a solids content of 12 to 15% by volume. The dry solids (DS) content of the concentrate is 27 to 28% by weight at 3,200 G. This can be compared to solids-ejecting or nozzle machines, which give around 15% by weight for bacteria. Longer residence time in the bowl or higher RCF gives still higher DS contents. At 10,000 G and similar residence time, a decanter can produce an *E. coli* concentrate of 32 to 36% by weight at a Q_{act}/Σ comparable to that in a disk bowl centrifuge.

The high DS capability of decanters is also used to concentrate solids from nozzle machines or from ultrafiltration units, working on fermenter broths in production of amino acids and antibiotics, respectively. The improvements are yield and less costly waste disposal (e.g., drying costs). Single-cell protein production from hydrocarbons or their simple derivatives is today a not-so-profitable business, but the technology exists. Mayer and Woernle presented a flow sheet for biomass downstream processing. Bacteria are preconcentrated in large nozzle machines to about 6% by weight DS and then flocculated. This biomass is then dewatered to about 30% by weight DS in decanters and goes after granulating to drying. The liquid phase from the decanter is clarified in high-speed separators and sent to sterilization together with the clarified liquid from the preconcentration step. From sterilization it is recycled back to the fermenter.

Mammalian cells

It is possible to use high-speed centrifuges in mammalian cell processes, in which the cells are discarded after removal. Usually, the inlet shear is not large enough to release DNA into the culture fluid. Care must be taken that shear and air contact in the centrate are avoided, because some protein substances may be destroyed. The centrifugation of intact cells for perfusion must be performed at very low pressure and shear forces. Special machines have been developed recently for this purpose.

Cell Debris

In those cases where the intracellular products do not form inclusion bodies, the cell debris often has to be removed after cell breakage. The separation is difficult because of the very fine particles produced. Moreover, the viscosity may be high. Dilution of the suspension reduces the viscosity, but the Q_{act}/Σ may still be 5 to 10% of that of the whole cell. It is necessary to use high-speed solids- ejecting machines for this step. In cases where it is possible to flocculate the debris, the Q_{act}/Σ increases by a factor of at least 10. It has been reported that borax-flocculated yeast cell debris can be removed to about 80% in a pilot-size decanter centrifuge at a Q_{act}/Σ of 4×10^{-4} m/s.

Inclusion Body Recovery and Washing

In cases where the substance forms inclusion bodies, it is advantageous to remove cell debris and other cell constituents before the next processing step. The inclusion body has a higher settling velocity than the cell debris, so that a classifying centrifugation will recover the inclusion bodies while leaving the cell debris in the centrifugate. There is a strong interaction between, on the one hand the homogenization method (i.e., the number of passes) and the resulting particle size distribution and, on the other hand, the ease of washing the inclusion bodies. There exists an optimum number of passes.

The classification split is not 100% sharp, making it necessary to repeat the separation a number of times. This is done either by resuspending the recovered inclusion bodies batchwise or by a continuous simultaneous feeding of the inclusion body concentrate and water into a well-mixed flow through feed tank, continuing until the required purity has been reached. Axial or preferably radial discharging solids-ejecting machines are applied at a Q_{act}/Σ of 1 to 2 x 10^{-8} m/s. To reduce the consumption of chemicals in the next stage, the final separation is often a dewatering step. This can be performed in a high-speed decanter at a Q_{act}/Σ of 2 x 10^{-6} m/s.

Extraction of Cell-Free Broth

Solvent extraction

For 50 years, liquid–liquid extraction has been used to concentrate and recover penicillin from filtered broth. This was often performed in multistage centrifugal extractors. Today, the trend for filtered broth is to use a two-stage, countercurrent system with mixers in between disk bowl machines, which are of the three-phase solids-ejecting type. Many other antibiotics need only a one-stage extraction, for example, erythromycin and tetracycline.

Two-phase aqueous systems

The polyethylene glycol (PEG)/dextrane or PEG/salt systems used in protein purification are characterized by a very small density difference between the phases, from 100 down to 14 kg/m^3. The viscosity of the polymer phase could be high, perhaps 100 mPa s, and a very low interfacial tension, leading to easy splitting of drops. Therefore, it is not surprising that the Q_{act}/Σ may be rather low, from 5 x 10^{-8} down to 1 to 2 x 10^{-9} m/s. The type of equipment used for the systems with low-density difference is radial solids-ejecting machines. Decanters may be used when the density difference is larger.

It is important for optimum performance that the interface between the light and heavy phase in the bowl be in the right position. Equation 4 shows the relation between the three liquid levels in the bowl. The interface should usually be placed just outside of the disk stack in the disk bowl machines. The two-phase aqueous systems can also be used for partitioning cell debris and other impurities from a dissolved protein product. Datar used a PEG 4,000/ potassium phosphate system. No density difference data were given. The Q_{act}/Σ was 2 x 10^{-7} m/s.

Precipitation

Precipitation is often used to isolate and concentrate antibiotics (e.g., rifamycin, erythromycin, and tetracycline). The suitable machine is the radial solids-ejecting disk bowl type. The methods and parameters involved in protein precipitation have been reviewed.

Whole-Broth Extraction

Solvent extraction

Whole-broth extraction of penicillin was the first development of this technique. The solids flow is so high that decanter centrifuges must be applied. Many of the constituents of the broth (e.g., macromolecules and the mycelium) have emulsion-stabilizing properties, so de-emulsifying agents must be added. After extraction, the rich solvent is separated to reduce the water content before further processing. The solvent-contaminated mycelium must be further treated before disposal.

Two-phase aqueous systems

The two-phase aqueous systems can be applied for extraction also in the presence of cell debris. Hustedt et al. reported on three enzyme recovery processes that use cross-current extraction. The phase separation is made in small solids-ejecting or nozzle disk bowl machines. Provided that the density difference between phases is large enough, extraction decanter centrifuges can also be used. Among the commercial uses of this technology on a large scale is the recovery of enzyme in a polymer–salt system, where the separation is performed in a three-phase radial solids-ejecting centrifuge.

INSTALLATION AND OPERATION

Plant Design

In modern biotechnology plants, the design philosophy must be documented as part of the validation procedure. These procedures are becoming common also in the food industry. Stringent guidelines for safe design of plants are being issued more frequently. In Europe, the draft standard for performance criteria for centrifuges in biotechnology deals with leak tightness, cleanability, and sterilizability.

Cleaning-in-Place

Automated CIP is today regarded as a quality guarantee. The validation of CIP procedures is one of the necessary protocols to obtain approval to manufacture bioproducts. An automated and logged CIP procedure is repeatable and traceable, less costly than manual cleaning, and safer for the plant personnel than manual cleaning is. Centrifuges and their peripherals are today designed with cleanability in mind. With use of modern programmable logic controllers (PLCs) for valve control, the CIP of a disk-stack centrifuge (or decanter) has been greatly simplified. A CIP cycle consists of several steps of rinsing with different cleaning liquids. The sequence could be cold water, warm (60-70°C) 4% caustic, 3% nitric acid if necessary, hot water, and finally cold water. It is often necessary to use some type of detergent to improve the cleaning effect, mixed with the caustic or the water. In the radial solids-ejecting machines and the modern pressurized discharge nozzle machines, the discharge mechanism is activated especially during the initial stages.

Containment, Asepsis, and Explosion Protection

These three areas are related in the sense that the installation designer must define a space around the centrifuge, including tanks and piping with controlled interfaces with the environment. In the case of containment, the task is to protect process operators from harmful process substances. The National Institutes' of Health biosafety level is an attempt to classify the risks involved. The interface with the environment is sterile HEPA filters. This is a case of "no leaking out." The reverse case, "no leaking in," applies for asepsis and explosion protection. The barriers are formed by HEPA filters and liquid seal pots, respectively. In both these cases, it is necessary to keep the centrifuge system under a slight, positive pressure relative to the environment, either by lowering the pressure in the room (asepsis) or increasing it in the centrifuge system. For explosion protection, this is done by filling the system with inert gas (e.g., nitrogen) and bubbling the inert gas through a water seal pot. Centrifuges in closed systems have to be modified in a number of ways: mechanical instead of labyrinth seals, cooling jacket around the bowl cover, and devices that can handle the pressure wave at discharge.

Sterilization

Sterilization with saturated steam at temperatures above 100°C is by many considered to be the reliable way for avoiding contamination, and it is prescribed for BL2-LS. For the centrifuge manufacturer, this means that in addition to the modifications described in the previous section, the bowl cover must be designed as a pressure vessel. But the most complicated changes concern the installation, because the bowl internal and external surfaces, the cover, the service liquid lines, etc., must all be sterilized. Toreach the desired temperature (above at least 121°C), condensate must be drained away. Krook et al. discuss the sterilizable system for a medium-sized machine with pressurized nozzle discharge in some detail. The sterilization efficiency for a radial solids-ejecting centrifuge of pilot scale was investigated. It was found that it was not possible to sterilize the whole system at once. By trying different combinations of open and closed valves, regimes were developed for heating different sections, and these were combined to sterilize the whole system. In the end, sterilization was validated with different chemical and biological indicators.

After the development of sterilizable microfilters, there is available an alternative to centrifugation in separating solids and liquids. A general rule is that with increasing particle size and increasing scale of operation, centrifugation becomes more interesting than microfiltration and vice versa. It is conceivable that *E. coli* on a 200-L fermenter scale is a microfiltration application. Many factors must be considered in the choice. What is becoming more apparent is that the choice done in pilot scale becomes the most difficult one, because the scale-up and validation procedures make it easier to keep same technology on the larger scale, even if it is not the optimal choice. In many cases, a combination of a filter and a centrifuge is the best answer.

CENTRIFUGES

With the emergence and growth of animal cell bioprocesses during the past 15 to 20 years, the need has arisen for unit operations that gently and effectively separate animal cells at the process scale. Centrifuge-based separation systems available today are especially well suited to this task. By its nature, centrifuge separation is inherently robust and delivers consistent separation performance. Centrifuge systems now incorporate a number of design changes, including some that are quite novel, aimed at tailoring their performance to the special requirements of animal cell bioprocesses. As a result, today we have a powerful array of options available for centrifuge-based unit operations for use in both the animal cell processes under development and the existing processes that are being improved for added efficiency and productivity.

Procedure

As biotechnology's commercial success has grown, it has created the need for highly regulated, effective, and, most recently, efficient production processes. Such biotechnology production processes (bioprocesses) have historically drawn heavily from other process industries, such as the chemical process industry for concepts, practices, and equipment, including centrifuges. Bioprocess has therefore become, in some cases, a hybrid field where established process techniques are intermingled with laboratory methods that have been multiplied manyfold for the required process scale. Nowhere does this appear to be more the case than for the bioprocess of animal cells. Although bacteria or yeast has been the more common choice for biological production to date, each can have such drawbacks as producing improperly formulated product or product that is unrecoverable as insoluble inclusion bodies. In such cases, animal cells have often been called on to overcome these limitations.

Unfortunately, animal cells do not share some of the more desirable attributes of these other organisms. By comparison, animal cells are slower growing; exhibit lower protein expression levels; are susceptible to infection by a large number of adventitious agents such as viruses, prions, and so forth; and, because of their lack of a cell wall, are much less tolerant of many traditional process operations. Numerous advances in animal cell biotechnology and fermentation have enhanced the productivity of animal cells. But the bioprocess equipment has often lagged. Animal cell bioprocess has therefore become a "force fit" of techniques and equipment originally designed and developed for other purposes. Nonetheless, the fact that animal cells are often the only practical way to produce what is required has generated a strong push by many in the industry to develop processes and equipment tailored to animal cell bioprocess. Centrifugation is no exception.

CENTRIFUGATION

Methodology

Centrifugation's role in animal cell bioprocess is primarily as a unit operation immediately adjacent to the bioreactor. In this role, the centrifuge is positioned to either (i) clarify the cell mass exiting the bioreactor, as in the case of a batch operation, or (ii) retain the cells for further processing, as in the case of a continuous operation. These two operating modes, however, differ somewhat depending on whether the product is intracellular or extracellular. Intracellular processes, where the focus is on harvesting the product-laden cells, require that the centrifuge concentrate the intact

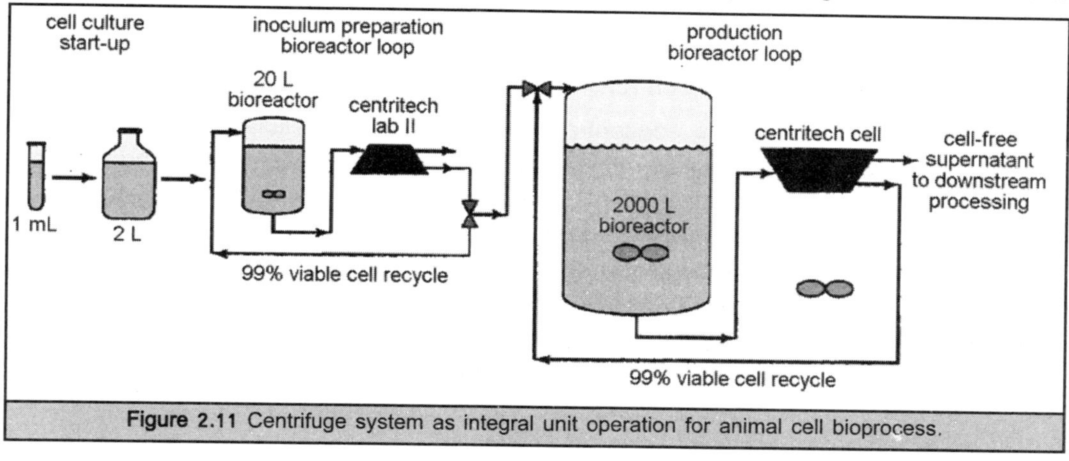

Figure 2.11 Centrifuge system as integral unit operation for animal cell bioprocess.

cells from the fermentation broth. Care is taken to avoid damaging the cells during this concentration step to maximize product recovery. Any ruptured or lysed cells also release agents into the process streams that can be harmful to the desired product, especially when that product is a protein. Occasionally, this type of care is also needed when the cells are to be stored or are otherwise required to remain viable for further processing. In many of these cases, aseptic operation is essential. Extracellular processes, on the other hand, require a different set of operating principles where the key focus is on keeping the cells viable.

Any cell damage that the centrifuge causes in these processes will have a direct deleterious effect on the process productivity. Not only would the producing cells themselves be lost, but here, too, harmful intracellular agents would be released. The situation is further aggravated by the fact that the extracellular product is relatively vulnerable to any harmful agent because both are present in the bulk fluid. It is clear that centrifugation processes that avoid cellular damage are preferable in these cases, especially when operating in a continuous or perfusion process mode. A further distinction in animal cell bioprocess that bears mentioning is that the producing cells can be either freely suspended or immobilized. The earlier discussion relates directly to suspension systems. For immobilized systems, where, for example, the cells require that they be tethered to induce growth or production, the need for centrifugation is either at the end of the process, whenever the tethered cells are released for further processing, or for clarification of debris released by the tethered cells, or both. When called for in this manner, the centrifugation considerations are similar to those for freely suspended cells.

Batch System

Perhaps the most prevalent bioprocess mode in use today for freely suspended cells is a simple batch system using carefully constituted media. But in the past decade or so, two more involved processes, fed batch and perfusion, have gained much of the attention. The fed-batch process tends to make more efficient use of the media expense and other operating costs, whereas the perfusion process, being continuous, tends to make more efficient use of the producing cells and the physical plant investment. The fed-batch process is currently the more prevalent of these two processes, partly because it is easier to operate in the laboratory and at the pilot and production scale. Because a batch culture is by its nature limited in its duration, process experience can be built up relatively quickly during the course of several runs. Here the need for processing the cells is confined to the end of the production run.

Centrifugation may be used to concentrate the cells for intracellular product or clarify the supernatant for extracellular product. In either case, as mentioned earlier, avoiding cell damage would likely lead to important productivity gains. Perfusion processes afford the benefits of continuous operation. To do so, however, perfusion requires an effective way to retain the cells throughout the production run with little or no cell viability loss. Centrifugation can deliver the needed performance provided it is continuous in both the liquid stream and the cell concentrate stream, can be operated aseptically, and does not damage the cells. The specific designs of centrifuges applicable to animal cell bioprocess are reviewed in the next section with reference to these important aspects.

The considerations of such processes are directly applicable to anchored cells when those cells are attached to microcarriers. Cells attached to microcarriers are, however, particularly sensitive to fluid shear forces. For example, cells that are not otherwise damaged by fluid shear are lost if the shear that is present separates them from the microcarriers and they do not reattach. In these cases, the bioprocess centrifuge needs to treat the process fluid especially gently. The emergence of insect cell bioprocess has added yet another dimension to the need for gentle handling of the cells. Although insect cells have several advantages over animal cells, they tend to be even less robust than animal cells are during rough handling in the process. Coupled with the fact that insect cells typically generate intracellular product (specifically so with baculovirus systems), the centrifuge operation must generate especially low shear forces to avoid cell rupture and the resulting productivity loss.

Design of Centrifuge

Basic approaches

Centrifugation is a very robust and versatile separation technique. Because it relies primarily on density differences, centrifugation can readily account for changing process conditions by simple modulation of the intensity or duration of the *g*-field as needed during operation. At lower speeds and *g*-forces, the centrifuge can be used to gently remove the larger and more dense cells from the process stream; at higher speeds and *g*-forces, the centrifuge can be used to spin down ever smaller cell debris to increasingly clarify the process stream to the degree required. The actual design of the centrifuge, however, depends on the various optimization trade-offs specific to the process application. Typical design trade-offs include high sample throughput versus instrument size, high centrifugal speeds versus large load volumes, versatility versus dedicated instrument specialization, batch versus continuous operation, and, for continuous operation, solids-retaining versus solids-ejecting design.

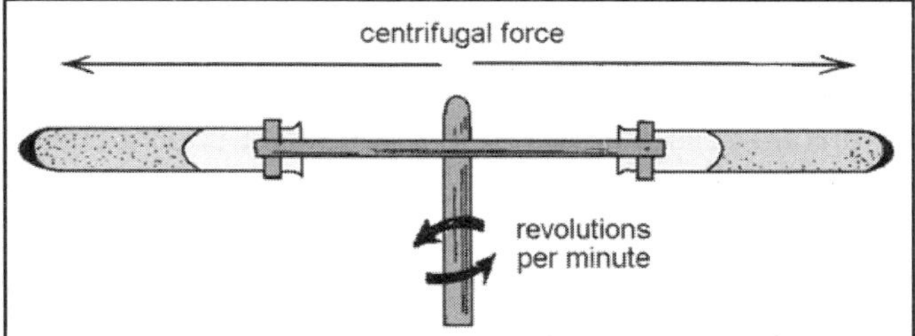

Figure 2.12 Centrifugation. Extremely rapid rotation of liquids in centrifuge tubes produces a force in the tubes analogous to that of gravity, but much more powerful. Thus, suspended particles more dense than the liquid in a centrifuge tube separate (sink) rapidly. Following centrifugation, the supernatant fluid in the tube can be decanted (poured off), leaving the pellet of dense particles in the bottom of the tube.

Laboratory-scale applications tend toward smaller, more versatile, batch or continuous solids-retaining systems, whereas pilot- to production-scale applications tend toward high throughput, specialized, continuous solids-ejecting systems. When applied specifically to the separation of animal cells, centrifugation typically involves the added considerations of avoiding high fluid shears and maintaining high aseptic integrity. Additional design trade-offs are therefore involved, including (i) increased volume throughput versus maintaining the shear forces below critical levels that can result in cell death or damage, and (ii) providing a totally closed aseptic system versus relying on rotating seals, a source of potential contamination and heat generation as well as generation of high shear stresses.

Design approaches

To date, the design of animal cell centrifuges has been approached from three distinct directions. The first two directions have been through engineering extensions of other centrifuge equipment. The third is a centrifuge design specifically engineered for animal cell bioprocess. The first direction has been to take centrifuge designs for larger process scale or bacteria and yeast and adapt them for animal cells. This approach has been more or less successful depending on whether cell damage was to be avoided or continuous operation was desired. Such centrifuges tend to spin the cells down into a paste or cake, which requires some type of cell removal process such as scraping or flushing out of the cells under high shear forces. The second direction has been to take centrifuge designs for blood component manufacture and adapt them for animal cells.

Blood component manufacturers have been making commercial use of continuous-flow centrifugation for many years now. Numerous such blood separation system designs have focused on cell concentration, supernatant clarity, throughput, and other issues that are also important to animal cell bioprocess. But there are several key differences between blood component manufacture and animal cell bioprocess. Blood cells, for example, flow much more readily than typical suspended animal cells. Animal cell centrifuges therefore require some type of low shear method to nudge the animal cells out of the centrifuge in a continuous mode. Also, formed elements of blood are more tolerant of shear stresses than are most animal cells in bioprocess. Flow channel designs in the centrifuge that work for blood therefore tend to be not at all suitable for animal cells. And the usual time frame for blood separation (on the order of a few hours) is much less than that necessary for animal cell bioprocess (on the order of days or weeks). The third direction was taken by Alfa-Laval Separation AB during the 1980s when a continuous- flow centrifuge design was developed specifically for animal cell bioprocess.

Combining some of the features of larger process centrifuges and those of the smaller centrifuges, this unique design focused on maintaining a continuous flow of both the clarified supernatant and the cell concentrate. Key engineering issues that had to be overcome were (i) gently coaxing the concentrated animal cells to smoothly exit the system without significant cell damage; (ii) sustaining continuous operation over days and weeks; and (iii) creating a totally closed system for inherently safe aseptic operation. These three approaches have had the common goal to overcome the limitations of laboratory-scale and even small production–scale centrifugation. At these smaller scales, research centrifuges have often been pressed into service. Several companies provide floor-model centrifuges that can process, for example, as much as 6 L at *g*-forces more than adequate for cellular separations. Kendro, through its Sorvall brand, has recently introduced a series of major software packages to their centrifuges to provide automatic process documentation for easier and more reliable process control and regulatory compliance.

Other suppliers are now providing software as well. These centrifuges are designed for batch operation only. Some of these companies offer continuous-flow models, but they tend to have small throughput capacity, do not eject the cellular material, and are typically difficult to set up and clean.

Because one of the keys to successful animal cell centrifugation is the gentle handling of the cells, there have been over the years several research groups exploring the various aspects of flow-induced shear stress on different animal cell lines. Although it has been shown that cell damage in bioreactors can be attributed to bubble breakup and other gas–liquid interface phenomena, other findings indicate that cell damage can also occur at shear stress levels at lower levels of energy dissipation. Shear stress susceptibility, moreover, varies significantly among cell lines, thereby adding shear sensitivity to the list of key criteria for selecting a particular cell line for process scale-up.

Current products

Several suppliers provide centrifuges for animal cell bioprocess that are based on proven designs for other applications. These designs typically provide high throughput and *g*-forces, clean-in-place (CIP) and steam-in-place (SIP) capability, and some method for removing the cellular material that accumulates in the centrifuge. More complete descriptions, including specific schematic depictions of how these systems operate, are available through the respective suppliers. At the small, batch end of the spectrum, Alfa-Laval's AS-26 SP Super Centrifuge is capable of high throughput rates in the range of 300 to 500 L/h. The AS-26, however, is strictly a batch centrifuge because it retains the solids that must later be removed by manual intervention. Also, the animal cells are removed as cell debris, so this centrifuge would be unsuitable for processes where cell viability after separation is desired.

At the *pilot scale,* Alfa-Laval's BTPX 205 system uses the classical disc-stack design to generate up to 12,800 x *g* and a maximum throughput of 1,200 L/h. The 205 has a solids-ejecting design that is tailored to produce very dry solids. It is therefore well suited for, but limited to, batch animal cell processes where cell viability is unimportant. Although its solids-ejecting capability would allow the 205 to be considered for use in a continuous animal cell process, its rough cell handling would result in unacceptable levels of cell damage. At the *production scale,* Alfa-Laval's BTAX 510 system delivers up to 15,000 x *g* and a maximum throughput of 4,000 L/h. The 510, like the 205, uses a disc-stack, solids-ejecting design. So the 510 process applicability is, at its larger scale, similar to that for the 205.

Westfalia Separator AG's CFA-01 separator is also a continuous-flow system based on a disk-stack solids-ejecting design. Specific investigation has demonstrated the effective use of this system for animal cell processing. This system, however, like the systems discussed earlier and the decanter designs discussed next, requires a rotating seal. To remove the frictional heat from its seal, the CFA-01 requires a sealing liquid that must be highly purified to maintain aseptic operation. These key considerations combine to significantly increase the complexity of the overall system. Later work with the Westfalia CSA-1 disc-stack centrifuge came to mixed conclusions and reported heating issues.

Decanter centrifuges, such as Alfa-Laval's Sharples Super-D-Canter systems, use a solids-ejecting design that differs from that of the disc-stack systems. The decanter design displaces the solids by pushing or scraping them out of the centrifuge using a type of helical blade arrangement. Using this design, high animal cell viabilities may be obtainable in the ejected solids. But these solids are not well suited to continuous processes where the cells must be resuspended, for example, to be introduced back into the bioreactor. *Carr Separations,* Inc., as a relative newcomer, has developed a different design specifically for bioprocess as well as for other pharmaceutical and food applications. The Powerfuge P6 Separation System uses a new solids separation design that can generate very dry solids. This new design also allows for easier CIP than traditional disc-stack designs, an issue that can be a critical drawback for the original design, especially in the larger-scale equipment.

The Powerfuge is capable of generating up to 20,000 x *g* and throughput rates of 60 L/h. Recently, Carr introduced a smaller, low-end bench-top version of the Powerfuge called the Pilot Separator. For animal cell process, the suitability of these systems is similar to that for decanter centrifuges. Kendro, through its Heraeus brand, offers a pair of bench-top centrifuges designed to operate with a continuous-flow rotor system that can generate 24,000 x *g* and a flow rate of 54 L/h. These units are compact and readily portable. They are continuous in the liquid phase but are not designed to eject the solids. Although popular for laboratory-scale work, these units can provide only qualitative support to any planned scale-up plans during process development. Baxter International, Inc. has modified the design of one of their blood component systems to focus on animal cell bioprocess at the laboratory and smaller production scales. Called the Cell Harvester, this system can be used to develop centrifuge-based cell separation protocols and to prove out process concepts. Because manual intervention is needed to remove the cells, the applicability of this system to continuous processes is extremely limited. One key advantage of the Baxter system, however, is its totally closed system design. This design allows for the use of a consumable to avoid the various issues that accompany the need for providing CIP/SIP.

The third, unique design approach mentioned earlier that was taken by Alfa-Laval in the 1980s, also uses a consumable insert. These disposable inserts are provided sterile so that, upon making sterile connections to the system, the user has created a totally closed, sterile flow path and separation zone for the process fluids. This design approach, called Centritech by Alfa-Laval, is now available through and supported by Kendro under its Sorvall brand. Key trade-offs of the Centritech design are very gentle cell handling and a totally closed, inherently aseptic system versus high throughput and *g*-force. Centritech technology is embodied in both a smaller bench-top laboratory-scale system and a larger floor-model pilot- to production-scale system. Depending on the application, however, the laboratory- scale unit has been and can be used for small production processes.

ALTERNATE BIOPROCESS UNIT

Filtration has been the traditional choice as an alternate bioprocess unit operation to animal cell centrifugation. The most common form of filtration used to separate the cellular material from the fermentation broth is tangential flow filtration (TFF). The foremost advantage of using TFF is that it is a technology that has been widely used to good effect. TFF does, however, have the disadvantage that, as for all filter-based systems, fouling and clogging can be a constant and often unpredictable problem if not sized properly. Filter systems also deliver performance that degrades over the course of the run, necessitating either that they be reconditioned or the run stopped to replace the filters. Even when reconditioned, filters do not usually return to their original performance, so operating them becomes a classic case of diminishing returns. Recently, issues such as documenting well-characterized products and other concerns have made the reuse of filters less desirable. In these cases, the only viable option is to stop the run, change the filters, resterilize the process, and start up again.

A further disadvantage of TFF is the inherent presence of shearing at the interface caused by the membrane. Material buildup can affect the way the system works, but a substantial shear remains across the boundary layer. Such shearing can run counter to the low shear requirements for animal cell processing and also requires that a recycle loop (with a pump) circulates the process fluid many times greater than that necessary for continuous (perfusion) operation. Another alternative to animal cell centrifugation is the use of spin filters. Spin filters have the very attractive aspect that as a cell retention device for perfusion processes, they can be integral to the bioreactor. Thus, a single unit operation is created. Cell damage is a concern, but evidence suggests that with proper design and operation this issue can be minimized or even overcome.

A disadvantage of the spin-filter systems is that they too, like TFF, are susceptible to fouling and clogging. What can be worse, however, is that when a spin-filter system integral to the bioreactor clogs unexpectedly, the necessary maintenance typically requires that the bioreactor containment be breached. The risk of infection from such a breach can be significant. Another disadvantage is that the scaleability of spin-filter systems is limited. They may not be a practical unit operations choice if larger scales are anticipated during process development or once in full production. Another device that has been used for cell retention in continuous processes is the inclined separator. The design of this device is mechanically simple, can be made to be relatively easy to CIP and SIP, and, because it relies on gravity (i.e., 1 x *g*), has many of the same process benefits as centrifugation. A drawback with this technology today appears to be its relative lack of throughput scaleability and, of course, that its separation efficiency is limited to that achievable at 1 x *g* (vs. centrifugation at 100 x *g* or more).

FATE OF CENTRIFUGATION

Going forward, centrifugation is likely to be as commonly used for bioprocess as it is for the comparable chemical process industry. The inherent advantages of centrifugation (such as robustness and versatility) continue to fuel creative new concepts for harnessing these advantages for animal cell bioprocess. Novel centrifuge designs and the unique continuous methodologies for bioprocess will likely combine to make the centrifuge a necessary unit operation in a large number of the new animal cell bioprocesses being developed in the near future. Moreover, recent changes in the FDA regulations are conducive to the accelerated adoption of new and more efficient process methodologies. Although in the past there were often heavy disincentives to consider the type of process improvements that, for example, continuous-flow centrifugation allows, many of the regulatory-based concerns have been reduced or eliminated. As a result, we are likely to see adoption of new, higher productivity process improvements in existing processes as well as the development of new ones.

3

BIOLOGICAL PROCESS MONITORING

Biological process monitoring plays a key role for most modern industrial production methods. The bioprocess systems are no exception to this rule. Effective methods of monitoring are required in order to develop, optimize, and maintain biological reactors at maximum efficiently. An optimized process should lead to streamlined performance, reductions in running and material costs, and improvements in quality control. A number of bioprocess monitoring instruments have been regular features of the industry for many years, in particular, sensors capable of measuring physical parameters. The challenge is to develop cost-effective devices capable of measuring a much wider range of parameters, providing process operators with a deeper insight into the process. Eventually effective monitoring will lead to better control systems, providing cost and quality benefits.

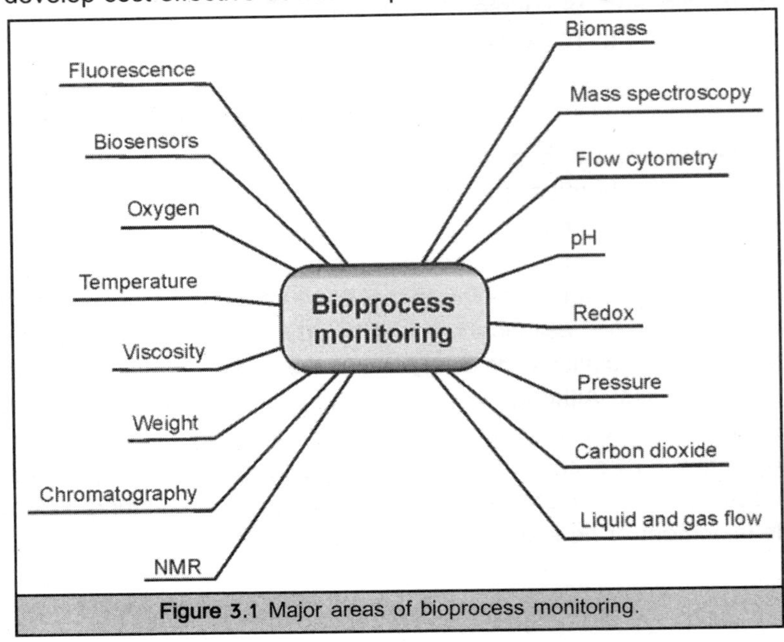

Figure 3.1 Major areas of bioprocess monitoring.

OPERATING CRITERIA

For any measuring technique, there are a number of criteria that the device must satisfy if it is to accepted by commercial bioprocess operators. The following list of desirable characteristic is not exhaustive, but does cover the main areas of concern.

Reliability

Reliability is a key issue for bioprocess monitoring equipment. Confidence in the ability of an instrument to maintain its credibility in terms of performance is a fundamental parameter. This encompasses a number of features relating to the operation of the instrument, for example, ease of use, maintenance, repair, and replacement. A successful monitoring instrument will have a low failure rate; again, this is related to reliability.

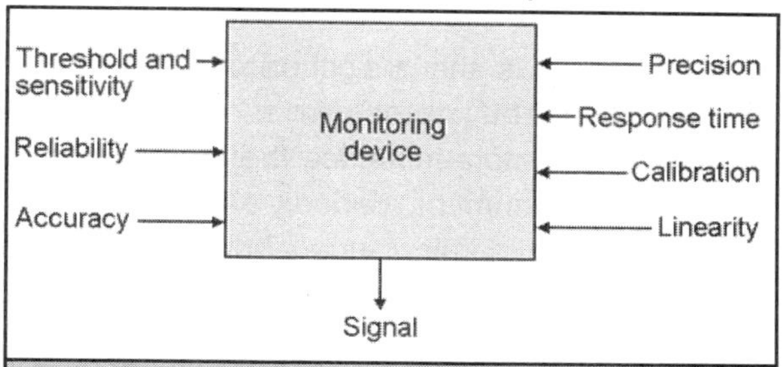

Figure 3.2 Main criteria for a bioprocess monitoring instrument.

Accuracy

Accuracy can be described as the relationship between the measured value (by the instrument) and the actual value of a bioprocess variable. An accurate instrument will achieve a low percentage error between these two values. Generally this can be stated as

$$\% \text{ error} = (\text{measured value} - \text{true value})/\text{true value} \times 100\%$$

Precision

Precision is a measure of instrument reproducibility, that is, the ability to obtain the same value with repeated measurements of a process variable (at a constant level). It follows that a precise instrument may not, necessarily, be accurate. Therefore it is important to distinguish between these two parameters.

Response Time

The measurement of any process variable will entail a time delay between change in the parameter and display of the measured value. This response time should not be detrimental to the progress of the bioprocess, particularly if the measurement is linked to a control action. Response times will be affected not only by the type of instrument but also by the method of measurement. Off-line sampling and measuring can involve a number of time-consuming steps. In contrast, in situ devices can provide a real-time measurement.

Calibration

In order to maintain the accuracy of a sensor it is (generally) desirable to carry out a calibration step. This is carried out using set standards of comparison. The effects of calibration can impinge on the process, in other words, it may be difficult to carry out calibration of an in situ sensor, whereas an on-line device could easily incorporate a calibration step(s) during routine running.

Linearity

Under ideal conditions, the output signal from a sensor would be directly proportional to the analyte concentration. However, this is not always the case, and other models have to be used to reach a true value. Despite this drawback, linearity is not an overriding prerequisite for a successful monitoring instrument. Developments in modern software can be adapted to compensate for nonlinearity.

Threshold and Sensitivity

Sensitivity is the magnitude of the output signal per unit change in the target analyte concentration. The lowest level of detection is related to the sensitivity and the signal-to-noise ratio. A number of factors influence the sensitivity of a monitoring device, including sensor design, the operating environment, periods of maintenance, and interfering noise levels. The sensitivity will influence the dynamic range above which the device becomes saturated; the highest level of detection will be the threshold limit for the device.

METHODS OF MONITORING

In addition to the variety of monitoring devices available, a number of methods (for carrying out the measurement exist. Sampling and sample handling is a vital issue, affecting both the accuracy and frequency of measurements. Broadly, the main approaches to sampling can be described in one of several ways.

Off-Line or At-Line Monitoring

Off-line monitoring involves taking a sample from the bioreactor and carrying out the measurement at a different site, usually under laboratory conditions. Off-line sampling can be detrimental in terms of cost, efficiency, and threat to asepsis. Manually removing and measuring a sample requires technical labor. The received signal will not be "real time" because of the delays. In order to obtain the sample, the sterility of the bioreactor must be maintained. Hence, provision must be made to achieve this. In some cases, decentralized equipment has become available that allows measurements to be made simply close to the production process (at-line), reducing some of the delays associated with laboratory analysis.

In Situ or In-Line Monitoring

In situ sensors are placed directly in the bioprocess vessel. The use of in situ sensors has long been established in the bioprocess industry, where "dip-in" devices are used to monitor a number of parameters such as pH and dissolved gas concentrations. A number of advantages are gained by operating sensors in this fashion, including real-time monitoring with the sensors operating continuously (any rapid change in the analyte concentration can be readily observed), and labor requirements and problems of contamination are both significantly reduced. However, in situ sensors can have a number of drawbacks, for example, the sensor system must be amenable to sterilization; if the lifetime is short, replacement may be difficult (during a process run); the surface of the sensor could become fouled by components of the growth medium, affecting the signal output.

On-Line Monitoring

Methods of on-line monitoring involve the automatic removal and measurement of sample or sample stream from the bioreactor. One example of this has been the development (since the early 1980s) of flow injection analysis (FIA). This is a liquid handling technique that has proved flexible in adapting to most chemical and biochemical reaction procedures, representing an effective compromise between the desirability of in situ monitoring and the technical ease of off-line measurements.

The main advantages of on-line methods include the following: sensor sterilization can be readily accomplished, sample pretreatment (e.g., gassing, dilution, removal of interferents), and sensor calibration can be built into the system. The main disadvantages are a need for an effective and reliable sampling system and the fact that the signal is discontinuous; frequency measuring rate is determined by the limitation of the overall FIA arrangement. All of these methods have their advantages and disadvantages. The choice of approach adopted depends on a number of factors, not least of which will be the availability of both sensor and system.

MEASURING DEVICES

pH Measurement

Monitoring the pH of a number of industrial biological reactions is an extremely effective way of determining their course, providing the operator with valuable information. Measuring pH, along with temperature, is one of the most common practices in bioprocess monitoring. Correcting action can be taken, either manually or automatically, to prevent unwanted increases or decreases in pH. The most common form of pH sensor used for fermentation monitoring is based on the electrode design introduced by Ingold in 1947. Under operating conditions, the probe must always be over-pressurized (with respect to the fermentation broth) to keep a positive flow of electrolyte out of the porous plug. A combined glass and reference electrode that is capable of withstanding repeated sterilization at temperatures of 121°C and pressures of 138 kN/m^2 is required. In this case, the term *combined* refers to the single unit containing both electrodes.

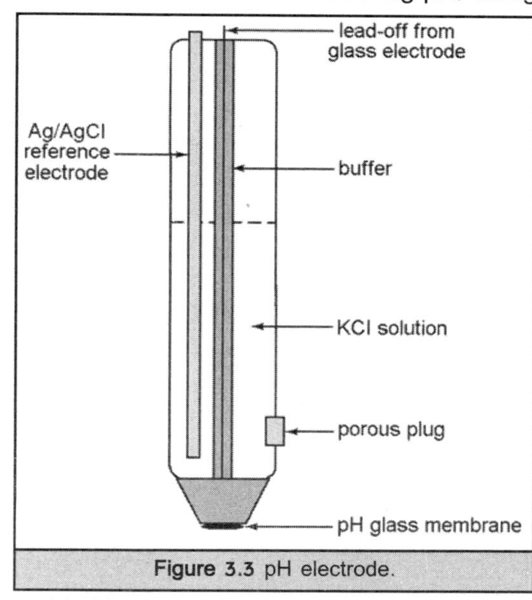

Figure 3.3 pH electrode.

Typically, the reference electrodes used are silver/silver chloride (Ag/AgCl) or calomel (Hg/HgCl) and maintained in a concentrated solution of potassium chloride in a buffer solution within the body of the sensor. Under these conditions an equilibrium is established with the electrolyte:

$$Hg/HgCl_2/KCL \rightarrow 2Hg + 2Cl^- \; Hg_2Cl_2 + 2e^- \qquad \ldots(1)$$

This leads to a situation whereby the reference electrode potential is dependent on the activity of the chloride ions.

Generally the sensor format is that of a single unit glass-reference design. It has been known since the early years of this century that certain forms of glass are electrically conducting.

Furthermore, the electrical potential at the glass-liquid interface changes with activity of hydrogen ions. This potential is directly related to the activity of the H^+ via the Nernst equation,

$$E = E^0 + (RT/\eta F) \ln aH^+ \quad ...(2)$$

where E is the measured potential, R the gas constant, T the absolute temperature, F the Faraday constant, and ln aH^+ the activity. With this type of pH electrode, the potential determining process is determined by interactions between H activity (in the sample solution) and sianol groups on the glass.

Autoclavable pH probes are available (for sterilization purposes), but this can lead to a reduction in useful probe life. Generally the sensor's response does degenerate over time with repeated use. Interactions with the medium constituents can lead to problems. However, proprietary cleaning agents are available to regenerate the probe. A recent advance in the field of pH monitoring has been the introduction of sensors using ion-selective field effect transistors (ISFETs). These are solid-state devices based on the use of semiconductor structures. The pH is determined by measuring the change in the voltage required to maintain a zero current between the gate and drain electrode as the transmembrane potential alters. Several commercial companies have marketed ISFET-based pH meters with potential applications in a number of areas. These sensors are robust, virtually shatterproof, and are easily cleaned. In addition, they require no filling solution and are easy to operate and maintain.

Temperature

Maintaining optimum conditions for any bioprocess will invariably involve monitoring and controlling the temperature of the broth. Generally, bioprocesses are monitored over a temperature range of between 0°C and 100°C (excluding the sterilization cycle). Furthermore, this may require a control regime operating within a narrow range of temperatures. A number of devices are available for obtaining an accurate measurement of temperature conditions during a bioprocess operation, based on a range of techniques.

Thermistors

Thermistors are semiconductor devices composed of agglomerates of polycrystalline metallic oxides (e.g., magnesium oxide, cobalt oxide, ferric oxide). The operating characteristic of these devices relies on a sensitive change in resistance, resulting from temperature fluctuations. This change is a function of the absolute temperature. The output response from thermistors is of a nonlinear temperature versus resistance curve, with the resistance decreasing as the temperature increases. Generally the response can be described by the following equation:

$$R(T) = R_0 \exp[B(1/T - 1/T_0)] \quad ...(3)$$

Thermistors tend to be cheap and stable, with rapid response times.

Resistance thermometers

Resistance thermometers are based on the changes in the electrical resistance of metallic conductors, with changing temperature. This relationship can be described by the following equation:

$$R = R_0(1 + \alpha T) \quad \ldots(4)$$

R is resistance. *T* is the temperature, and *a* the temperature coefficient of linear expansion.

Platinum is the preferred metal for these devices, but other metals can be used (e.g., nickel, tungsten, or copper). The advantage of platinum is that it can be obtained in a highly pure form, hence the electrical properties of the metal can be well characterized. A platinum wire of 100-Ω resistance at 0°C is typically used. Generally, calibration of the sensor is carried out at this temperature. Readings can be obtained using a Wheatstone bridge circuit. Overall, these sensors tend to be precise, accurate (± 0.25%), reliable, and sensitive. Platinum sensors are stable in both sterilization and fermentation conditions.

Thermocouples

Thermocouples are based on the See-beck effect, first postulated in 1821. It was noted that a circuit consisting of two dissimilar metals joined together at junctions maintained at different temperatures will generate an electrical current. The electromotive force generated will be determined by the choice of metals and the respective temperatures. For practical applications, one of the junctions (termed the *cold junction*) is maintained at a constant temperature and used as a reference point. The second junction is in contact with the media to be monitored; fluctuations in the temperature can be studied by monitoring the current response. The response (*EmV* versus *T*[K]) approximates to a parabolic curve. Generally it is preferable to measure over the linear range of this response. One advantage of therm o- couples is that they do not require an external electrical supply. However, they are susceptible to cold junction problems, with the need to maintain a constant temperature.

Mercury-in-glass thermometers

Mercury-in-glass thermometers can be used directly in small batch fermenters. However, the fragility and potential toxicity of these thermometers restricts their use. In order to overcome the problem of using mercury, alcohol can be substituted. Generally these sensors are not very accurate (especially over a wide range) and provide only an indication of the temperature. However, these thermometers have the advantage of being both portable and providing a direct reading.

Bimetallic thermometers

Bimetallic thermometers usually consist of a bimetallic helical coil surrounded by a protective tube or wall. The coil winds or unwinds with changes in temperature and causes movement of a fixed pointer. By affixing a pen to the fixed point, a temperature profile of the bioprocess can be obtained. Despite being more robust than the mercury thermometer, they are less accurate (the range determined by the metals used) and, again, are limited to giving only an indication of temperature.

Pressure

Many bioprocesses operate under conditions of overpressure. Monitoring the magnitude of this pressure is an important factor, both in terms of safety and optimization of the process. Industrial and laboratory fermenters are designed to operate up to a safe working pressure. Increasing the applied pressure above the upper limit can be dangerous. Furthermore, maintaining a positive reactor head pressure can prevent contamination of the bioreactor. In order to facilitate effective sterilization, fermenters need an accurate pressure monitor. Pressure will also affect the solubility of gases (such as O_2 and CO_2). Several different approaches have been used in the development of pressure gauges.

Bourdon tube pressure gauge

The Bourdon tube pressure gauge is composed of an arched or spiral-shaped tube. Part of the tube is of an elliptical shape, made of either metal or quartz. Under pressure, the elliptical section tends to become circular. The interior of the tube is influenced by the process pressure, whereas the exterior is under atmospheric pressure. The overall effect of the change in cross-sectional area is to straighten out the curved shape of the tube. One end of the tube is fixed to an inlet from the reactor. The other end is sealed and allowed to move; this is connected (via a geared section and pinion) to an indicator needle. Pressure can be read directly from the gauge. These devices tend to be robust and economical.

Strain gauges

Strain gauges are based on the phenomenon that when a wire is placed under strain, the electrical resistance changes. This is due to both the alteration in geometry and resistivity changes. The electrical output can be related to changes in pressure. Typically the sensors are constructed using either a metallic or semiconductor element, usually in a grid structure composed of a number of conductor threads. Generally the change in resistance to pressure can be related by the following equation:

$$\Delta R/R = \Delta L/L + \Delta\rho/\rho - \Delta S/S \qquad \ldots(5)$$

R is resistance, L is length of thread (a multiple of the number of threads), ρ is the resistivity of the material, and S is the cross-sectional area of the material.

Piezoelectric manometers

Certain materials, such as quartz crystal, posses an asymmetrical electrical charge distribution. When an alternating electric field is applied across a piezoelectric material, small mechanical deformations (to the crystal) are generated. Alternatively, if the crystal is subjected to pressure on one face, an electrical charge will be generated on the opposite face. This effect (the piezoelectric effect) forms the basis of operation for piezoelectric manometers. By attaching electrodes to one face of the crystal, changes in pressure can be relayed as an electrical output.

Diaphragm-type and pressure bellows sensors

Diaphragm-type sensors incorporate a membrane that deforms as a result of pressure from either a liquid or gas, forming a pressure difference across the membrane. This movement can be monitored and a corresponding readout obtained, indicating the pressure of the matter under investigation. Generally, the membranes used can be made from materials such as stainless steel, palladium, or silicon. When in operation, these sensors can suffer from the deposition of material on the surface, leading to erroneous measurements. In order to prevent this, periodic checks should be made.

Pressure bellow sensors operate, as the name implies, by monitoring the movement of a bellows when subject to a pressure. Changes in the bellows extension length can be related to pressure. The bellows can be linked directly to a variable transformer that is capable of recording the physical movement and displaying a readout.

Viscosity

Information on the rheology, or viscosity, can help in ensuring the efficiency of a biological process. An example of this would be the effectiveness of oxygen transfer throughout a medium and determining the degree of branching of filamentous microorganisms. In addition, viscosity can effect pumping, mass transfer, and mixing. Viscosity is the apparent shear resistance between adjacent layers of liquids or gases moving at different speeds. Typically, in fluids this is the result of molecular cohesion; rising temperatures lead to a decrease in viscosity. In contrast, viscosity increases for gases under conditions of rising temperature. This results from an increased measured molecular activity.

Newtonian fluids

Consider two parallel surfaces with area *A* and distance *Y* apart. If a shearing force *F* is applied to one side, it will move with a velocity *V*:

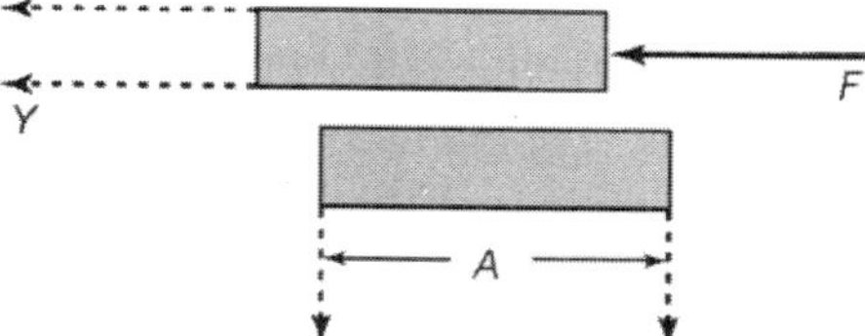

It is found that for ideal or Newtonian fluids the shear stress $F/A(\tau) = dV/dY$ multiplied by a constant (μ). This constant is called the coefficient of absolute (or dynamic) viscosity. The term dV/dY is called the *shear rate* (γ). Plotting τ against γ gives the flow curve characteristics of a particular fluid. Newtonian fluids display a straight-line response, passing through the intersection of the axis. The slope of this gradient is equal to μ.

Typical units of viscosity are Newtons per second per square meter (NS m^{-2}) and centipoise. Examples of Newtonian fluids include all gases, liquids with low molecular weight (notable exceptions are the colloidal suspensions and polymeric solutions with long molecules, e.g., specially developed mineral oils), water, and some oils.

Non-Newtonian fluids

Non-Newtonian fluids are classed according to their flow curves, which may be of various shapes according to the nature of the fluid. For each fluid the ratio between shear stress and shear rate is not constant.

Bingham plastic

For these fluids a certain minimum shear (τ^0) is required before viscous flow takes place, after which the relationship is linear. Examples include some fats and plastic melts, some greases, and detergent slurries. Perhaps the best everyday example is toothpaste, a certain amount of pressure must be applied to the tube before the paste begins to flow.

Pseudoplastic fluids

Apparent viscosity decreases with shear rate, tending to a constant value at high shear rates. This behavior is typical of suspensions of long-shaped particles of high-polymer solutions such as cellulose derivatives. The process has been explained as the result of progressive alignment of particulars or molecules as shear rate increases. Examples include starch suspensions, biological fluids, and rubber solutions.

Dilatant fluids

In contrast to the pseudoplastic fluids, dilatant fluids display an increasing apparent viscosity with increased shear rate. Examples include sugar solutions, starch, and wet cement aggregates.

Casson plastic

These fluids display a decreasing apparent viscosity with increasing shear rate after a yield stress (τ^0) value has been exceeded. Examples include blood, printing ink, and some fruit juices.

Generally, bioprocess media display non-Newtonian behavior, that is, a nonlinear response to shear rate. These fluids do not have a constant viscosity, it varies according to the shear rate. In practice, because of the problems of maintaining sterility, most viscosity measurements are made off-line. There are a number of commercially available instruments (viscometers), applicable to the measurement of bioprocess rheology properties. Generally, these instruments can be classified into three different types:

1. *Cone and plate viscometer.* This instrument consists of a flat plate in contact with an inverted cone, the fluid to be measured fills the gap between the cone and the plate. In operation, the cone is rotated in the fluid, and the angular velocity and torque are measured. These values can be related to the shear stress and rate.
2. *Coaxial cylinder rotary viscometer.* This instrument consists of two cylinders, one located inside the other. The fluid to be measured fills the gap between the two cylinders. Again, the angular velocity and torque are measured when one of the cylinders is held stationary and the other rotated.
3. *Impeller viscometer.* This device is similar to the coaxial cylinder viscometer. However, the fluid measurements are made using a rotating impeller inside the stationary cylinder.

Oxygen

Generally, oxygen measurement falls into two main categories: dissolved oxygen and exit gas analysis. For most aerobic fermentations, an accurate determination of the dissolved oxygen concentration is of particular importance, for example, in maintaining the concentration above a specified minimal level. In addition, oxygen transport processes can be determined only by using an accurate method of measurement. The exit gas analysis for oxygen, usually in conjunction with CO_2 measurement, can be used to determine the metabolic state of aerobic microorganisms and their oxygen uptake.

Dissolved oxygen probes

A wide range of dissolved oxygen probes based on the use of electrochemical determination are available. Typically these fall into two main types: galvanic (potentiometric) or polarographic (amperometric or Clark). In essence these electrodes measure the activity (or partial pressure) of oxygen present, rather than the concentration. The range covered by these devices is approximately 0.001 to 1000 ppm. Both the galvanic and polarographic electrodes use an oxygen-permeable membrane (e.g., Teflon, polyethylene, and polypropylene) to separate the electrode components from the fermentation media. Similarly, the measurement for both types is based on the electrochemical reduction of oxygen at a cathode.

Polarographic electrodes

The principal components of the polarographic oxygen sensor include a precious metal cathode (e.g., gold or platinum), a silver/silver chloride (Ag/AgCl) anode, and an electrolyte solution (e.g., sodium chloride). During operation the following reactions take place at the respective electrodes:

$$\begin{aligned} &\text{Cathode: } O_2 + 2H_2O + 2e^- \rightarrow H_2O_2 \\ &\qquad + 2OH^- \rightarrow H_2O_2 + 2e^- \rightarrow 2OH^- \\ &\text{Anode: } Ag + Cl \rightarrow AgCl + 2OH^- \end{aligned} \quad \ldots(6)$$

Typically, polarographic (or amperometric) sensors operate at a fixed voltage, 0.6 to 0.8V, between the two electrodes. This voltage range represents the diffusion-controlled plateau of oxygen reduction. In effect, the current recorded can be directly related to the oxygen concentration.

Galvanic electrodes

In contrast to the polar graphic oxygen electrode, the galvanic sensor employs a non–noble metal (zinc, lead, or cadmium) anode with a noble metal cathode (silver or gold). The reactions occurring at the electrodes are

$$\begin{aligned} &\text{Cathode: } O_2 + 2H_2O + 4e \rightarrow 4OH^- \\ &\text{Anode: } 2Pb(e.g.) \rightarrow 2Pb^{2+} + 2e^- \end{aligned} \quad \ldots(7)$$

In contrast to the polar graphic sensors, galvanic electrodes require no external voltage source. The combination of a sacrificial anode and noble metal cathode is sufficient to provide an adequate electromotive force, resulting in a spontaneous reduction of oxygen at the electrode.

Despite their differences in operation, both types of electrodes are subject to similar constraints. During steady state, oxygen flux to the cathode is dependent on a number of transport-related steps including movement of oxygen through the bulk of the fermentation media and diffusion across the membrane and through the supporting electrolyte. In relation to the fermentation media, the viscosity and local hydrodynamic characteristics (in the vicinity of the electrode face) will play an important role. An example of this would be the incorrect placing of the electrode in a highly viscous media displaying non-Newtonian characteristics. If the electrode is situated in an area of quiescent fluid, the received signal may not be an accurate representation of oxygen partial pressure throughout the bioreactor. The design of the sensor will determine its operating characteristics. These include the membrane thickness, the membrane diffusion coefficient, and the surface area of the cathode. Temperature can also have a significant influence on the sensor response.

Generally a change in temperature will affect both the solubility and diffusion of oxygen through the membrane. In order to compensate for these effects a built-in thermistor is generally included in commercial devices. There are several problems encountered by both types of sensor during bioprocess operations. Fouling of the sensor membrane surface by components of the bioprocess media can lead to erroneous signals. With continued use, this could include the growth of microorganisms on the membrane surface. The effect of a buildup on the membrane would be to change the diffusion characteristics of the sensor (equation 8). Dissolved oxygen electrodes are also susceptible to signal drift during operation. In addition, gas bubbles and restricted flow toward the sensor can affect the received signal. Other important considerations are the effect of sterilization on the sensor, and calibration. Typically, calibration is carried out after sterilization, using medium gassing to obtain the full range of sensor values. In order to avoid sterilization problems, the sensor can be withdrawn into an isolating lock where it can be sterilized using milder conditions. The actual measurements can be expressed in one of several ways. One approach is to use percent of saturation on a scale of 0 to 100% oxygen saturation. Alternatively partial pressure can be used,

$$pO_2 = (p_{atm} + \Delta p - pH_2O)0.2095 \qquad ...(8)$$

where $(p_{atm} + \Delta p)$ is the absolute pressure and pH_2O is the partial water pressure. Finally a direct concentration can be obtained using Henry's law coefficient:

$$c = pO_2/H \qquad ...(9)$$

The coefficient value is subject to alteration through interactions with media components such as sugars, products, salts, and so forth. Hence, these factors must be taken into consideration when determining oxygen concentration.

O_2 *Optodes*

A recent development for measuring oxygen, and indeed a range of other analytes, has been the introduction of sensors based on the use of photometric transducers: optodes. The driving force for the development of these devices was linked to the evolution of immobilization technologies and fiber optics. The optical fibers (used to transmit information over long distances: 2-100 m) can be sited in harsh environments (e.g., high temperature and extremes of pH), allowing in situ real-time monitoring. A fiber is constructed using an optically transparent material called the *core* (this is the guide path for the light) and a cladding, surrounding the core. Light is totally reflected at the cladding-core interface, allowing it to be transported over long distances without significant loss. The light interacts with a sensitive element or layer, generating a signal. Several spectroscopic techniques have been used in the construction of optodes; of these, fluorescence has been the most widely used in the construction of O_2 sensors. Fluorescence is the emission of light from an absorbing species that has been excited by light of a different wavelength. Optical sensors for oxygen are constructed using an immobilized flurophore that undergoes dynamic quenching. Increasing concentrations of the quencher lead to a decrease in the fluorescence. The relationship between fluorescence intensity and extent of quenching can be summarized as

$$I_0/I = 1 + K_{sv} \times [Q] \qquad ...(10)$$

where I_O and I represent the measured fluorescence intensity in the presence and absence of the quenching agent, K_{sv} the Stern-Volmer proportionality constant, and Q the concentration of the quenching agent.

For a number of applications, dyes, whose fluorescent properties depend on the presence of a target analyte (e.g., oxygen), can be used. The dye can be attached directly to the surface of the fiber or onto a membrane. With the most common configuration, where sensing takes place at the end of the fiber light is returned after acting with the sensitive layer. This can occur through the same fiber or through a parallel fiber in a bifurcated device. The magnitude of the interaction is determined at the other end of the fiber using a detector system. These sensors have a number of advantages compared with electrochemically based systems including that they do not require a reference signal, they can be miniaturized, multisignaling from a range of analytes measured at the same time can be achieved, and the fibers can be steam sterilized. Despite these advantages there are a number of drawbacks to these devices, including interferences from ambient light, photobleaching and washout of the sensitive indicator, and the limited range of some devices. A number of fiber-optic-based O_2 optodes have been described, using a range of indicator molecules. Their use in measuring oxygen concentrations during bioprocess applications will indubitably become more widespread as the technology develops.

Tubing methods

An alternative approach to measuring dissolved oxygen content is to use the tubing method described by Philips and Johnson. This method is based on the use of a polymeric tubing (e.g., permeable Teflon) immersed in the fermenter media. A stream of helium or nitrogen gas is passed through the coil. Oxygen in the fermenter medium diffuses through the tubing into the carrier gas stream. The oxygen concentration in the inert gas is then determined using a suitable detector, for example, a paramagnetic gas analyser. Despite the longer response times for this method (2 to 10 minutes for a reading) compared with the electrode approach, it is a reliable technique. In addition the tubing is robust and can withstand repeated sterilization cycles.

Exit gas analysis

There a number of methods available for determining the oxygen concentration of exit gas from a bioprocess reactor. Several of these are based on exploiting the strong affinity shown by oxygen to a magnetic field. The paramagnetic gases, such as oxygen, display a positive magnetic susceptibility. There are two types of detectors based on this phenomenon: the magnetopneumatic and thermomagnetic analysers.

Magnetodynamical analyzer

The magnetodynamical analyzer is based on a dumbbell-shaped di-magnet held in a nonuniform magnetic field. The dumbbell structure is free to rotate about an axis. In the presence of oxygen, the dumbbell is rotated (under the action of the increased magnetic force) in proportion to the oxygen partial pressure. By applying a known corrective electrostatic force to return the dumbbell to its original position, the current required can be directly related to the partial pressure.

Thermomagnetic analyzers

Thermomagnetic analysers are based on the observation that changes in temperature affect magnetic susceptibility (χ). Generally this is described by the Curie law,

$$\chi T = \text{constant} \qquad \text{...(11)}$$

whereby χ is inversely proportional to the absolute temperature. The instrument consists of a flow-through ring element divided up into two measuring chambers. The detection site is composed of a thin-walled glass tube containing resistors that heat the incoming gases.

These resistors form part of a Wheatstone bridge circuit and are used to detect changes in resistance resulting from flow- rate variations. When the gas is heated, oxygen loses a high proportion of its magnetic susceptibility. This gas, in turn, is displaced by incoming cool oxygen, generating a convection current or “magnetic wind.” The flow of current (or heat transport) causes a change in resistance, imbalancing the Wheatstone bridge. In effect the convection current can be related to the oxygen concentration.

Carbon Dioxide

For many bioprocesses, the measurement of CO_2 is an important feature. Increased levels of dissolved carbon dioxide can inhibit growth and reduce the production of secondary metabolites.

Electrochemical probe

Steam-sterilizable electrochemical probes for determining dissolved CO_2 partial pressure are commercially available. These function in a similar way to oxygen probes and are based on a pH sensor immersed in a saturated bicarbonate buffer, separated from the bioreactor fluid by a hydrophobic membrane. Dissolved CO_2 gas molecules diffuse from the bioreactor media, through the hydrophobic membrane, into the carbonate buffer. The bicarbonate buffer equilibrates with the dissolved CO_2 gas molecules:

$$CO_2 + H_2O - HCO_3^- + H^+$$

$$aH^+ = K(aCO_2)/(aHCO_3) \qquad \text{...(12)}$$

Here, *a* is the activity. Hence, increases or decreases in pH can be directly related to the concentration of dissolved CO_2. This logarithmic relationship requires a two-point calibration, which is usually carried out by measuring the pH response, following substitution of the carbonate buffer with a reference solution.

CO_2 Optrodes

Optical sensors for the detection of CO_2 have also been described. Generally, these sensors are analogous to CO_2 electrochemical systems, whereby changes in pH resulting from the formation of bicarbonate ions are detected. One example described the use of a fluorescein dye as a pH indicator. All of the advantages and disadvantages already outlined for the O_2 sensor can be equally applied to these devices. Future developments in this field should lead to a more widespread use of these sensors.

Exit gas analysis

With a well-mixed aerobic fermentation, the concentration of dissolved CO_2 can be estimated from analysis of the exit gas. One particularly important approach used for determining the exit gas CO_2 concentration is infrared (IR) analysis. Particular molecules absorb IR radiation at characteristic wavelengths. This absorption is related to the chemical structure of the molecule, that is, the type and number of function groups present. When these molecules absorb IR radiation, the vibrational and rotational states are altered. Observing this absorbance (in either the gas, solid, or liquid phase, or combinations of these), the concentration of the target analyte can be determined. In the gas phase, CO_2 absorbs strongly at wavelengths $<15\ \mu m$. The main components of an IR analyzer consist of a suitable radiation source, a method of selecting the range of wavelengths required, a sample and reference cell, a detector, and a means of displaying the information in terms of CO_2 concentration. Generally, CO_2 is monitored using a positive filtering method. With this method, the optical system detector is filled with CO_2. The system detects the reduced radiation energy of the measuring beam reaching the detector. Comparing the reference and sample responses provides a value for the CO_2 concentration present. One of the main problems encountered with this approach is conflicting absorbance patterns from "interfering" molecules.

Redox Potential

Monitoring the redox potential of a bioprocess medium can provide information about the equilibrium between oxidizing and reducing species (electron acceptors and donors, respectively) present. Measurement of redox potential is achieved using a combined metal-reference electrode system. Typically the metal electrode can be gold, iridium, or platinum; with platinum being the usual choice. The reference electrode is either Ag/AgCl or calomel. The redox sensor is linked to a pH meter that is fitted to provide a readout in millivolts. During operation, the redox potential (measured between the metal and reference electrodes) varies as the logarithm of the ratio of oxidizable and reducible components in the media. The redox value varies linearly with the pH of the media and with the logarithm of dissolved oxygen tension. This can lead to complications when trying to interpret the readout from the sensor. A full description of the use and problems associated with the redox sensor was outlined by Kjaergard.

Chromatography

Chromatography has proved to be a valuable tool for numerous applications in diverse areas of analysis. The principle is based the separation of a mixture of compounds by means of a suitable matrix. Target analytes are separated by their characteristic interaction with the matrix. Following separation, the compounds can be detected using a range of detectors including flame ionization (used for gas chromatography applications), UV detectors, electrochemical detectors, infrared and refractive index detectors, conductivity detectors, and mass spectrometers. A number of chromatographic methods are available, several of which can be used for bioprocess monitoring. Overall chromatographic analysis is carried out off-line. However, recent developments have shown that on-line methods of analysis can be developed.

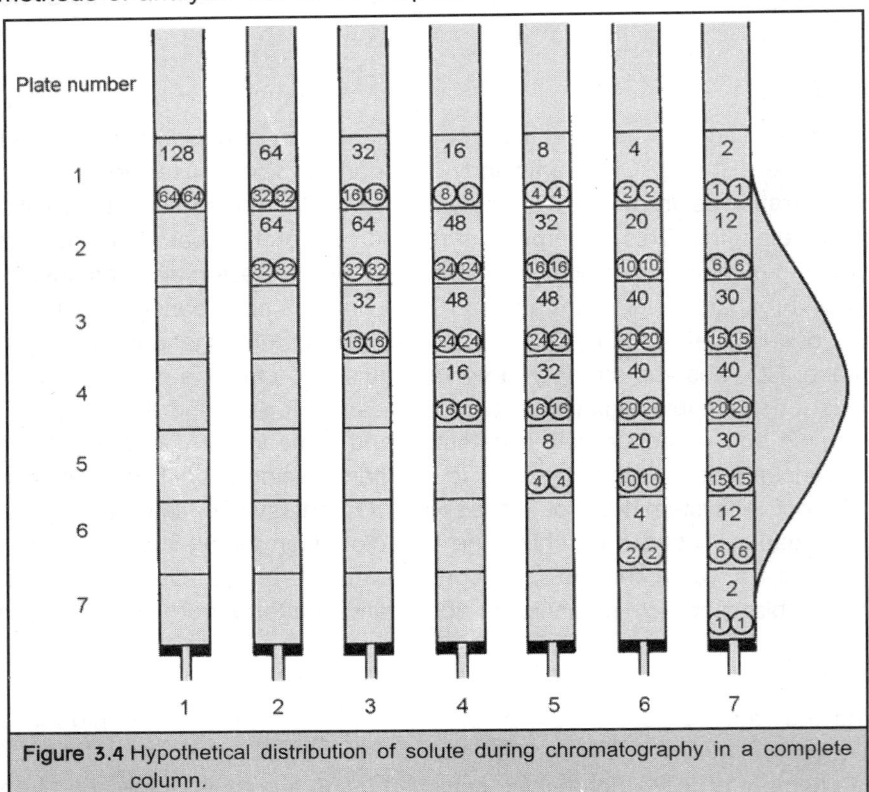

Figure 3.4 Hypothetical distribution of solute during chromatography in a complete column.

Liquid chromatography

For the bioprocess industry, liquid chromatography is used in sample analysis as a preparative method for the large-scale purification of biomolecules and as recovery technique for high-value products such as therapeutics and pharmaceuticals. One important development in this field has been the introduction of high-pressure liquid chromatography (HPLC). This approach to liquid chromatography involves the use of well-characterized homogeneous columns that contain small (3–5 μm-diameter) particles. In operation, these columns can withstand high pressures (up to 6,000 psi). The overall effect is to achieve a fast, accurate separation of the components in a sample. Liquid chromatography permits the separation of a sample solution into consistent components through the interaction of two phases. One type incorporates a solid stationary phase in conjunction with a fluid mobile phase containing the sample of interest. During their passage over the stationary phase, sample components are separated by their degree of solubility or, alternatively, their affinity for the solid matrix. The result is a separation profile indicative of the component mixture of the sample. There are a numerous examples of HPLC systems being used on-line to monitor bioprocess production. Typically, on-line monitoring is carried out using a sampling system based on membrane filtration.

Liquid–liquid chromatography

With the liquid–liquid chromatography method, the stationary phase is coated with a suitable fluid. The sample is carried in the liquid phase under pressure. Partitioning between the two liquids leads to a separation of the sample components. Detection of the separated components can be carried out using detectors such as UV absorption, conductivity, or refractive index. Liquid-liquid chromatography can be further divided into normal-phase type, in which the carrier liquid is less polar than the stationary fluid phase, and reverse-phase type, in which the carrier liquid is more polar. The reverse-phase method is particularly useful for separating organic compounds in an aqueous carrier solvent.

Ion exchange chromatography

Ion exchange chromatography is used to separate ionic compounds; typically from an aqueous carrier. This approach is particularly suited to measuring low concentrations of ions. The method is based on the ionic interaction between cations or anions in the sample solution, with the corresponding (oppositely charged) functional group on the column matrix.

Gas chromatography

Gas chromatography (GC) employs gas as the mobile phase and is principally used for separating and measuring the composition of volatile compounds such as aldehydes and alcohols, for example, in a bioreactor headspace. Again, this is a chromatographic approach that has been widely used for bioprocess monitoring. Connected to a suitable sampling system, GC can be used in an on-line format. An important development in the field of analysis has been the introduction of Fourier transform infrared spectroscopy (FT-IR) instrumentation. Combined with chromatography, this approach can provide a powerful tool for the accurate, sensitive analysis of gas mixtures. This technique will undoubtedly become more important over time.

Gel filtration

Gel filtration is a method of separation based on molecular size and is carried out using an appropriate polymeric matrix. By selecting the correct pore-size exclusion limit, molecules in a sample solution can be separated. Small molecules, relative to the gel "sieve," pass rapidly through the chromatography column. However, larger molecules diffuse into the polymer matrix, where their passage is impeded. Hence, elution time from the column is determined by molecular size.

Mass Spectrometry

The use of on-line mass spectroscopy (MS) has developed since the first report of Reus et al. A wide range of gasses, both free and dissolved (e.g., CO_2, O_2, CH_4), can be measured. In addition,

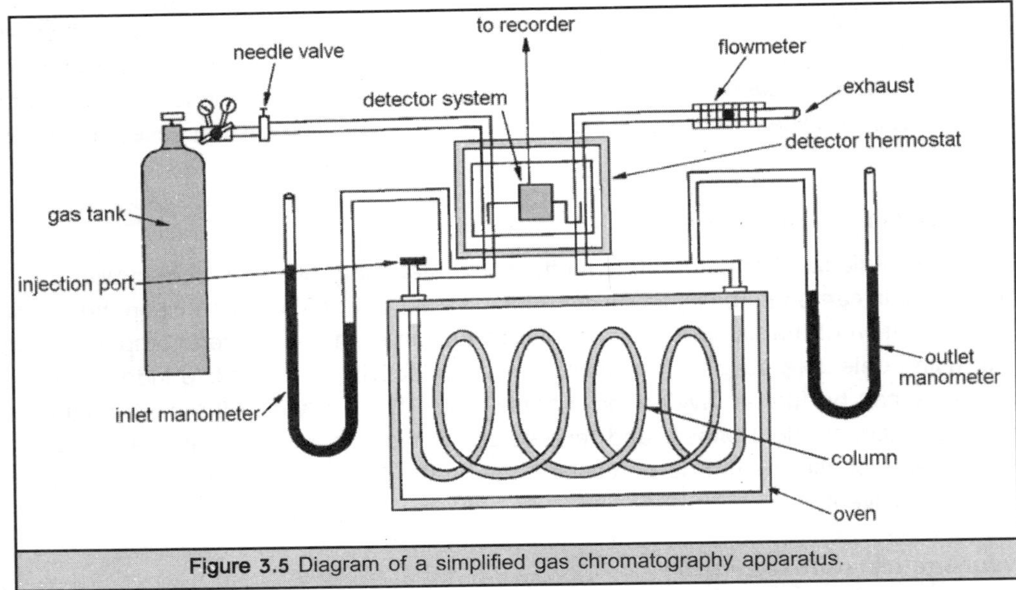

Figure 3.5 Diagram of a simplified gas chromatography apparatus.

volatile organic compounds such as methanol, ethanol, acetone, and simple organic acids can be monitored. The technique is based on the rupture of molecules by a high-energy source into corresponding ions. Magnetic separation of the charged ions may then be achieved, based on their mass: charge ratio. Each compound produces a characteristic display pattern. Typically an instrument consists of an electron bombardment source, the magnetic ion separation system, and the detector. There are several types of mass spectrometer available.

The quadrapole MS is based on the use of a radio frequency system to separate the ions. A combination of applied voltage and frequency range produces a change in the trajectory of the ions. The term *quadrapole* comes from the four rods used to produce the radio frequencies. There is an alternative MS instrument that is potentially more precise and stable. This instrument is based on the use of a magnetic field to separate ions. Two types of magnetic instruments are available. One system uses a fixed magnetic field and several detectors. The other uses a variable controlled magnetic field with one detector system. Mass spectrometers offer a number of advantages for process monitoring including speed, specificity, precision, and stability. The major drawback to the use of mass spectrometers is cost; in addition some peaks can be difficult to resolve using the less elaborate instruments. Despite these drawbacks, MS monitoring of volatile compounds has been widely reported.

Weight

Monitoring fluctuations in the weight of a vessel used for bioprocesses is an effective way of measuring the contents and flow rate of additions. There are a number of load cells available for this task. The principle of a load cell is based on measuring the compressive strain that is placed on the device (e.g., a solid or tubular steel cylinder) when under an axial load. Electrical resistance strain gauges can be included in the device structure. A proportional electrical resistance to the applied load can then be measured; this varies in relation to changes in load. Obviously, a temperature compensator must be included in the instrument to account for heat effects on the resistors. These devices need to be rugged and have long-term stability. For reasons of safety, it is also desirable for the device to be accurate. A load cell designed to detect tensile forces can be used for weight measurements of suspended vessels.

Liquid Level Measurement

Liquid levels in a vessel can be determined using a number of techniques. Determining the level can assist in formulating mass balances and measuring nutrient additions. Hydrostatic pressure can be measured using a single sensor positioned at the bottom of the vessel. Alternatively, two sensors spaced at the top and bottom of the vessel (differential design) can be employed. These sensors are electromechanical devices based on the deformation of a spring; the resulting signal is displayed as an electrical signal (capacitance or reactance).

Conductivity sensors

Conductivity sensors are particularly suited to monitoring levels of foam (in a number of fermentations), where in excess it can cause problems. The sensor consists of a stainless steel probe, insulated except for the tip. The device can be used with noncorrosive conducting fluids. If the liquid or foam level rises to contact with the tip, an electric current is passed through the sensor; the foam acts as an electrolyte and the vessel as a ground. The sensor can be coupled to a control device that dispenses antifoam to counter the increased level.

Capacitance sensors

Capacitance sensors operate by detecting changes in the relative dielectric constant of the media, compared with air. Measurements are made by detecting variations in electrical capacity brought about by changes in the liquid level.

Acoustic sensors

Acoustic sensors operate via transducers that generate and detect ultrasonic waves. Two formats are available: a single device that both transmits and receives the signal, and two separate transducers with one function. By monitoring the time it takes for the sound wave to travel, the liquid level can be determined. With the single device the sound wave is directed onto the liquid surface. When the level rises, the time delay is shortened.

Temperature probes

Liquid levels can be measured using a series of thermistors sited vertically through the vessel. During operation, an electrical current is applied to the sensors, raising their temperature above the ambient. When the liquid (or foam) level reaches a sensor positioned lower down the vessel, the sensor cools. This causes a temperature difference and, hence, a change in electrical resistance. The resistance can be displayed as an output signal indicating the liquid level.

Flow Measurement and Control

Flow measurement is an important feature for many bioprocesses, for both gases and liquids. These measurements are carried out on both the influent and effluent streams.

Gas flow measurement

The most common form of gas measurement is carried out using a variable area flow meter, or rotameter. This device consists of vertically mounted (usually conical) tube enclosing a free-floating body that is able to move up (floating in the gas flow of interest) and down a tapered bore running through the tube body. Typically, the tube is constructed of glass or metal and the float is a ball or hollow thimble shape. Because the position adopted by the flow rate is dependant on both the gas flow and the viscosity of the medium, the instrument requires careful calibration. Another important instrument is the thermal mass flow meter. The measuring principle is based on the use of a heating device situated in the gas flow, which is used to heat the gas. The mass flow rate can derived from the heat balance,

$$Q = H/C_p(T_2 - T_1) \quad \text{...(13)}$$

where Q is the mass flow of the gas, H is the enthalpy, C_p is the specific heat of the gas, T_1 is the temperature of the gas before heat is transferred to it, and T_2 is the gas temperature after the transfer of heat. The mass flow detected using this approach is independent of pressure and temperature.

Liquids

A number of techniques are available to measure liquid flow rates. The choice of method will be determined by the nature of the liquid (e.g., viscosity, conductivity, etc.). For nonsterile liquids, instruments such as the turbine flow meter can be used. Fluid flow over a turbine blade array is proportional to the rotation speed of the turbine, hence a direct correlation can be made. A common approach to measuring fluid flow in ancillary pipe-work is to use a differential pressure device, which utilizes the pressure drop caused by a constriction in the pipeline. This approach is based on the general Bernoulli equation for fluid flow. The constriction causes an increase in the velocity of the fluid, which in turn results in a corresponding pressure drop across the constriction. The differential pressure is a function of the flow velocity and density of the fluid. Measurement of the pressure difference can be used to determine fluid velocity. Measurement of fluid flow under sterile conditions can be carried out using a magnetic-inductive flow meter. This device can operate with a fluid that includes particulate matter in suspension. The outside of the device comprises two windings that, in operation, are supplied with an alternating current. This induces a magnetic field. A pair of electrodes is placed in conjunction with the device. The average flow velocity is proportional to the electrical potential induced in the field. Potential differences in the field are detected by the electrode arrangement. Hence, accurate flow measurements can be made.

Biomass

The monitoring of biomass concentration can be carried out using a number of techniques. Conventionally, biomass concentration is measured off-line using labor- intensive, time-consuming methods such as dry weight cell, plate or microscopic cell count, and measuring the optical density of diluted samples. However, a number of more rapid methods have been developed. Of particular interest is the development of real-time on-line methods. Generally, methods for determining biomass concentration can be divided into two classifications: direct and indirect. The former is based on determining the physical properties of the cell and its components. In contrast, indirect methods measure factors related to the cell and its activity (e.g., respiration, electrochemical behavior, and nutrient fluctuation).

Bioluminescence and chemiluminescence

The use of bioluminescent techniques is based on determining levels of adenosine triphosphate (ATP) concentration. Generally, levels of ATP remain constant for living cells, decreasing when the cells die. ATP concentration can, therefore, be related to biomass. A particular enzyme that has proved useful for this method is luciferase, which catalyzes the following reaction:

$$\text{ATP} + \text{luciferin} + O_2 \xrightarrow{\text{luciferae}} \text{oxyluciferin} + \text{AMP} + \text{light} \qquad \ldots(14)$$

By detecting the light produced by this reaction, ATP concentrations, and hence, biomass, can be determined. This method can detect cell numbers a low as 10^5 cells ml^{-1}; it can be automated, and the assay time is fairly rapid. However, the method does suffer from several drawbacks, for example, the extraction of ATP may be incomplete, there may be free ATP present from other sources, and there maybe degradation of ATP by the extraction reagents. Chemiluminescence is based on the detection of light produced by the protein-catalyzed oxidation of luminol in the presence of hydrogen peroxide.

Acoustic resonance densitometry

Acoustic resonance density is based on determining the change in a resonant frequency that results from changes in cell density. This is a noninvasive method that does not require contact between the instrument and the culture medium. The method incorporates an oscillatory circuit, amplifier, and test cell. Theoretically, the fluid density of the sample can be calculated from the square of its oscillation.

Recent reports have described using this approach to monitor cultures of hybridomas and human lymphoma cells. Further advantages for this method include independence from flow rate and viscosity. However, there are disadvantages including the need for a filtration system (to enable the resonance density of the medium to be measured in the absence of cells), with the inherent problems associated with such systems, including poor sensitivity and problems caused by bubbles and particulate matter.

Capacitance, conductivity, and electrochemical methods

The electrical properties of cells have been exploited in the development of a number of techniques. Changes in media capacitance have been related to biomass concentration (e.g., a decreasing capacitance coupled with an increasing biomass concentration). In addition, it has been suggested that cell viability is linked to capacitance measurements. Several in situ commercial instruments are currently available for determining biomass concentration based on capacitance measurements. The Biomass Monitor measures the radio frequency conductance and capacitance of a cell suspension using a constant voltage, four-terminal, phase-sensitive detector system. Furthermore, the probes can be inserted directly into bioreactors using a standard 25 mm-diameter (Ingold-type) port. The probes are fully sterilizable and can be cleaned in situ during operation.

Fluorescence

Fluorescence is a characteristic possessed by a number of important biological compounds including proteins, enzymes, and coenzymes. Simply defined, it is the absorbance of light energy at a particular wavelength, followed by reemission at a longer wavelength. After the passage of light energy, the compound returns to its ground-state level. In the field of bioprocess monitoring, fluorescence has been used, in particular, to monitor reduced nicotinamide adenine dinucleotide (NADH) and reduced nicotinamide adenine dinucleotide phosphate (NADPH) concentrations. These compounds are irradiated at 340 nm and emit at 450 nm.

Both compounds are present in living organisms and are vital components of metabolic processes. By monitoring the intensity of fluorescence at the characteristic emission wavelength, it should be possible to calculate the total biomass concentration. Unfortunately, the use of fluorescence as a marker for cell concentration does suffer from a number of drawbacks including that the fluorescence signal can originate from changes in metabolic state and not from cell concentration levels, sensitivity may be affected by the presence of compounds that can quench both the excitation and emission wavelengths, and other compounds may fluoresce at the same wavelength. Nevertheless, the use of this approach has been reported in a number of papers.

Flow cytometry

Flow cytometry is a measuring technique based on the irradiation of a sample solution (containing a cell population) with a suitable light source, followed by monitoring of the scattered or absorbed light. In addition, fluorescence can be used as the measuring parameter. This technique can be used to ascertain a number of cellular features, such as the accumulation of cellular components (e.g., DNA, RNA, and proteins), and cell dynamics (e.g., cell size distribution). Furthermore, flow cytometry can be used to differentiate and quantify a range of species populations present in a mixed medium.

Light scattering and turbidity

Optical techniques based on either the scattering of light (nephelometry) or the degree of transmitted light or optical density (turbidity) have been developed for determining biomass. By monitoring the degree of light scattering using nephelometry, both cell numbers and mass can be determined. This approach is particularly suited for bioprocesses that involve low cell concentrations, where the background (compared) level is near zero. Turbidity measurement can be used for both online and off-line determination of biomass. Light scattering by a turbid medium results from a number of factors including particle size, shape, and number. Quantification is based on the Lambert-Beer law.

Biosensors

One possible answer to the problem of monitoring metabolites, both in situ and on-line, may be solved in the near future by the use of biosensors. Since their conception in the 1960s, these devices have generated considerable interest. This has spread to a diverse range of fields including clinical diagnostics, environmental protection, bioprocess monitoring, and defence applications. In general the operation of a biosensor is characterized by three functional steps: recognition, physiochemical signal generation, and signal processing. The biological component (e.g., enzyme, whole cell, antibody, and cell receptor) imparts a high degree of selectivity on the biosensor. Coupled to the biological component; the transducer is designed to respond to the changing physicochemical parameters caused by the specific interaction of the biological component with the substrate. A range of transducers have been used in the development of biosensors and include the electrochemical (potentiometric and amperometric), optical, calorimetric, piezoelectric, and thermometric.

Most of these types have been adopted in the development of biosensors intended for bioprocess monitoring. For a more detailed insight into both the fundamentals and potential for biosensors, a number of specialized publications are available. Enzyme-modified field effect transistors (FETs), whereby the biologically active layer is positioned on top of the ion electrode membrane, have also been used for bioprocess monitoring. An off-line biosensor system has been available for a number of years. The instrument can be used to determine a range of metabolites (e.g., glucose, lactate, and ethanol). Recently the company has introduced an on-line instrument. Indeed, on-line biosensor systems have been demonstrated using a variety of biological-transducer systems. A wide range of metabolites have been monitored using various bioprocess regimes. Commercially, a significant number of biosensor devices are available for measuring a range of analytes. However, at present, the majority of these devices are aimed at the medical diagnostic market. The goal is to develop cheap, reliable sensors that can operate under a range of bioprocess conditions for extended periods with a minimum of maintenance.

Nuclear Magnetic Resonance

Nuclear magnetic resonance is based on the detection of the response from particular nuclei when exposed to a magnetic field and electromagnetic radiation. Following absorption, the resonance frequency is shifted in a characteristic response pattern, according to the environment of the sample under detection. This approach can be used to determine a range of intracellular factors such as ATP, ADP, sugar phosphate, and polyphosphate, as well as pH. In these examples, concentrations of ^{31}P are used to characterize the compounds. This is an off-line monitoring system that is currently expensive; hence, its use is primarily in the research field, not in routine production environments.

Artificial Intelligence

The data obtained using the various measuring instruments already discussed are invariably used to control and optimize the bioprocess being carried out. Recent developments in the field of artificial intelligence have led to investigations into the use of such systems for improving bioprocess control, based on the received measurement output signals. This has included the use of both knowledge-based expert systems and neural networks during bioprocess operation. A recent report described the successful use of a neural network as a tool for evaluating the received measurement from an enzyme (penicillin-G amidase) pH-FET sensor linked to a flow injection system. Undoubtedly, the adaptation of such "intelligent" systems will develop over the coming years and will play an important role in the precise control of bioprocess applications. Although a diverse range of monitoring equipment is available, only a relatively narrow range of the more reliable instruments is used routinely in practice. As culture techniques become more elaborate and high-value-added products are produced, conventional methods will either prove insufficient or require supplementation with a range of sensors able to directly monitor key process parameters. Despite a clear demand for new sensors (e.g., lactate, glutamine, and glutamate for animal cell culture), cell cultivation represents only a modest market for analytical instrument manufacturers. Hence, while progress is to be expected, it may be slower than might be wished.

4

BIOPROCESS VALIDATION

Bioprocess validation for pharmaceuticals has been defined by the U.S. Food and Drug Administration (FDA) as "establishing documented evidence which provides a high degree of assurance that a specific process will consistently produce a product meeting its pre-determined specifications and quality attributes". In spite of the apparent clarity of this definition, pharmaceutical companies vary widely in their approaches to process validation. Operational definitions of process validation are often developed by firms to suit their own validation philosophies. Subcategories of process validation such as formal process validation, process characterization, process development, process evaluation, and the like can result, and the meaning of these phrases is highly specific to a particular company. For the purposes of this article, the term *bioprocess validation* refers to any activity consistent with the FDA definition. The discussion that follows is necessarily subjective, and it does not reflect the view of any particular company or the industry at large. Rather, it describes concepts used by those who practice this craft in industry. This article is concerned with processes that produce the active ingredient in a biopharmaceutical product for use in humans as therapeutics or vaccines.

Biopharmaceuticals, protein agents derived from recombinant gene expression or hybridoma production systems, include various biotechnology products such as monoclonal antibodies, enzymes, hormones, soluble receptors, growth factors, and immunogens. Pharmaceutical manufacturing processes are divided into two subprocesses. The primary process produces the bulk biological substance (BBS), also referred to as the active pharmaceutical ingredient (API). The secondary process transforms the BBS into the final product, which is what is actually sold. Secondary processing steps include formulation, sterile filtration, filling, lyophilization (in some cases), capping, and packaging. This article is limited to the validation of the primary processes which produce biopharmaceutical BBS. Pharmaceutical manufacturers typically classify validation activities into the following broad categories: process validation, equipment and facility validation, cleaning validation, analytical methods validation, and computer validation.

Equipment and facility validation is further subdivided into design qualification, installation qualification (IQ), operational qualification (OQ), and performance qualification (PQ). Bioprocess validation is viewed by some experts to be an element of PQ. Process validation is a regulatory requirement to enable product licensure. Limited process validation is also critical for clinical trial materials to ensure product safety. Process validation regulations originated within the FDA, which was created in 1906, initially to prevent the adulteration of products. The sulfanilamide disaster led the Food, Drug, and Cosmetics (FD&C) Act in 1938 to ensure product safety. The thalidomide disaster in 1962 led to the Kefauver–Harris Amendment to the FD&C Act, which required demonstration of product safety and efficacy. It was in this revision that the concept of good manufacturing practices (GMP) originated. In 1978 the act was further revised to include the concepts underlying current GMPs (cGMPs) and good laboratory practices (GLPs). The origins for process validation derive from FDA draft guidelines and workshops that occurred in the late 1970s and early 1980s. *Guideline on General Principles of Process Validation*, published in 1987, represented the first (and still only) regulatory document focusing on this topic.

The definition for *bioprocess validation* quoted earlier was taken from this FDA document and has gained wide acceptance by other regulatory agencies around the world. This guidance was specifically intended for the production of finished drug products. Today, the regulatory basis in the United States for process validation is contained within cGMPs for finished pharmaceuticals (drugs). The *Code of Federal Regulations* states that "control procedures shall be established to monitor the output and validate the performance of those manufacturing processes that may be responsible for causing variability in the characteristics of in-process material and the drug product." It also states that "appropriate written procedures, designed to prevent microbiological contamination of drug products purporting to be sterile, shall be established and followed. Such procedures shall include validation of any sterilization process." Until recently, these FDA validation requirements and guidelines only applied explicitly to drug products and did not apply to bulk biologics. Nevertheless, there was a trend for these regulatory principles to apply implicitly to areas more broadly than originally defined.

There are numerous publications by industrial scientists, but currently no umbrella regulatory document exists on the validation of bulk production processes for biopharmaceuticals. Important aspects to this subject are captured within numerous other regulatory documents that focus on other aspects of manufacturing and control. Regulatory requirements that pertain to process validation are changing and are likely to continue to change for the foreseeable future. The most current information on process validation and many other regulatory subjects may be accessed on the Internet. Proposals published recently by the FDA pertain to the cGMPs for finished pharmaceuticals and include content on process validation. How these proposals will be extended to bulk biological products has become clearer with the recent publication by the FDA of a draft guidance for the manufacturing of APIs and the guidance for submission of the chemistry, manufacturing, and controls (CMC) section of a biologics license application (BLA) for most *biopharmaceutical products*. In addition to being a regulatory requirement, bioprocess validation provides important benefits to the manufacturer. Biopharmaceuticals are among the most expensive products to manufacture, costing hundreds to thousands of dollars per gram to produce. The assurance of process consistency therefore enables a manufacturer to maximize plant productivity and obtain favorable cost efficiency by reducing the incidence of process failures, rejection of lots, and reworking of rejected but salvageable lots. In today's manufacturing environment where plant capacity is closely matched with product demand, unexpected process failures can lead to product shortages, with negative financial and company product image impacts, particularly for life-saving products.

Process validation is a key means to minimize the occurrence of unexpected process failures. Process validation can also provide a manufacturer with operational flexibility to respond to inevitable production problems (e.g., power failures, equipment breakdown, and scheduling conflicts). To the extent that they can be anticipated in the process validation program, ranges can be established for operational parameters that, if maintained, will ensure that the products meet specifications. These important benefits only accrue after considerable investment because process validation is costly and time consuming. It occupies the full period from producing supplies for the first clinical trial through the start of production in the commercial plant and continues throughout the commercial lifetime of the product. During product development, for example, more resources from process development, operations, quality control, and quality assurance groups can be consumed in validating the process, equipment, and facility than were used to develop the process and to produce clinical supplies. Nevertheless, careful planning and sound scientific strategies make process validation cost effective and meet regulatory requirements. There are several different classes of process validation.

Clearance validations demonstrate the capability of a process to remove or inactivate contaminants to and beyond the degree necessary to ensure product safety. Of greatest concern are contaminants that have toxic, pharmacological, or infectious potential in humans. Typical contaminants in biopharmaceutical processes with these potential liabilities include DNA and proteins derived from the recombinant host cell line, certain components of culture media (e.g., methotrexate and other drugs used to maintain selection pressure, insulin, bovine serum albumin, growth factors, and animal-derived components of culture media), viruses derived from the host cell line or raw materials, additives for viral inactivation (e.g., guanidine), and affinity chromatography ligands (e.g., monoclonal antibodies, protein A). *Validation of operating ranges* involves the determination of a range for critical operating parameters through which excursions can occur without affecting the quality of the product. The usage of the term *operating parameters* here is taken very broadly to include not only compositions, temperature, pH, and related factors but also holding times of in-process intermediates, variations in raw materials, cycle limits on multiuse resins, age of raw materials, and age limit of cells used in the process. *Validation of process consistency* must be established through the characterization of multiple lots of product produced by separate production runs operated and controlled under the same conditions. Each of these classes of validation are described in greater detail in this article.

STRATEGY AND DOCUMENTATION

Bioprocess validation generally starts after a process is defined in development and evolves with the progression of clinical trials. Process validation for phase I and II trials is limited to addressing potential product safety concerns. Validations for the clearance of model viruses and retroviruses and host cell DNA are sufficient. More extensive validation activities occur in parallel with production of phase II and III clinical supplies and the conducting of the phase II and III clinical studies. At this stage, the virus clearance validations are expanded to include a greater number of model viruses spanning a greater range of virus characteristics. The full range of clearance and operating range validations are undertaken in a pilot plant or development laboratories. Process consistency validation, however, is derived from the pilot and commercial plants. License applications can be filed with consistency data derived from a pilot plant. However, the awarding of a license is contingent on the filing of process and product consistency data derived from a pilot plant or commercial manufacturing facility. In the United States, three qualification lots are usually sufficient; the European Union usually requires five lots. Process validation continues after product launch for the length of time the product is manufactured. Validation maintenance programs often call for revalidation of specific aspects of the process at regular intervals. Trending or control charting of process variables can lead to refinements of operating range limits.

Revalidation is also required when the process is modified because of improvement, equipment alterations, raw material changes, process scale-up, substitution of a processing step with an alternate, resequencing process step, and so forth. Validation is subdivided into prospective, concurrent, and retrospective according to the timing relationship between the validation work and production. Prospective validation studies "are conducted prior to the [commercial] distribution of a new product, or product made under a revised manufacturing process, where the revisions may affect the product's characteristics". Concurrent validation studies employ data derived from production runs for product in distribution. How prospective and concurrent validations are conducted is often defined in preapproved validation protocols. Retrospective validation studies demonstrate, "in some measure, the adequacy of the process by examination of accumulated test data on the product and records of the manufacturing procedures used.

Retrospective validation can also be useful to augment initial premarket prospective validation for new products or changed processes". Retrospective validation is generally not conducted according to a preapproved validation protocol, but this practice may change in the future. In reality, most comprehensive process validation programs involve all three types of validation throughout the lifetime of a product. Consistency validations are most often run prospectively in full-scale production equipment for clinical supplies or commercial purposes. Concurrent and retrospective validations can be used to set new operating ranges not covered by previous validation work or to modify (usually narrow) operating ranges based on manufacturing experience using statistical process control, control charting, or related techniques. A final precaution should be taken regarding recent regulatory concerns on retrospective validation. Retrospective validations are acceptable as an adjunct to a well-defined and comprehensive prospective validation program, but they cannot substitute for it.

Another technical consideration to be defined is the scale at which process validation studies are to be conducted. One could assume that the most authoritative results would be obtained by conducting validation experiments at the scale intended for commercial manufacturing. This assumption normally leads manufacturers to validate process consistency using the equipment to be licensed. It is not unusual, however, to identify 50 operating parameters for which low and high operating range limits are sought. This could involve 100 experiments (if interactions among parameters were disregarded) and up to 1×10^{15} experiments (if interactions are included), which is clearly not feasible in full-scale equipment.

The infectivity and limited quantity of viral reagents used in clearance validations prevent these studies from being executed at full scale or in a GMP facility. GMP prohibits the intentional introduction of infectious agents into processing areas. In other clearance validations, limitations on the quantity of reagents used for spiking (e.g., host cell proteins, host cell DNA, and affinity ligands) prevent these studies from being conducted in full-scale equipment. One approach to these problems is to scale down processing steps. Scale-down factors from 10- to 10,000-fold have been employed in industry. In some cases, validation experiments can be run in parallel using multiple identical scaled-down units (e.g., fermentors), which enable statistical design of experimental techniques and a significant shortening in the overall time expended for validation experiments. In other cases a hybrid approach can be taken in which samples of in-process streams taken from the full-scale process are used as feed streams to scaled-down unit operations to assess performance. Virus clearance studies that employ live viruses and retroviruses are conducted in this manner. The use of scaled-down systems for virus clearance validations also facilitates adequate biological containment.

If scaled-down models are to be employed, then it is necessary to establish that the models are predictive of full- scale performance. A useful scaled-down model may be developed by attending to three considerations: performance, design, and quality (PDQ). The performance of the scaled-down model should be identical to or at least predictive of the full-scale process. Typical performance characteristics are titer, cell density, viability, processing rates, and step yields. In cases where there is a performance difference between the scales, it should be quantified and the basis for the difference should be understood from technical and scientific principles whenever possible. The degree of variation in the model should not be greater than variation at full scale to ensure that process changes are distinguishable from natural variation in the model. During the design of the scaled-down model, it is important to observe the same technical principles used to scale up a unit operation. Typical examples are maintaining constant culture split ratios in cell culture and to preserve chromatography bed heights and linear flow velocities in purification during scale-up or scale-down.

The quality of the intermediate produced by each scaled-down unit operation should be identical or similar to its full-scale counterpart in its analytical attributes. Finally, it is important to document the qualification of scaled-down models used for process validation. As the definition of *process validation* states, documentation of validation work is essential. The validation master plan defines the overall validation philosophy, the systems to be validated, the general testing scheme to be employed, the responsibilities of various groups, and the approval mechanisms of the program. It may also define the evolution of validation throughout the life cycle of the process, from its origins in process development through validation maintenance during commercial manufacturing. Another class of document is the validation protocol, which prospectively defines the details for testing of each validation stage. The validation protocol is defined by the FDA as "a written plan stating how validation will be conducted, including test parameters, product characteristics, product equipment, and decision points on what constitutes acceptable test results".

The final document is the process validation report, which includes the validation protocol, deviations from the protocol, the results from the validation experiment, and the conclusion regarding the validation status for the processing stage. An important phrase in the definition of process validation is establishing documented evidence. Validation experiments must be documented in reports to be acceptable to regulatory agencies. Copies of the most critical validation reports and summaries of the others are included as part of the license application. All other process validation reports must be available to the regulators for inspections as part of the licensing process or during routine postlicense inspections. Failure to submit essential process validation reports can lead to a refusal to file determination by the FDA in the United States. A failure to produce appropriate validation reports during regulatory inspections can lead to a rejected license application or other regulatory actions such as the issuance of an FDA 483 form.

RECOVERY AND PURIFICATION PROCESS

Clearance validations are generally concerned with the capacity of the recovery and purification process to remove process contaminants. Of particular concern are contaminants with toxic, pharmacological, or infectious potential in humans; studies of these risks are thus among the first to be undertaken in a validation program. The overall approach to this form of validation can be illustrated using the clearance of DNA. DNA contaminants could contain genes that encode harmful proteins (e.g., toxins and oncogene products), viruses, or retroviruses endogenous to the host. If these gene contaminants could transfect patient cells (resulting in retention and duplication of the gene), then expression of the gene product is a theoretical possibility. The World Health Organization (WHO) has established a 100 pg/dose limit to DNA contaminants in biopharmaceutical products. Proposals are under consideration to increase this limit to 10 ng/dose.

One approach to validating adequate clearance of DNA by the process is to measure the DNA level in the BBS over the full range of process operating conditions. If these measurements fall below the 100 pg/dose limit, then the process is validated for this clearance capability. This approach is taken for contaminants whose assays are sufficiently sensitive to quantify levels in the BBS. Often the assays for DNA and many other process contaminants are not sufficiently sensitive to quantify the contaminant in the bulk product, and an alternate validation approach must be used. Clearance factors are established for upstream unit operations for which DNA levels can be measured in the in-process streams. For downstream steps for which DNA levels are immeasurable, host cell DNA is added (spiked) to the feed stream for each unit operation at a level sufficient to detect it in the output stream.

Clearance factors for the downstream steps can then be calculated. The clearance factors are then summed together to give a total clearance factor for the process. This total may be adjusted downward to eliminate double counting steps that have similar mechanisms of clearance. If, for example, a process includes two anion exchange steps, only one would be included in the analysis of clearance. This is based on the notion that virus particles capable of being removed by this mechanism (i.e., anion exchange) would only be removed by the first step. No further clearance could be expected in the second or additional anion exchange steps. The total clearance factor is then applied to the range of DNA levels present in unprocessed bulk harvests. If the calculated DNA level in the BBS is below the 100 pg/dose level, then the process is validated with respect to DNA clearance. Spiking studies frequently use a hybrid of the full-scale process and scaled-down process models during validation. During full-scale operation, samples of in-process streams are collected and stored. Portions of these samples are spiked with the appropriate contaminant(s) and are then processed over a scaled-down model. In some cases, the spiked species may be labeled with a radioactive label to facilitate quantitative detection in in-process streams. This approach is often used for measuring DNA clearance.

When animal cells are employed during production, clearance of putative viruses constitutes the largest group of clearance validations. The general approach is similar to that described for DNA clearance. The product safety concern is to minimize the risk of viruses or retroviruses contaminating the product. The viruses can originate as active or latent viruses associated with the host cells (i.e., endogenous viruses) or can be introduced with raw materials (i.e., adventitious viruses), particularly those derived from animals. In most cases, if viruses exist at all, they are present at levels below detection in inprocess streams, making spiking–clearance studies necessary for all steps in purification. For infectious contaminants such as viruses, there are two modes of clearance. The first involves removal by each purification step, similar to the manner in which DNA is removed. The second involves inactivation, where additives or conditions are used to disable the virus without removing it. The determination of the viral load to the process depends on the specifics of the host cell system. For Chinese hamster ovary (CHO) cells, the load can be estimated by measuring the level of retrovirus-like particles (RVLPs) present in the unprocessed bulk harvest. CHO cells are known to produce RVLPs, which resemble intracytoplasmic type A retroviruses and cell membrane budding type C retroviruses. Even though these particles are defective in their ability to infect other cells, they nevertheless provide a basis for estimating a retroviral load in the process. Levels of 10^5 to 10^7 RVLPs per milliliter of unprocessed bulk are typical. The level of viral clearance to target depends on the perceived risk to the patient. One target for retroviral clearance in processes involving CHO cells is to allow no more than 1 RVLP per million doses of final product.

The total process retroviral clearance and the level of RVLPs in the unprocessed bulk are used to calculate an estimate of the RVLPs in the final product. The process is considered to be validated for retroviral clearance when the estimated level of RVLPs in the final product is less than or equal to the target. Process clearance factors of 15-20 logs are typical for murine retroviruses. The FDA has suggested that a minimum of 3 logs of clearance is necessary based on the initial level of contamination entering the purification process. For most viruses other than endogenous retroviruses, the validation exercise becomes more academic; there is no reasonable basis for establishing the virus load in the process because there should not be any present. In these cases, measured clearance factors are reported to provide some assurance that if these types of viruses were inadvertently introduced into the process, there would be a defined capacity to clear them.

The spiking studies use model viruses that are representative of the most likely contaminants and when grouped together cover a broad spectrum of virus characteristics. Clearance studies are often performed for a number of viruses to test the ability of the process to clear viruses in general. The choices of model viruses depend on the cell line employed and the viruses could contain. For CHO cells, typical model viruses include a small, non-enveloped virus (e.g., simian virus [SV40], human polio virus 1, animal parvovirus), a medium-to-large enveloped RNA virus (parainfluenza virus, influenza virus, Sinbis virus), a medium-to-large DNA virus (e.g., herpes simplex virus 1 and pseudorabies virus), and a murine retrovirus (e.g., murine leukemia virus). The first three viruses are chosen because they cover a range of virus characteristics, and the murine retrovirus is chosen to model one that could infect the CHO host cell line.

VALIDATION STAGES

The purpose of validating process operating ranges is to demonstrate the robustness of the process to excursions in operating parameters that are likely to be encountered in a practical manufacturing environment. The process can be subdivided into a series of validation stages or modules. In some cases, a stage corresponds to a specific processing step (e.g., production fermentation or a chromatography step). In other cases, a stage maybe a series of steps (e.g., media preparation or seed expansion). The choice of defining validation stages depends on the specifics of the process and the manner in which the process is controlled.

Once the stages are identified, inputs and outputs to each stage are defined. Typical inputs include the operating parameters, properties and storage duration of the feedstock derived from the previous upstream step, and quality attributes of critical raw materials used in the step. Typical outputs include the performance attributes for the stage (e.g., titer and yield) and the properties of the product stream (product homogeneity, purity, contaminant levels, and chromatogram shape). For chromatographic steps, the shape of the chromatogram has been used to indicate step performance. Suggested quantitative measures of chromatogram shape have been proposed, such as SPEW, for the retention times (or chromatogram distance) when the product collection starts, peaks, and ends and the width of the elution peak. In general, all in-process controls must be included within the list of process step outputs. It is usual to prioritize the inputs to be validated based on the

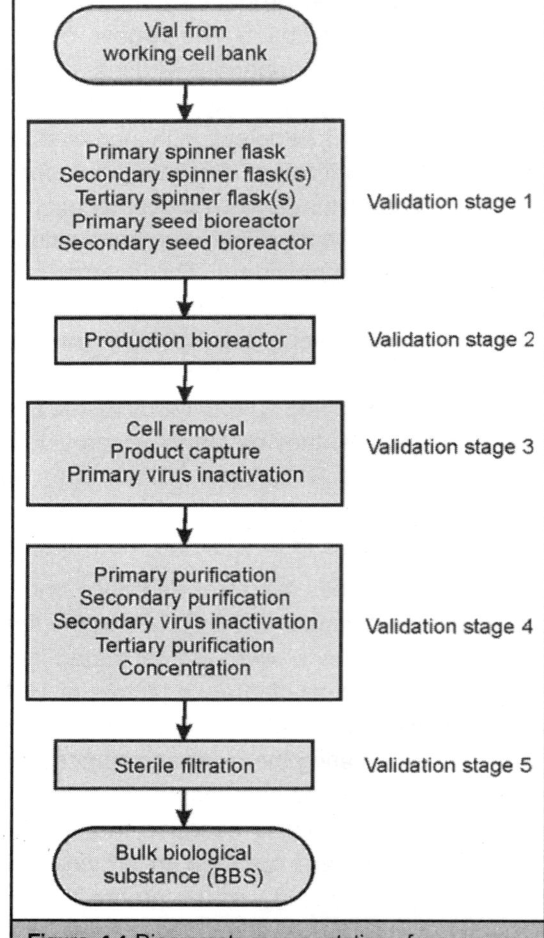

Figure 4.1 Diagramatc representation of a sequence of validation stages, some of which include multiple processing steps.

(i) sensitivity of process outputs to variations in each input, (ii) proximity of the validation stage to the final product, and (iii) technical difficulty in controlling an input to within a specified range. Assessment may be based on an analysis of all data that are available and documented, including process development, monitoring of clinical manufacturing, and any experiments specifically designed to gauge the sensitivity of the process to variations. The most important stage inputs must be included in the operating range validation program.

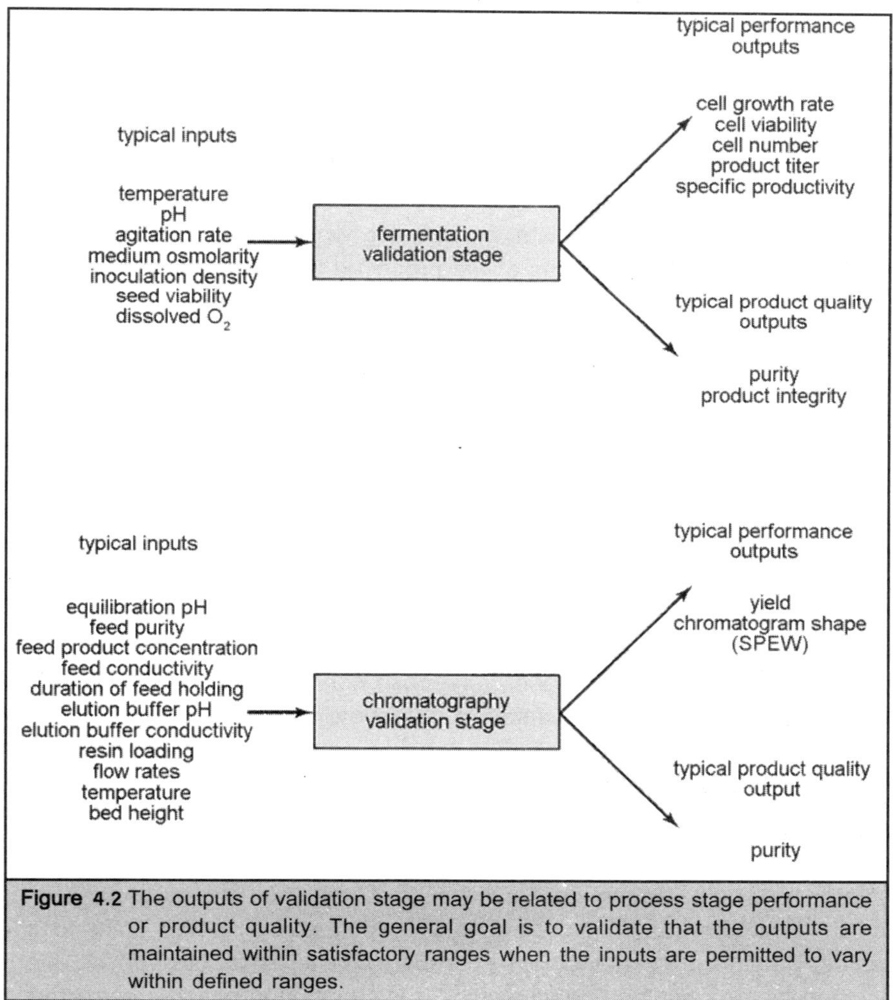

Figure 4.2 The outputs of validation stage may be related to process stage performance or product quality. The general goal is to validate that the outputs are maintained within satisfactory ranges when the inputs are permitted to vary within defined ranges.

Having identified the parameters to be validated, the next step is to define the criteria on the outputs that will be used to determine whether a validation experiment is successful. These criteria are typically a mixture of process performance results (e.g., titer, yield, and concentration) and quality of the product intermediate (e.g., protein purity, product homogeneity, and contaminant levels). The need for both types of criteria is based on the assumption that no amount of testing of the end product can ensure consistency without also showing the process to be controlled. These criteria are based on a variety of considerations and experience with the process. In some cases, these criteria are based on the output range results from operational history of the process during development or clinical supply production.

In other cases, they may be based on equipment limitations. For example, a minimum titer from the production fermentation tank may be established by the resin bed size and lower capacity limit for the first chromatography step in a process. In still other cases, output criteria may be derived from product specifications. If, for example, product purity is primarily determined by the performance of the first purification step of a five-step process producing a monoclonal antibody, then the product purity specification can be used to define a purity output criterion for the first step. It is then necessary to define an overall experimental paradigm. Usually, the experimental program begins with a set of proposed operating ranges for the inputs to be evaluated for each validation stage. These ranges may be based on equipment capability, theoretical considerations, process development data, operating history, the need for flexibility in manufacturing, or other considerations. Each validation stage is then run using inputs at the limits of the proposed operating ranges. These experiments can be run to investigate one input at a time or in parallel using statistical design of experiment (DOE) techniques.

The outputs from these experiments are obtained, and the proposed operating ranges become proven acceptable ranges (PARs) if the output criteria are met. If the output criteria are not met, then the input ranges are narrowed and the experiments are repeated until the criteria are met. This approach is taken for each validation stage. However, it maybe necessary to explore interactions among the validation stages. For example, the performance of a purification step or the quality of the product it produces often depends on the quality of the input feed stream. Because the input feed stream is an output from the previous purification step, one way to handle these interactions is to include the input stream properties as other inputs to the validation stage. Input streams of high and low quality are then required. One approach to obtain these streams is to operate the upstream purification step under best- and worst-case conditions. Alternatively, by knowing the function of the purification step to be validated (e.g., endotoxin removal, host cell protein removal, DNA removal, or removal of product isoforms), one can prepare input streams of high and low quality by spiking in the appropriate species to be removed.

The pair-wise validation of processing steps has been referred to as forward linkage validation. *Worst case* and *most appropriate challenge* are terms that describe the validation of a process run with its critical operating parameters set at the limits of their operating ranges. PARs are generally used to establish maximum operating ranges (MORs) bounded by action limits for the important process variables. When a process variable exceeds the outer limits of an MOR, then the impact on the process and the product may be unknown. This event often requires corrective action to be taken by operations staff and various QA systems. Depending on the importance of the variable, the duration and extent of the excursion beyond the MOR, and other operating data from process operations, a variety of follow-up actions may be appropriate. These actions can include relatively minor responses such as requiring a more detailed batch record review or more analytical testing than is normally required for product release when appropriate analytical systems and databases are available. Other QA responses include launching an investigation to look for the cause of the problem, requiring the demonstration of lot stability before release, requiring reprocessing of the lot, or rejecting the lot.

PROCESS CONTROL

From the standpoint of process control, it is necessary to determine the extent to which the process and product vary naturally under normal operating conditions. This knowledge is important in defining the normal operating range (NOR) bounded by alert limits for each important process variable. When a NOR is exceeded, operations staff may have to take corrective action. NORs must fall within the MORs established in most cases by the PAR studies. When NORs exceed MORs, the process is not sufficiently controlled to provide "a high degree of assurance that a specific process will consistently produce a product meeting its pre-determined specifications and quality attributes".

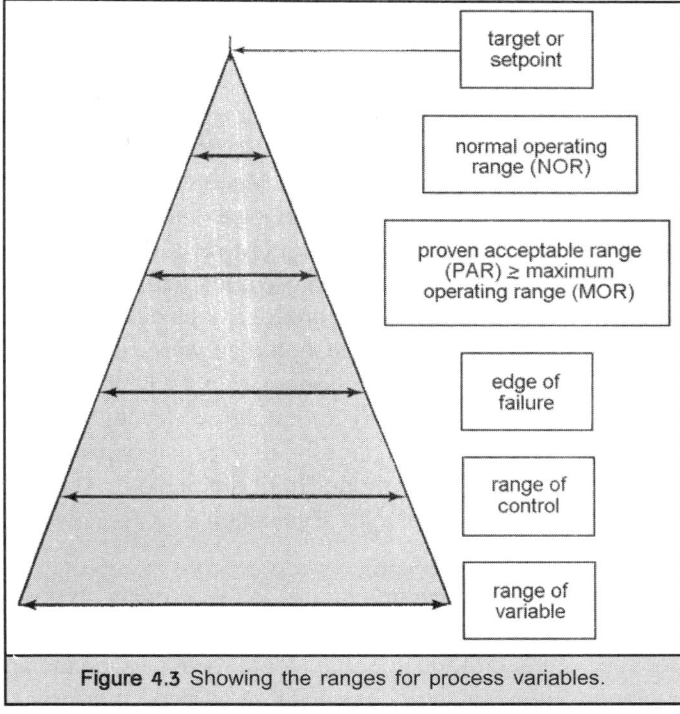

Figure 4.3 Showing the ranges for process variables.

NORs are typically calculated from replicate runs of the process that produces product that meets specifications. The calculation of a 99% or 95% confidence interval is a typical means of establishing NORs. Process development or optimization often establishes a desired value known as a target or setpoint (e.g., 37°C) for a process variable. Under normal operating conditions, however, a process variable varies within an NOR (e.g., 37 ± 0.5°C). Validation studies can determine a PAR (e.g., 37 ± 2.0°C), which can be used to set the MOR (e.g., 37 ± 2.0°C). The MOR falls within a broader range bounded by the edge of failure (e.g., 32-40°C). A process variable that exceeds the edges of failure will cause the process or product to fail, but the edges of failure are not typically determined during process development or validation except by accident. The ranges over which the variable can be controlled (e.g., 20-80°C) and for which it can exist (e.g., ≥ –273°C) are still broader.

LICENSING

Process validation is a requirement for product licensing. However, it continues throughout the lifetime of the product. Experience with a process in the commercial manufacturing facility is often very limited when the facility files license applications. Experience may mean production of as few as three or five lots to meet the qualification lot requirements for filing in the United States or Europe, respectively. These numbers of lots are insufficient to generate meaningful process or product specifications. As a consequence, initial process validation data are generally derived from pilot plant experiences obtained during the production of supplies for clinical trials or the development of the process. Regulatory authorities accept this information as sufficient for licensing, but many of the MORs, NORs, and product specifications may be refined during actual commercial production experience. In most cases, changes to these ranges are reported to regulatory authorities in annual updates, particularly when the change results in a narrowing of a control range or an improvement in product quality. In other cases, however, the change may need to be proposed to the FDA and regulatory approval gained before implementation.

SIGNIFICANCE

The significance of process validation has grown from its origins in the mid 1970s to the present time. Of major importance to the biopharmaceutical industry has been the total absence of medical incidents traceable to process control problems, largely because of the application of validation principles. More recently, regulatory authorities have reconsidered their former positions that most changes to the manufacturing process, including relatively minor ones, led to a product that might be different. In the past, clinical studies were required to establish the safety and efficacy of the new product. Today, many changes can be justified using analytical data and process validation studies without the need to undertake a clinical trial. More recently, the FDA has dropped the requirement for filing an establishment license application (ELA) for "therapeutic recombinant DNA-derived products and/or monoclonal antibodies". The FDA has also dropped the requirement for the FDA Center for Biologics Evaluation and Review (CBER) to release test each lot of product with this classification.

These changes are due largely to the increase in powerful analytical methods that are available to test these types of products and the greater sophistication of validation to characterize the process. Because of the growing importance of process validation and the absence of all but the most general guidance from regulatory agencies, it is likely that process validation will become the subject of future regulatory focus by the FDA and the European Union. Most cGMP regulatory requirements are specified for final product, and efforts in the United States to define cGMPs specifically for the production of bulk pharmaceutical chemicals (also known as APIs) are in progress. It is in these guidelines that we can expect more practical directives on process validation to be detailed. Given the expense of process validation work, guidance that leads to improved definition of requirements and elimination of process validation studies of marginal significance will be welcome. A reexamination of the balance among prospective, concurrent, and retrospective validation approaches will also be useful. Industry will need to contribute to these debates to ensure that validation costs and their impact on the overall health care system cost are directed toward areas with the greatest impact on the safety and efficacy of biopharmaceutical products.

5

BIOFILTRATION

Biofiltration is the biological removal of air contaminants from waste gases by means of aerobic microorganisms that are immobilized on a porous solid support. In a biofilter, the waste-air stream is passed through a biologically active packing material; organic or inorganic air pollutants are degraded by microorganisms into harmless end products such as water, carbon dioxide, mineral salts, and new microbial mass. Clearly, biofiltration technology is limited to non-toxic gases with good biodegradability and water solubility, because these gases are used by microorganisms as their sole source of energy and carbon. The idea of employing microorganisms to eliminate contaminants from waste gases is very old. Initially, biofiltration was used mainly for the treatment of odorous gases (such as hydrogen sulfide and ammonia) emanating from sewage-treatment plants.

The first reports on actual applications of this technology, dating back to the late 1950s and the early 1960s, dealt with soil-bed installations in West Germany and the United States, where it was demonstrated that biodegradation, rather than sorption, was responsible for most of the odor abatement. After the first publication of Bohn's extensive investigation on soil-filter beds, several additional full-scale applications of biofilters in West Germany and the Netherlands followed. These biofilters were employed to control the odors from wastewater- and thermal-sludge-treatment plants, composting facilities, rendering plants, carcass incinerators, and others. Jager and Jager, who compared several methods for off-gas purification from the composting plant of Heidelberg, first demonstrated the economic convenience of biofiltration. Process and technology development was attempted by Thistlethwayte et al., who utilized a trickling filter column packed with river gravel or glass balls and seeded with activated sludge. Waste air contaminated by H_2S, C_2H_5SH, $(C_2H_5)_2NH$, and C_4H_9CHO was purified, in countercurrent with a nutrient solution, with removal yields ranging from 40% to almost 100%. In addition, applications in which organic impurities were eliminated from waste gas, or air in beds inoculated with biologically active material, have been reported together with the first criteria for correct process design and operation. A further development of this technology was marked by a biofilter installation with continuous regeneration of the bed, which was obtained by removing active material at the bottom of the reactor and recharging it after regeneration at the top of the reactor.

A regeneration process by microbial, thermal, and chemical methods was also reported by Helmer, who used dry- and wet-soil filter beds to extract products of anaerobic fermentation of organic materials (e.g., NH_3, H_2S, sulfides, alcohols, and aldehydes). The importance of a sufficient moisture content in the filter bed and of a good distribution of the gas supplied to the biofilter was demonstrated for the first time. Leson and Winer estimated that in 1991 the total number of biofilter and soil-bed installations used in the United States and Canada mainly for odor control was less than 50. There was greater confidence in biofiltration in Germany and the Netherlands, where over 500 biofilters were operating in the same period. In the last two decades, the number of installations has significantly increased in Japan, growing from about 40 in the 1980s to 90 in the 1990s. On the contrary, in other European countries, including Italy, Switzerland, and Austria, the number of biofilter installations has been quite limited. Although biofiltration has developed mainly with the primary aim of eliminating odorous compounds from exhaust gases, since the early 1980s this technology has been of increasing importance in air pollution control, and the number of applications of biological filters has strongly increased due to an increasing concern about environmental quality and more stringent air-emission quality standards.

The application field of biofiltration has widened thanks to better knowledge of the biodegradation process, in particular the control of the process conditions, as well as to improved biofilter construction and packing material composition. In particular, special packing materials having optimal structural properties prevent aging phenomena, decrease the pressure drop over the packed bed, and increase the long-term operational stability of the filter, as well as its microbial activity. Furthermore, the considerable research carried out on the isolation, selection, and construction of strains or mixed cultures of microorganisms (mainly bacteria) has led to the extension of biodegradation to both anthropogenic (i.e., manmade) and xenobiotic compounds (chemicals that do not occur naturally). Compounds that sufficiently resemble structures of biological origin are eliminated rather quickly, whereas biodegradation of xenobiotics is very slow (recalcitrant compounds) or even impossible (persistent compounds). Nevertheless, the continuous adaptation of microorganisms to new substrates is at the basis of the increasing number of microbes capable of degrading xenobiotics.

As a consequence, biofiltration is now a well-recognized technology that is successfully applied to a wide range of industrial facilities, waste-disposal and food-processing activities to control odors, volatile organic compounds (VOCs), and air toxics (both inorganic and organic), especially those that are readily biodegradable. Biofiltration development is the result of the implementation of health and safety guidelines, enforced in many countries to protect people and the environment, combined with several advantages that this technique has demonstrated over traditional chemical and physicochemical treatment methods, such as water washing, chemical scrubbing (ozonation and chlorination), activated carbon adsorption, incineration, and catalytic oxidation. Besides the overall low energy requirements (moderate temperature and pressure conditions), easy maintenance, simple control, and low capital and operating costs, it must be stressed that biological filters do not transfer the pollution problem to another environmental compartment, as occurs with many other alternative control technologies. Because the pollutants are converted into harmless oxidation products (which are part of a natural cycle), they do not create any secondary pollution and thus can be considered environmentally safe.

BIOTECHNOLOGIES FOR WASTE-GAS TREATMENT

Biofilter Systems

There are three basic types of biological reactor systems used to treat waste gases: bioscrubbers, biotrickling filters, and biofilters. These can be grouped into those employing microorganisms dispersed freely throughout the liquid phase (bioscrubbers), and those using microorganisms immobilized on a packing or carrier material (biotrickling filters and biofilters). Moreover, in bioscrubbers and biotrickling filters the water phase is continuously moving, whereas in biofilters it is stationary. A bioscrubber consists of a scrubber unit and a regeneration unit. In the scrubber (absorption column), water-soluble gaseous pollutants are absorbed and partially oxidized in the liquid phase (the culture medium containing the microorganisms), which is distributed from the top of the unit. The contaminated water is subsequently transferred into an aerated stirred-tank reactor (regeneration unit), similar to an activated-sludge unit, where the contaminants are fully biodegraded. The regenerated suspension is continuously recirculated to the top of the scrubber section, thereby enhancing process efficiency.

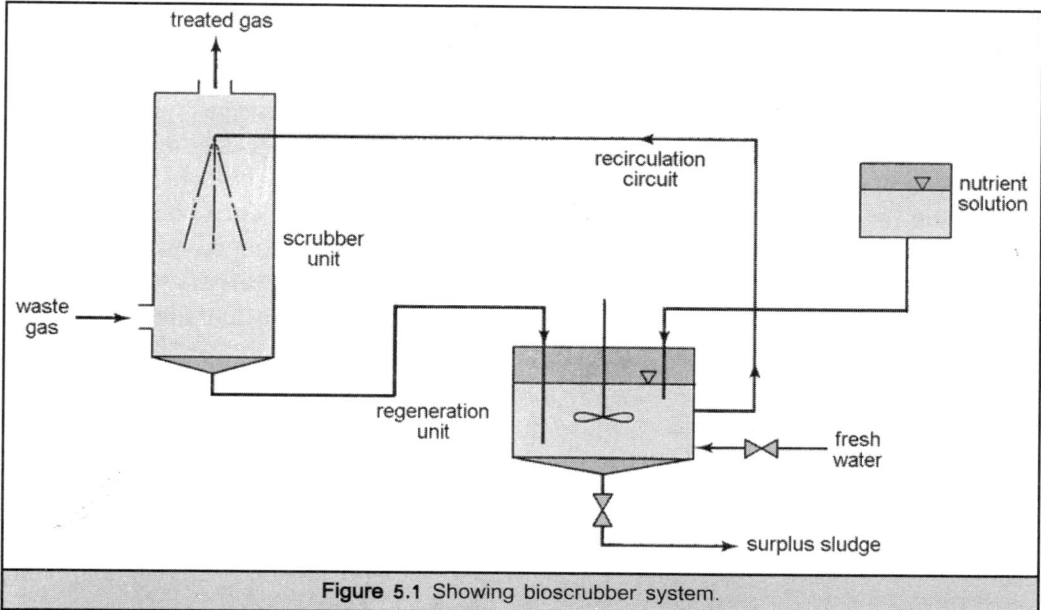

Figure 5.1 Showing bioscrubber system.

Biotrickling filters and biofilters are different from bioscrubbers in that gaseous pollutant absorption and biological degradation occur simultaneously in the same compartment, resulting in more compact systems. The polluted air flows through a biologically active bed, where microorganisms are attached in the form of a biofilm. As the gas diffuses through the packed bed, the pollutants are transferred to the biolayer and degraded. As illustrated, in order to ensure optimal operation of a biofilter, the inlet gas usually requires pretreatment processes such as: (i) particulate removal in order to prevent possible clogging and/or sludge build up, (ii) load equalization in case the waste-gas concentration is subject to strong fluctuations, (iii) temperature control, and (iv) humidification.

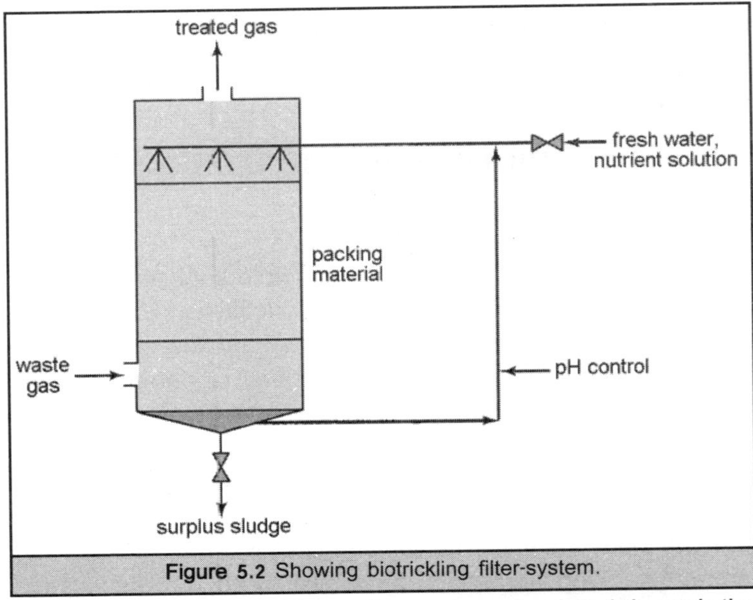

Figure 5.2 Showing biotrickling filter-system.

In biological trickling filters the packed bed consists only of inert materials (glass, ceramics, and plastics), while the liquid phase, containing inorganic nutrients, flows downward over the packaging material in countercurrent with the contaminated gaseous stream and is continuously recirculated through the bioreactor. Bioscrubbers and biotrickling filters are applicable mainly to the treatment of waste gases containing good or moderately water-soluble compounds, whereas biofilters, due to the large surface area available for mass transfer, are also suited to treat poorly water-soluble compounds. Moreover, due to their high reaction selectivity, biofilters are particularly suitable for treating large volumes of air containing easily biodegradable pollutants with relatively low concentrations, typically below 1,000 ppm. Compared with the other biological systems, biofilters have the widest application because they are easy to operate, simply structured, and imply low installation and operating/maintenance costs. Also, the reliability of biofilter operation is higher than that of bioscrubbers, where the risk exists of washing away the active microorganisms. Moreover, the presence of a large amount of packing material with a buffering capacity diminishes the sensitivity of biofilters to different kinds of fluctuations. Because the major disadvantage is the difficult control of parameters such as pH,

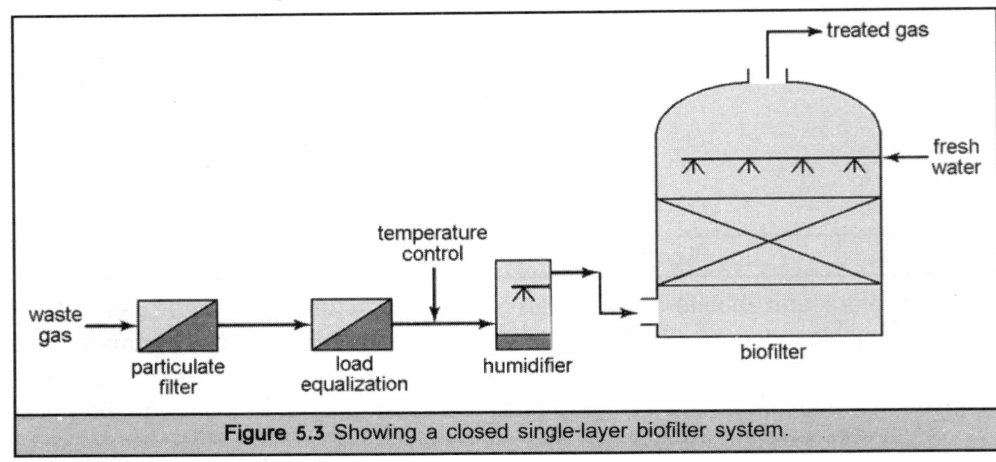

Figure 5.3 Showing a closed single-layer biofilter system.

temperature, and nutrient supply, biofilters may be unsuitable for degrading halogenated compounds (as acid metabolites are produced) and treating gas streams containing high concentrations of VOCs, unless long residence times or large bed volumes are applied. Biotrickling filters and biofilters are currently utilized mainly in compost-production plants, sewage-treatment plants, and agriculture, whereas biofilters and bioscrubbers are preferred in industrial applications.

Varieties of Biofilters

Among the different types of biofilters described in the literature for a variety of applications, open biofilters, consisting of single beds of compost or porous soils, commonly 1-m deep, are used mainly for odor and VOC abatement. Because they are in direct contact with the open air, their performance is strongly influenced by weather conditions (rain, frost, temperature fluctuations, etc.). The simple design and low cost of these systems are counter-balanced by difficult monitoring and control, as well as by large space requirements. All the other real-scale configurations are closed biofilters containing mixtures of organic materials and bulking agents. Among these, simple single-layer systems, often accurately controlled and monitored, but scarcely flexible and requiring large space, can be used for VOC treatment.

Multiple-layer systems, characterized by increased complexity and cost but less space requirements, consist of separately supported layers that allow optimal growth conditions for different microbial populations in each stage; a better control of the bed environment during the treatment of pollutant mixtures is also achieved. More sophisticated and very promising configurations have been successfully tested on bench- or pilot-scale projects for the treatment of VOC mixtures, BTEX (benzene, toluene, ethyl benzene and xylene), and reduced-sulfur compounds, among which there are multistage, modular, and step-feed systems, that in general are characterized by better control and flexibility, but higher costs and operative complexity.

BASIC PRINCIPLES

Biofiltration is a very complex process that involves mass transfer and reaction and is also influenced by the fluid dynamic of the gaseous stream through the reactor. Furthermore, it is necessary to maintain an environment as close as possible to the optimal conditions for the microorganisms. Therefore, several important process parameters, such as filter material structure, moisture content, temperature, and pH, must be kept under strict control and within an optimum range.

Filter Material

The choice of filter material is of great importance in order to maintain the efficiency of a biofilter, in that it must (a) guarantee optimum environmental living conditions for the microorganisms; (b) constitute at the same time a nutritious reserve, a humidity reservoir, and a mechanical odorless support; and (c) provide the structural stability of the bed. The choice of the material is strongly influenced by the need to minimize the overall volume necessary for the reactor, optimize the removal efficiency, keep energy consumption to a minimum, and minimize maintenance. In addition, the characteristics of the carrier material impact directly on the microbial growth and activity, thus in turn affecting biofilter performance. Biofilter beds have the advantage of immobilizing the microflora on the packing material, as a result of which these organisms, forming a biolayer, are not drained from the system, as is often the case in freely dispersed systems. Small particles of natural organic materials, such as compost, peat, soil, or mixtures of these materials with bark, leaves, wood chips, heather branches, humus earths, or brushwood (less than 10 mm in diameter), are widely used as packing media in biofilters because they provide a high specific surface area (from 300 to 1,000 m^{-1}), favorable living conditions for the resident microbial population (ensured by high retention capacities of water and nutrients), and favorable immobilization for the microflora involved. In practice these packing materials have shown the common disadvantage of being strongly subject to aging phenomena, resulting in bed shrinkage.

This phenomenon disturbs the homogeneous flow distribution of the gas and may cause a considerable increase in the pressure drop, a decrease in the specific area, and shorter filter lifetime. Moreover, the natural unhomogeneity of the structure of these natural materials prevents the uniform distribution of the gas flow, provoking short-circuiting channel development in the filter bed, with an increase in the flow resistance and a decrease in the biological degradation capacity of the bed. In order to prevent these problems, inert materials, such as polystyrene spheres, lava particles, glass beads, porous clay, and ceramic, are usually added to the natural filling materials. This combination, improving the uniform distribution of the gas flow, lowers the head losses up to 100–150 mm of water gauge, thus contributing to reducing the power requirement necessary to convey the gas through the filter and, consequently, the operating costs. Moreover, the addition of porous materials (e.g., granular activated carbon) with high internal porosity and hydrophilic properties, increases the buffering capacity of the filter, (which is very favorable, particularly in the treatment of gas streams with strongly fluctuating pollutant concentrations).

The selection of the most suitable material for a specific application depends on the nature of the pollutants and on the magnitude of the concentration fluctuations in the gaseous stream. In order to buffer the fluctuations in the concentration so that a constant supply of contaminants to the biofilter can be achieved, the adsorbent should have both good adsorbing power and reasonable desorbing properties. This combination provides the adsorbent with the capacity of adsorbing at high concentrations and desorbing at low concentrations and allows a significant reduction in the required filter volume. Compared to soil, compost has the advantage of providing lower resistance to the gas flow and consequently contained pressure drop, which should not, in general, exceed 250 mm of water gauge. Peat is the material with highest water-retention capacity and constitutes the optimal substrate for the microorganisms; nevertheless, as it implies a higher pressure drop, it is often mixed with other materials in order to improve the structural stability.

Humidity

The good performance of a biofilter depends greatly on an optimal moisture content of the filter bed. Moisture, the most critical operational parameter, is, in fact, essential for the survival and activity of the microbial flora (which can absorb and degrade substances only in the aqueous phase) and contributes to the filter buffer capacity. A lack of moisture causes cracking of the bed mainly at the gas inlet side, where the microbial activity is stronger due to the higher air-pollutant concentration. This results in (a) a considerable decrease of the microbial activity, (b) development of short-circuiting channels that cause bypass of the gas flow, and (c) contraction of the filter bed with consequent breakthroughs. On the contrary, an excess of water content promotes the development of anaerobic zones in the bed, due to occlusion of the pores, and provokes the formation of odorous volatile metabolic products, which are released in these zones and transported by the gas flow. Moreover, due to the higher gas-flow resistance, these zones are bypassed by the gas flow, resulting in a decrease in the mean residence time of the gas and, hence, in the efficiency of the process, and causes a 10% increase in the power requirement. A moisture excess causes problems also in the oxygen transfer due to a reduction of the gas–water interface per unit biofilm volume, and in the drainage of the filter components, with production of low-pH and high-load leachate that requires disposal.

The optimal moisture content of the filter bed should then range between 40 and 60% by weight, depending on the material. In order to preserve the microbial activity, it should exceed at least 40%. Water consumption by a biofilter is generally low, depending on the temperature and the relative humidity of the input gas and, for open filters, on precipitation. Due to evaporation of the aqueous phase, freshwater must be continuously supplied to the system in order to ensure a degree of saturation of more than 95%, which is necessary for optimal living conditions. This is normally realized by humidifying the inlet gas stream in a water-spray scrubber upstream of the filter unit and by periodically spraying water onto the surface of the packing material by means of surface sprinklers, spray nozzles, and so on. This periodical water supply is necessary not only because it is not always possible to humidify the inlet waste gas up to the required level, but also because heat is produced by the microbial oxidation of the pollutants. Moistening equipment must be operated in such a way that the moisture content stays within the indicated limits at any point of the medium. This operation must be realized in a very controlled way, otherwise the structure of the bed may be disturbed, which results in compaction of the filter bed in the long run and in increased gas-phase pressure drop.

pH

Maintaining an optimal pH in the filter material is a very important operational requirement for maximizing the biofiltration process because the microbial activity is strongly dependent on the pH. The effect of pH on biological systems is the indirect result of its action on the enzymatic reaction rate, which usually falls outside an optimal value. In biofiltration, as well as in most aerobic biological processes, the range of pH within which the biological systems can operate is typically between 5 and 9, with an optimum range of 6–8. Because in biofilters the aqueous phase is stationary, problems of package acidification may arise during the degradation of pollutants such as sulfur- or nitrogen-containing compounds and halogenated hydrocarbons, which can generate acidic metabolites such as sulfuric, nitric, or hydrochloric acid.

As a consequence of their accumulation, the microbial activity may decrease or even stop, seriously compromising the overall elimination capacity of the biofilter. These problems can be prevented by addition of buffering agents, such as lime, dolomite, limestone, marl, phosphates, or other water-insoluble alkaline materials, even if the neutralization products can sometimes reach inhibitory levels. Besides the above adverse effect on the process, acidification of the biofilter, as well as the acidic nature of the waste gas, filter material, or leachate, can corrode the air distribution and leachate collection pipes. In such cases, the use of materials resistant to corrosion, together with a continuous pH monitoring, are suggested.

Thermostatic Effect

Temperature remarkably influences the microbial growth through its effects on both bacterial enzyme activity and on various metabolic processes. Because the microorganisms generally applied in biofilters are mesophilic, temperature should be maintained between 20 and 40°C, with an optimum of 35°C for most aerobic microorganisms. Suboptimal temperatures can slow down the degradation capacity. The bed temperature should not rise above 45°C, although the resident microflora can gradually adapt itself to temperatures as high as 50°C. Microbial activity slows dramatically above 65°C. The effect of low temperature on biofilters depends largely on the inlet gas temperature. In fact, although low bed temperatures, especially below 10°C, decelerate biological oxidation, this effect can be partially counterbalanced by increased gas sorption by solids. However, the microbial activity will recover again upon preheating the influent exhaust gas. Fortunately, microorganisms have shown a tremendous ability to rapidly adapt themselves to drastic variations of temperature.

Reaction and diffusion rates typically increase with increasing temperature; nevertheless this effect is offset by a decrease of the solubility of the compounds to be removed, as well as a decrease in the physical adsorption capacity of the packing material at higher temperatures. Operation of the biofilter at high temperature shifts the microbial population towards thermophilic microorganisms. In order to prevent the death of the microorganisms, it is often necessary to control the influent gas temperature by precooling through heat recovery, water injection, or wetting of the packing material, thus offering additional economic advantage and the possibility of raising the gas relative humidity. Moreover, the heat liberated by aerobic microbiological activity causes a significant increase in temperature within the filter material, which must be taken into consideration at the design phase, particularly in biofilters that treat waste gases containing high concentrations of volatile organic compounds. For good operation of the biofilter, it is obviously necessary to continuously monitor the filter-bed temperature.

METHODOLOGY

Physical Model

In a biofilter, the microorganisms adhere to the surface of the porous solid material used as the support and grow, thus forming a biofilm. All models commonly used to describe the phenomena taking place in a biofilter under steady-state conditions are based on the principles of heterogeneous catalysis. According to the macrokinetic model proposed by Ottengraf, the support particle is surrounded by a biofilm; both the pollutant and oxygen molecules dispersed in the gaseous phase continuously transfer to the active biofilm region. Due to the metabolic activity of aerobic microorganisms, a concentration gradient forms that is responsible for the continuous mass transfer of the pollutant from the gaseous to the liquid phase. At the same time, the products of the aerobic degradation (carbon dioxide, water, inorganic compounds, etc.) continuously counter-diffuse toward the gaseous phase, where they are removed by the upward flow and are finally released from the biofilter. The metabolic activity is made possible by the continuous diffusion of inorganic nutrilites (mainly nitrogenous and phosphoric compounds) from the support to the biofilm. Nutrilite consumption is only partially counter-balanced by the mineralization of dead cells, therefore periodic addition or support renewal is necessary. Because the small particle size and the low solubility of the organic pollutants in water make the mass-transfer resistance in the gas phase negligible, pollutant concentrations at the biolayer interface and in the gas-phase bulk can reasonably be assumed in equilibrium and related by Henry's law.

Degradation Kinetics

The degradation macrokinetics in biofilters were reviewed by Ottengraf. Assuming gas flow through the biofilter to be plug flow, and that degradation follows typical Monod kinetics law, two different situations may occur: first- order or zero-order kinetics, respectively. Several experimental works demonstrated that zero-order kinetics are typically observed in biofilters for most of the volatile compounds, even at very low concentration levels.

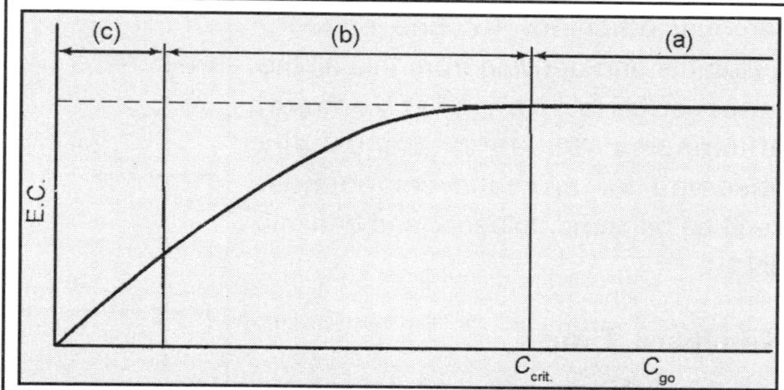

Figure 5.4 Dependence of the elimination capacity of a biofilter on the inlet off-gas concentration. (a) Reaction limitation; (b) Diffusion limitation; (c) 100% conversion.

Reaction limitation

The process is reaction-limited above a critical pollutant concentration (C_{crit} in the off-gas). Due to there being no diffusion limitation, the wet biolayer is fully active (e.g., the biofilter works at its maximum elimination capacity), and pollutant elimination is limited by the reaction. Under these conditions, the degree of conversion (η) is given by the ratio between the maximum elimination capacity, K_o which corresponds to an apparent zero-order kinetic constant, and the bed load per unit volume ($U_g C_{go}/H$):

$$\eta = 1 - (C_{ge}/C_{go}) = HK_o/U_g C_{go}$$

So, at the macroscopic level, the pollutant concentration in the filter bed linearly decreases with the height, and the compound is completely consumed when HK_o equals $U_g C_{go}$.

Diffusion limitation

Below the critical concentration, diffusion limitation occurs and the rate of reaction tends to exceed that of diffusion. As a consequence, the depth of penetration becomes smaller than the layer thickness and the biolayer is not fully active. The pollutant is nearly completely consumed by the biofilm before having entirely crossed it, and the conversion rate, which decreases with decreasing pollutant concentration in the waste gas, is controlled by diffusion. Under these conditions, the elimination capacity of the biofilter is influenced by both the reaction and the diffusion rates and the compound is completely removed when $HK_o \geq 2U_gC_{go}$. The applicability of this model to multicomponent waste gases is limited by the theoretical complexity of the system because the various compounds are not biodegraded independently. Industrial application regarding mixture of pollutants requires pilot testing for accurate sizing of a full-scale system.

MAIN CRITERIA

The main criteria for biofilter design and management are the degree of conversion or removal efficiency, the residence time, the superficial gas flow rate, the mass loading, and the elimination capacity.

Removal Efficiency

The *removal efficiency* (η), which provides information on the biofilter effectiveness, is given by the fraction of removed pollutant mass ($m_o - m_e$) with respect to total mass in the influent gas (m_o). The most common way of designing and managing biofilters consists of selecting the desired removal efficiency for one or more pollutants and deriving from this all the other criteria. The best removal efficiencies (95–99%) are usually observed for aromatic compounds, such as benzene, toluene, and benzoic acid.

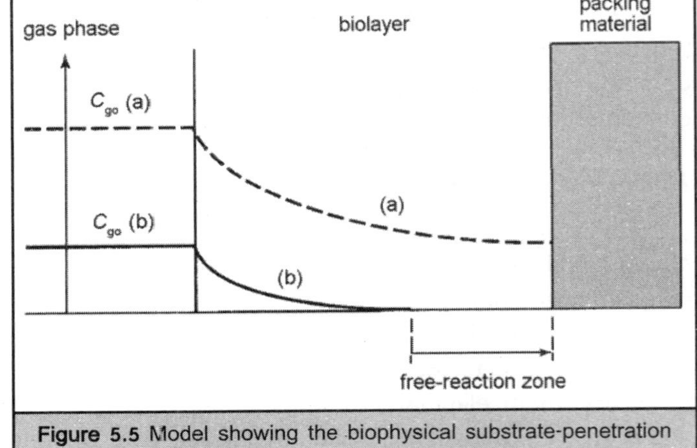

Figure 5.5 Model showing the biophysical substrate-penetration

Residence Time

The *residence time* of a gaseous pollutant in a biofilter (τ), given by the ratio of the empty bed volume of the filter (V_e) to the waste-gas flow rate (Q_g), is a fundamental parameter that cannot assume values smaller than a critical level if satisfactory removal efficiencies are desired. However, due to the difficulty in estimating empty-bed volume, residence time is often referred to total-bed volume. According to the pollutant degradability, this parameter typically ranges between 15 and 60 s. For liquid pollutants it is necessary to take into account their partition coefficients between liquid and gaseous phases.

Superficial Gas Flow Rate

The *superficial gas-flow rate* (U_g), defined as the ratio of the gas flow rate (Q_g) to the total cross-sectional area (A), is a parameter increasing with flow rate and decreasing with residence time; it usually ranges, depending on pollutant type, from 50 to 200 $m^3m^{-2}h^{-1}$. Since a superficial gas-flow rate increase obviously implies a reduction in removal efficiency, a maximum threshold of about 200 $m^3m^{-2}h^{-1}$ should not be exceeded.

Mass Loading

The *mass loading* (L_v) is defined as the ratio of the gas-flow rate (Q_g) to the bed volume (V) multiplied by the inlet pollutant concentration in the off gas (C_{ge}); for VOCs it should not exceed values of 3,000 ÷ 5,000 mg m^{-3}. In a biofilter, load increase could be ensured by either increasing waste-gas concentration or decreasing residence time. Although increasing gas concentration should, in general, accelerate biodegradation, excess concentrations above a certain threshold can inhibit microbial activity, so that decreasing residence time should be preferred. In addition, since the removal efficiency is affected by an increase in mass loading, and because excess organic load may obstruct the biofilter with the possible formation of toxic and/or acidic intermediates, a value between 10 and 160 g $m^{-3}h^{-1}$ should be employed to achieve satisfactory removal yields.

Elimination Capacity

The *elimination capacity* (EC) is defined as the ratio of gas flow rate (Q_g) to the bed volume (V), multiplied by the difference between the pollutant concentration in the influent and the effluent waste gases ($C_{go} - C_{go}$), it indicates a measure of the biofilter ability in removing the pollutants. From this definition it is evident that, because removal efficiency is expected to be close to 100%, the elimination capacity also ranges between 10 and 160 g m^{-3} h^{-1}. In addition, this parameter also is higher close to the inlet side, as happens for mass loading. The elimination capacity increases with mass loading and inlet pollutant concentration and decreases with residence time, the limiting value depending not only on the pollutant degradability, but also on the support material and the operating conditions. Beyond this value it is nearly constant.

APPLICATIONS OF BIOFILTRATION

Bench-Scale Results

Although many experimental works on biofiltration, both at laboratory and full-scale size, have been published during the last two decades, a critical comparison of results is difficult due to various reasons, among which are the use of German and Dutch as the most common languages in this research field, the use of different criteria to express the performances, the utilization of different volatile organic and inorganic compounds, the different types of biofilters used, and so on. Nevertheless, the work of Ottengraf can be taken as a useful reference basis to provide a complete overview on biofilter performance, as well as on the various effects influencing the process macrokinetics. For most organic pollutants, the steady-state continuous elimination follows zero-order kinetics, so that pollutant concentration decreases with the column height.

In the presence of mixtures of pollutants characterized by different biodegradability, a nearly complete removal of the most easily degradable compounds takes place at the bottom of the filter, whereas the most recalcitrant are mainly removed at the top. This suggests that different microbial populations could be responsible in the filter bed for the degradation of different organic components. The path explains another typical behavior of biofilter performances besides the one illustrated earlier. At low organic loads (corresponding to low concentrations for a given residence time), the elimination capacity linearly increases with this parameter, which means that nearly complete pollutant conversion is achieved. Over a critical load, the elimination capacity reaches a maximum value (corresponding to the apparent zero-order kinetic rate constant, K_o) and stays nearly constant. Maximum thresholds of 21, 27, 32, and 32 g m^{-3} h^{-1}, calculated for a mixture of toluene, ethyl acetate, butyl acetate, and butanol in a five-stage system simulating a biofiltration column with variable height, seem to reflect a direct dependence of this parameter on the pollutant degradability.

Pilot-Scale and Real-Scale Applications

Biofilters of different sizes (areas ranging from 10 to 2,000 m^2) are currently used in many industrial sectors, mainly in Europe, to remove a variety of mixtures of volatile organic and inorganic compounds from off-gases characterized by variable flow rates (1,000–150,000 m^3 h^{-1}) and low pollutant concentrations (usually less than 1,000 ppm). As one can see, that the main pollutants removable by biofiltration are listed, a relatively variable degradability characterizes not only the different classes of pollutants, but also the various pollutants within the same class. While for easily biodegradable substances the microorganisms suited for their degradation are usually present in the natural packing materials, inoculation and enrichment with cultivated organisms is often necessary for the slowly biodegradable ones.

MICROORGANISMS

Another factor strongly influencing biofiltration performance is the composition of the heterogeneous microflora, which primarily depends on the composition of the polluted gaseous stream. As a consequence of the progressive adaptation of microorganisms to the organic pollutants contained in the off-gas, which typically takes about 10 days, the population distribution shifts towards strains that naturally catabolize these pollutants.

Microbial Populations of Biofilters

Microbial population distribution along biofilters has recently been reviewed by Benedusi et al. The purification of waste air containing a limited number of pollutants is effectively achieved by a population restricted to a few microbial species and generally requires, to shorten the start-up operation time, inoculation of the biofilter with pure cultures or activated sludges previously adapted to each pollutant. This practice is particularly suited to the degradation of very recalcitrant or complex organic pollutants, such as halogenated and aromatic hydrocarbons, which often require cometabolism with more easily degradable substances to accelerate the process. On the other hand, gaseous emissions simultaneously polluted by several compounds, such as those coming from water works, compost plants, and so on, require as inoculum a more heterogeneous population possessing a wider variety of metabolic pathways, which is available in the microorganisms naturally present in activated sludges and compost.

Microbial loads of about 10^5, 10^4, and 10^3 colony-forming units per gram of support (CFU/g) have been reported for oligoheterotrophic, copioheterotrophic, and autotrophic metabolic types, respectively in heather-peat biofilters treating septic-tank emissions. Although Bohn demonstrated that enriching compost-based biofilters with microorganisms is not advantageous in terms of removal yields, probably because of the unfavorable competition with the indigenous population, the addition of activated sludges to medium devoid of microorganisms is a common practice. Finally, the growth and activity of the microorganisms are strongly influenced by the availability of oxygen and nutrients, the eventual presence of toxic substances in the exit gas, the degree of moisture, temperature, pH, and so on. A great advantage of the heterogeneous population present in the biofilter is the excellent ability to survive long periods (up to two months) without activity, provided that periodic aeration of the bed is ensured, even if microbial activity seems to be hardly affected. As revealed by plate counting in nutrient agar, most of the microorganisms growing in biofilters treating organic pollutants are heterotrophic eubacteria, actinomycetes, and fungi, which utilize the organic compounds contained in the gaseous streams as carbon and energy sources. Even when the presence in the off-gas of inorganic substances, like thiosulfate or hydrogen sulfide, should favor the growth of chemio-litoautotrophic microorganisms, such as *Thiobacillus* sp., by utilizing these substances as the energy source and carbon dioxide as the carbon source, the heterotrophic population becomes prevalent and is probably responsible for most of the degradation activity.

The bacteria more frequently detected in biofilters are soil bacteria, such as *Bacillus cereus* var. *mycoides* and several strains and species of actinomycetes belonging to the genus *Streptomyces*. Helmer, who did a microbiological characterization of compost-filter population, observed abundant growth of eubacteria (10^9 CFU/g), actinomycetes (2×10^6 CFU/g), and fungi (10^6 CFU/g) along the whole filter height, with higher density of fungi at the bottom. The highest bacterial densities are usually detectable near the inlet side, where substrates are more concentrated and more easily degradable, while different strains well-adapted to low substrate levels, as well as to more recalcitrant compounds, can be found at higher depth. The growth of several bacteria and fungi, whose cell densities decreased according to the sequence: *Actinomyces globisporus, Penicillium* sp., *Cephalosporium* sp., *Mucor* sp., *Micromonospora albus, Micrococcus albus,* and *Ovularia* sp., was also reported for similar supports used for mixtures of halogenated and aromatic hydrocarbons.

An opposite bacteria distribution has been reported by Benedusi et al. along a pilot heather-peat filter used to remove hydrogen sulfide and ethyl mercaptan from septic- tank emissions. They observed maximum bacterial densities just near the biofilter outlet side, with oligoheterotrophic population (2.4×10^8 CFU/g) predominating over all the others: copiochemioheterotrophs (5.7×10^6 CFU/g), sulfur-oxidizing chemioautotrophs (2.6×10^5 CFU/g), and nitrifying chemioautotrophs (7.2×10^5 CFU/g). Bacterial densities at biofilter bottom vary between 1% and 3% of these values, probably due to the hard pH decrease (up to 2.5–3.5) consequent to hydrogen sulfide oxidation (to H_2SO_4), as well as to the inhibiting effect of high concentrations of this pollutant on microbial activity. Other bacteria detected in biofilters are *Micromonospora vulgaris* and *Proteus vulgaris.* Among the main soil fungal taxa able to degrade polysaccharides naturally present as symbionts or saprophites in heather or peat (*Geomyces pannorum* var. *pannorum, Penicillium citrinum, Oidiodendron griseum, Penicillium frequentans, Pestalotiopsis guepinii,* and *Trichoderma koningii*), only *G. pannorum* var. *pannorum* and *T. koningii* seem to be able to develop in the bed under regular and irregular moisture conditions, respectively.

Other taxa, initially absent in the support material but normally living in polluted waters and soil, appear as a consequence of the air flowing and progressive acclimation, under conditions either of irregular (*Mucor* sp.) or regular moisture (*Mortierella bainieri, Drechslera australiensis,* and *Trichotecium roseum*). Other fungi detected in filter beds are *Circinella* sp., *Cephalotecium* sp., and *Stemphilium* sp. Although the contribution of fungi in biofiltration is not so clear, it is likely that an important role is played in the degradation of organic nitrogen, sulfur, and potassium, such as takes place in the soil. As far as the influence of the support structure on microbial density inside the bed is concerned, it was demonstrated that compost-based filters usually contain much higher population densities of these microorganisms than does soil. On the other hand, it seems that concentrations of bacteria cells and fungi spores in the treated emissions are only slightly higher than those observed in open air. Nevertheless, Liebe and Herbig advised against the risk of relevant fungi-spore emissions during biofilter maintenance and support-replacement operations. A great number of other bacterial strains have been inoculated and tested in bench-scale biofilters to treat waste gases polluted by specific compounds, most of which appear to be particularly promising for future development of this technology.

Metabolic Pathways

Different metabolic pathways are employed by microorganisms to degrade a variety of organic pollutants.

Aliphatic hydrocarbons

Several bacteria (mainly belonging to the genera *Nocardia, Pseudomonas,* and *Mycobacterium*), and some yeast and fungi, which are capable of utilizing saturated aliphatic hydrocarbons for their growth, are the main candidates for new applications in bioscrubbers and biofilters for the removal of such air pollutants. The oxidation of the terminal carbon to an alcohol, catalyzed by a monoxygenase, is responsible for the introduction in the hydrocarbon structure of only one of the atoms of molecular oxygen, which acts as a cosubstrate, while the other is reduced to water:

$$R\text{-}CH_3 + NADH + H^+ + O{:}O \xrightarrow{\text{monoxygenase}} R\text{-}CH_2\text{-}OH + NAD^+ + H_2O$$

Further oxidations by NAD^+-dependent alcohol and aldehyde dehydrogenases lead to a carboxylic acid that can be decomposed to acetyl-CoA by the β-oxidation route:

$$R\text{-}CH_2\text{-}OH + NAD^+ \xrightarrow{\text{alcohol dehydrogenase}} R\text{-}CH = O + NADH + H^+$$

$$R\text{-}CH = O + NAD^+ \xrightarrow{\text{aldehyde dehydrogenase}} R\text{-}COOH + NADH + H^+$$

$$R\text{-}COOH + ATP + CoA \xrightarrow{\beta\text{-oxydation}} CH_3\text{-}CO\text{-}CoA + AMP + PPi$$

On the other hand, polychlorinated hydrocarbons can be cometabolized in the presence of isoprene-degrading bacteria.

Aromatic compounds

The aerobic degradation of aromatics is only possible if the aromatic ring undergoes an enzymatic cleavage. Benzene can be degraded by several microorganisms by two divergent pathways after preliminary dioxygenation to catechol: the so-called *ortho*-pathway, utilizing the intradiol cleavage catalyzed by the catechol-1,2-dioxygenase, and the so-called *meta*-pathway, following an extradiol cleavage. Mixtures of benzene, toluene, and xylene are preliminarily dioxygenated, decomposed mainly via the *meta*-pathway, and finally metabolized through the citric acid cycle. The presence of an alkyl group in the benzene ring gives the microbe the opportunity to attack the compounds either on the side chain or the ring. In arenes with short side chains (up to seven carbons), both alternative routes are possible, mainly due to the widespread presence of transformable plasmids.

In particular, the plasmid TOL, containing catabolic operons for both alkyl-group oxidation and ring cleavage via *meta*-pathway, gives several strains of *Pseudomonas* the ability to degrade toluene, *m*- and *p*-xylenes, and other monoalkylbenzenes. On the contrary, if the side chain is sufficiently long (more than seven carbons), its metabolization by initial side-chain attack via ω- and β-oxidation, provides the cells of different species with enough energy for growth, and then the ring cleavage is not necessary. Of the dialkylbenzenes, *m*- and *p*-xylenes can be biodegraded by certain strains of *Pseudomonas* containing the TOL plasmid by preliminary oxidation of one of the methyl groups to a carboxylic group and methylcatechol, followed by *meta*-cleavage.

Members of *Nocardia* are able to cometabolize *m*-and *p*-xylenes via the *ortho*-pathway, and *o*-xylene via the *meta*-pathway in the presence of alkanes as carbon sources. More recently, it has also been demonstrated that *o*-xylene can be metabolized as the sole carbon and energy source, via 3,4-dimethylcatechol and subsequent *meta*-cleavage, by some strains of *Pseudomonas stutzeri* and *Corynebacterium*. There are reports of different bacteria (*Pseudomonas* sp., *Xanthobacter* sp., etc.) able to grow on styrene and methylstyrenes as sole carbon and energy sources, during which oxidation of the aromatic nucleus and subsequent *meta*-cleavage probably take place. In several bacteria, after a preliminary monoxygenation to a catecholic structure, phenol is degraded through the *meta*-pathway to directly give intermediates of the citric acid cycle. While for 4-chlorophenol degradation the chlorinated lactonic intermediate is directly dechlorinated before its introduction into the cycle, for 4-methyl-phenol a specific enzyme catalyzes the isomerization of the related intermediate into a more easily metabolizable lactone.

Biodegradation and filtration

Although the biodegradation of biphenyl, naphthalene, and polycyclic aromatic hydrocarbons (PAH) is a subject of less relevance for biofiltration applications given their solid state at ambient temperature, their high toxicity has brought lively interest in the actual possibility of purifying low contaminated off-gases by utilizing biofilters inoculated with liquid suspensions of strains specifically adapted.

Apart some subtle differences, biphenyl is biodegraded through a pathway common to many different bacteria belonging to the genera *Pseudomonas* and *Nocardia*, including a preliminary dioxygenation of the ring in the C2 and C3 positions to give a catechol-type intermediate, which is subsequently *meta*-cleaved and catabolyzed via 2-oxopenta-4-enoate and benzoate. Naphthalene is degraded mainly by members of the genus *Pseudomonas* via dioxygenation to 1,2-dihydroxynaphthalene, followed by extradiol cleavage of the ring between C1 and C9, whereas PAHS, which are potential carcinogens, mutagens, and tetragens, are dioxygenated and cleaved in two different ways, according to their phenanthrenic or anthracenic structures. Less detail is available for higher molecular weight PAHs.

ECONOMIC VALUES

Comparing cost data quoted in the literature for biofiltration systems is very difficult because of the use of different currencies, the fluctuations of their reciprocal rates, the cost variability with time, the different units employed, and so on. Nevertheless, there is no doubt of biofilter economic convenience when compared with other traditional abatement techniques, due to the following advantages: minimum operational input, reduced production and disposal of by-products, easy start-up procedures, operational stability at steady state, lcw operative temperature, low cost of materials, simple building technologies, simple control systems, and low power requirements. From the comparison of different biofilters in use in Europe, total operating and maintenance costs ranging from \$0.2 to 0.5 per 1,000 m^3 of waste gas can be estimated, which includes depreciation and interest, water consumption, replacement of filter materials and personnel, wastewater treatment, and energy costs for compressors.

Costs reductions showing a half of these amounts for United States installations do not include expenses for support replacement and reflect the lower cost of electricity. Capital costs are often negatively influenced by several factors, among which high pollutant concentration in the waste gas, recalcitrant nature of the components, necessity of gas pretreatment, reduced waste-gas availability, lack of space near the polluting source, and transport. Since the transport costs are strongly dependent on the country and the nature of the site, a cost comparison is only possible if this factor is not considered. On these bases, capital costs for open single-bed filters are in the range between \$300 and \$1,000 per m^2 of filter material in Europe and between \$600 and \$1,000 per m^2 in the United States.

These costs would increase up to \$3,800–5,700 per m^2 if lack of space would force the build-up of the filter on the plant cover, because of the additional costs of personnel, piping, and compression, and up to \$1,000–5,000 per m^2 if particular meteorological situations (snow, frequent atmospheric precipitation, etc.) require an upper covering or enclosed systems. Multiple-bed systems seem to be twice as expensive as the single-bed ones. The filling material may consist, according to the pollutants, of heather, peat, and earth, often mixed with other inert materials necessary to minimize head losses. Because of these variables and the market situation, costs may fluctuate remarkably from \$60 to \$700 per m^3. Taking odor abatement as a basis for comparison of different techniques, biofiltration is by far the most convenient process in terms of space availability, resulting in, as a function of waste-gas composition, total cost savings of 15-30%, 45-70%, 50-75%, and 80% with respect to bioscrubbing, chemical scrubbing, adsorption, and ozonation, respectively. The success of biofiltration appears to be the result of the very low incidence of running costs to total costs as compared to other techniques. Extending the above comparison also to incineration, it has been estimated that biofiltration allows total cost savings of 60-75%, and investment and operational cost savings of 40-75% and 70-80%, respectively.

FURTHER DEVELOPMENTS

Pollutants' degradability often depends on their origin and source: although xenobiotics, which are characterized by unnatural structures, cannot be degraded (persistent compounds), or can be

Table 5.1 Xenobiotic compounds and microorganisms able to degrade them in pure cultures

Pollutant	*Microorganism*
Chlorinated hydrocarbons	
Methyl chloride	*Hyphomicrobium* sp.
Dichloromethane	*Pseudomonas* DM1
	Methylobacter DM111
	Hyphomicrobium sp.
1,2-Dichloroethane	*Xanthobacter* GJ10
Vinylchloride	*Mycobacterium* L1
Epichlorohydrine	*Pseudomonas* AD1
Chlorobenzene	*Pseudomonas* WR 1306
1,2-Dichlorobenzene	*Pseudom onas* GJ 60
1,3-Dichlorobenzene	*Pseudom onas* sp.
1,4-Dichlorobenzene	*Alcaligenes* A175
Arenes	
Benzene	*Pseudomonas* sp.
	Acinetobacter calcoaceticus RJE 74
Toluene	*Pseudomonas* sp.
	Pseudomonas putida
	Bacillus sp.
Monoalkylbenzenes	*Pseudomonas* sp. NCIB 10643
	Pseudomonas fluorescens
	Pseudomonas putida RE204
	Acinetobacter lwoffi
m-, *p*-Xylene	*Pseudomonas* sp.
	Nocardia sp.
o-Xylene	*Pseudomonas stutzeri*
	Corynebacterium sp. C125
Styrene	*Xanthobacter* sp 124X
	Pseudomonas putida
α-Methylstyrene	*Pseudomonas* sp.
Biphenyl	
	Pseudomonas sp. NCIB 10643
	Nocardia sp. NCIB 10503
	Pseudomonas cruciviae
	Pseudomonas putida
	Pseudomonas pseudoalcaligenes
	Pseudomonas aeruginosa
	Escherichia coli
(Fused-ring compounds)	
Naphthalene	*Pseudomonas* sp.
	Pseudomonas fluorescens
	Pseudomonas putida
PAH	*Pseudomonas* sp.
	Pseudomonas paucimobilis
	Pseudomonas vesicularis
	Algaligenes denitrificans

degraded only with difficulty (recalcitrant compounds), the biogenic ones are easily biodegradable. For this reason, much research effort is being spent on the isolation, selection, and construction of different microorganisms able to grow on difficult carbon sources in pure cultures, at rates comparable with those usually encountered on common substrates. For these reasons, it is evident that future developments and new applications of this technology strongly depend on the actual possibility of utilizing these microbial strains in pure cultures or selected consortia under the peculiar conditions, as present in biofilters.

SUMMARY

As pointed out here, biofiltration has become by now a well-recognized technology for the control of odors and the elimination of air contaminants from waste gases, and represents a new generation of air pollution control technologies. Over the past two decades, thanks to the considerable progress made in the microbiological and process-engineering fields, biofiltration has gained the confidence of many of industries and has found increasingly widespread application. The suitability of this technique for the treatment of a wide range of inorganic and organic pollutants has been proven. Removal efficiencies as high as 90% or more are usually achieved for common pollutants, such as alcohols, ethers, aldehydes, ketones, and monoaromatic volatile chemicals.

The operating cost savings, the technological simplicity of the plants, the reduced maintenance and energy requirements, and the minimal generation of by-products that must be disposed of, make biofiltration, with respect to other air pollution control technologies, an attractive cost-effective alternative, particularly if applied to large volumes of gaseous-waste streams containing readily biodegradable contaminants in relatively low concentrations. Although its application to different compounds has been demonstrated, considerable research is being undertaken to find appropriate strains able to degrade recalcitrant compounds, in particular, most of which are xenobiotics. In addition, new strategies should be developed to improve the control of essential operating parameters affecting biofilter performance and to optimize reactor design and performance, in order to widen the application field of biofilters.

6

SEDIMENTATION THEORY AND APPLICATION

By definition, the bioprocessing industry uses biological cells as little factories to make a variety of products. Once they are presumed to have finished their product synthesis step, the cells need to be separated from the liquid medium containing them. During the course of research involving biological cells or during the isolation and purification of products derived from cells, cell separation is like the mid- game of a chess match. Correct moves at this stage provide an excellent chance to obtain a product of the desired characteristics. At least five general approaches are available for cell separation: centrifugation, cyclone separation, sedimentation, filtration, and microfiltration. In addition, cell flocculation can often be used to enhance the efficiency of these separation techniques. Sedimentation as a tool for cell separation is the focus of this article. It is also noted how centrifugation and flocculation, too, draw on sedimentation theory.

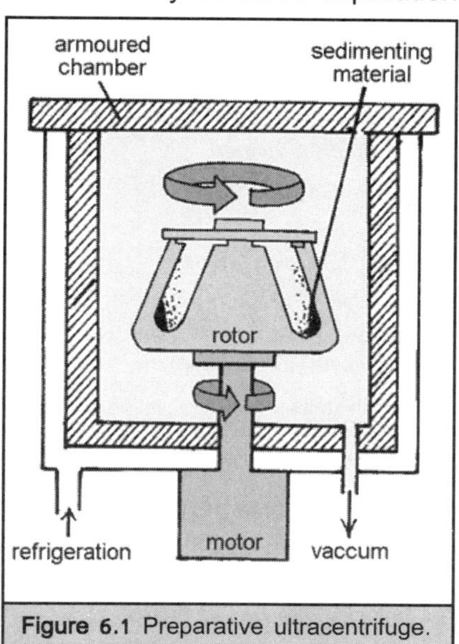

Figure 6.1 Preparative ultracentrifuge.

BASIC PRINCIPLE

The starting point in selecting a cell separation strategy is to locate the product of interest in a given suspension of cells. Is it the cells themselves? Is it all the cells or a specific cell type? Is the interest actually in a product inside the cells (i.e., intracellular) or in a product outside the cells (i.e., extracellular)? If the product is intracellular, cell removal may need to be achieved without significant cell damage. If the product is extracellular, the liquid phase must be separated while the solid content of the liquid phase is minimized, and the cells along with the other solids are merely the solid waste to be removed and disposed of. If the cells are the product, they must be separated without damage and their functionality kept intact. Sedimentation may be a useful tool for any of these situations. Sedimentation is relevant in a variety of biotechnology/bioprocessing situations:

1. Production of therapeutic or diagnostic proteins via mammalian cells
2. Achievement of high cell concentrations via cell recycle in continuous fermentors
3. Biological waste or water treatment
4. Production of beer
5. Study of different cell types involved in immunological response
6. Harvesting of cells in microcarrier or suspension cell culture

Table 6.1 gives specific examples of research in which sedimentation has been used as a cell separation tool. Sedimentation is also utilized extensively in other fields such as water or solids removal in mineral recovery and processing.

Table 6.1 Some examples of sedimentation as a cell separation tool in research

Cell and cell culture type	*Sedimentation application*
Mammalian, suspension	Sedimentation chamber as a separation device in a perfusion system for hybridoma cell culture for antibody production
Mammalian, microcarrier	Sedimentation separator pipe as a separation device for microcarrier cell culture
Mammalian, suspension	Gravitational sedimentation used to separate cells aggregated with the help of high-frequency, low-energy ultrasonic resonance fields
Bacteria	Sedimentation rates of cells of different shapes and sizes used to develop a method for enumeration and quantitation of bacteria
Yeast, *Saccharomyces cerevisiae*	Gravitation sedimentation used to recycle yeast cells for alcohol fermentation
Yeast, *S. cerevisiae* and yeast–mouse/mouse hybridoma cells	Sedimentation chamber used to separate cells from a chemostat culture
Starch endosperm of barley, *Hordeum vulgare L.*	Endosperm protoplast isolation method utilized gravity sedimentation to get good preparations
Spleen B-lymphocytes	Hybridoma-forming immune spleen B-lymphocyte cells separated based on their size and density using gravity sedimentation
Yeast, *Candida rugosa*	Sedimentation velocity measured to compare cells grown at different temperatures
Bacteria (*E. coli*) and yeast (*S. cerevisiae* and baker's)	Differences in flocculation and sedimentation properties utilized to achieve selective separation of cells carrying a certain plasmic
Sperm cells	A migration–sedimentation technique used to study whether it is a useful method to separate sperm cells leading to improved fertilization
Mammalian cells, erythrocytes	Utility of the erythrocyte sedimentation rate studied as a possible indicator of clinical disease or health
Sulfate-reducing bacteria, *Desulfovibrio desulfuricans*	Gravity sedimentation utilized for cell recycle in a system for microbial reduction of sulfur dioxide in an anaerobic digester

EQUIPMENTS REQUIRED

In processing applications, the sedimentation step is usually carried out in a typical processing tank. Such tanks are designed to have the capabilities of a feed line, an output or an overflow line, and a line to allow the removal of the sedimented material. However, equipment designs other than a processing tank have been shown to be very effective in special situations. One such design uses an inclined, parallel-plate sedimentation device shown, which can be used for cell recycle in continuous fermentations or cell culture. The inclined sedimentor design not only allows the fermentation or the cell culture to achieve high cell densities, it also offers an efficient means of selective cell retention.

A CelSep apparatus meant primarily for mammalian cell separations at the 1-L scale was commercially available. Although the unit is no longer on the market, the autoclavable unit worked well and allowed gravity separations under a continuous density gradient.

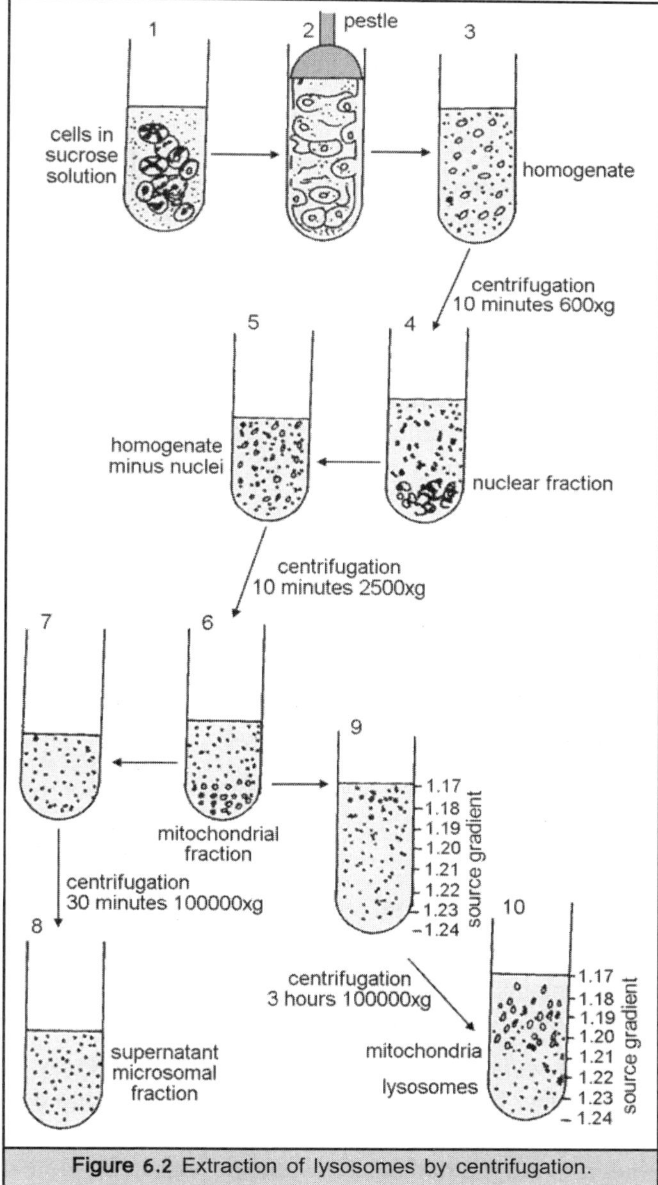

Figure 6.2 Extraction of lysosomes by centrifugation.

METHODOLOGY

1. Sedimentation as a separation tool is a mild technique. This is important in cell biology, where the functional capabilities of the isolated sub-population of cells must not be altered by the separation step.
2. Sedimentation lends itself to aseptic handling if needed. Although other techniques may allow cell separation without contamination, sedimentation devices for cell separation are typically easier to assemble, clean, and sterilize. Hence aseptic handling is more easily performed with sedimentation than with other options.
3. From a practical viewpoint, "unaided" sedimentation can be too slow a process if the sedimentation rates

are low. In industrial-scale cell separations, where microbial growth or enzymatic degradation can be a problem, long processing times are generally not acceptable. Such situations call for centrifugation, which "enhances" the sedimentation rates via the centrifugal force, because the problems of microbial growth or enzymatic degradation are less dominant for short processing times.

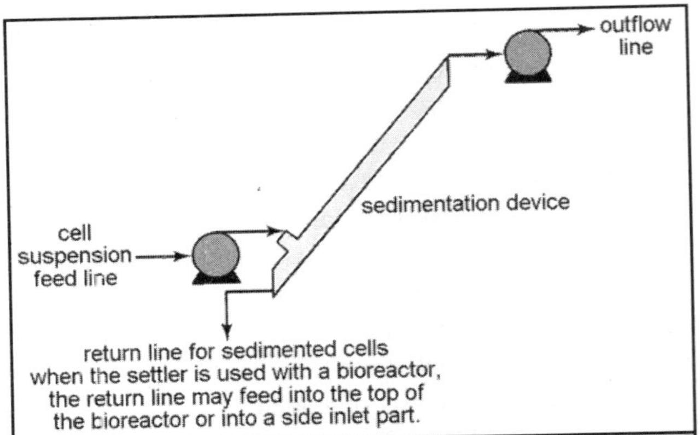

4. Sedimentation is a simple and a highly reliable separation technique; unlike other options, such as centrifugation; it does not require extensive training of the user. In its simplest form, sedimentation may be the lowest-cost technique for cell separation.

Figure 6.3 Conceptualized representation of an inclined, parallel-plate, cell sedimentation device used for cell recycle in a continuous fermentation or a perfusion bioreactor.

Overall, sedimentation principles have broad applicability in cell separations and extend beyond simple gravity sedimentation. Sedimentation principles also apply to centrifugation and ultracentrifugation.

THEORY AND APPLICATION

To consider the practical issues in cell separation problems, it is useful to understand the principles behind gravitational sedimentation. Several factors affect the sedimentation of cell in a liquid medium:

1. The *difference* between the density of a cell and that of the surrounding liquid medium
2. The density of the liquid medium
3. Cell size
4. Cell shape
5. Cell concentration
6. The viscosity of the liquid medium

As a practical rule of thumb, the particle size needs to be greater than about 1 μm for gravitation sedimentation to be relevant as a separation tool. Other approaches such as centrifugation, ultra-centrifugation, and membrane separation are usually needed for smaller particles.

The variables just listed are linked together by what is known as the modified *Stokes law,* named after the Ireland-born British scientist George Stokes (1819–1903), who is credited with having developed much of the general theory on the movement of a particle in a flowing fluid. Stokes was a gifted mathematician who also explained the existence of fluorescence and the variation of gravity at the surface of the Earth—all during the brief span of 1845–1854. Although Stokes' law is not directly used in cell separations, it helps us understand how the various parameters affect the settling operation. For a simplified case of gravitational settling when the velocity relative to the surrounding medium is relatively modest (often expressed in chemical engineering literature via a dimensionless parameter called the Reynolds number being set at < 0.2), the Stokes' law modifications suggest that

$$V_s = \frac{g(\rho_p - \rho)D_p^2}{18\mu}$$

where V_g is sedimentation rate (ft/s or m/s); g is gravitational acceleration constant (ft/s^2 or m/s^2), ρ_p is density of the particle (or cell) (s/ft^3 or kg/m^3), ρ is density of the liquid (s/ft^3 or kg/m^3), D_p is diameter of the particle (or cell) (ft or m), and μ is viscosity of the liquid [lb/(s ft) or Pa s&rsqb. Many of the factors in this equation may be beyond one's control in a given practical situation; it is not used very much for actual numerical calculation in cell separations. However, we can learn how cell sedimentation is affected under various conditions by examining this equation.

Density

In many large-scale industrial cell separation situations, the cell sedimentation velocity is too slow for sedimentation to be a practical tool. The sedimentation velocity is slow because biological cells and cell aggregates (if cell aggregates are present) tend to have a low density. The Stokes' law equation teaches us that the greater the *difference* between the densities of the cells and the liquid medium, the greater the cell sedimentation velocity. This property can be usefully exploited by a controlled increase of the liquid medium density. For illustration purposes, let us assume we need to separate a cell mixture containing two different cell densities in a given liquid medium. If the liquid density is increased via the addition of agents like Ficoll, fetal calf serum, or albumin, the *relative* ratio of sedimentation rates of the two cell types is increased, leading to a better separation. A refinement of this idea is to use a density gradient in the liquid medium through which the sedimentation is taking place. A technique based on this idea has been called iso-kinetic sedimentation.

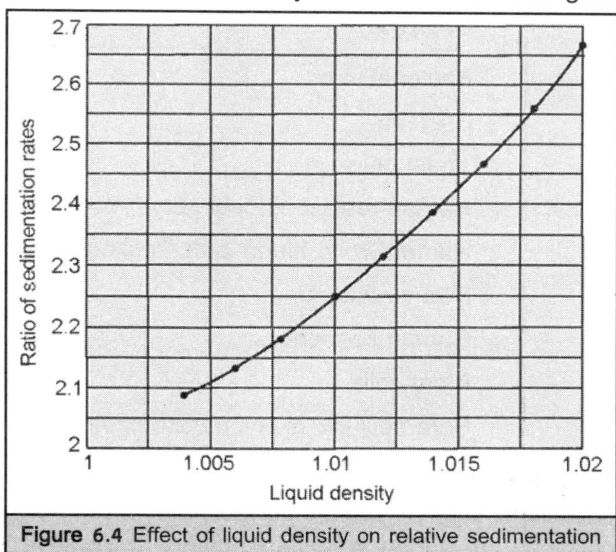

Figure 6.4 Effect of liquid density on relative sedimentation rate.

Cell Size

Table 6.2 shows the distance traversed in 1 h by hypothetical spherical particles of specific gravity 2.0 and different cell sizes, and settling in a fluid with a specific gravity of 1.0. In practice, the specific gravity of cells tends to be well below 1.2; the example merely illustrates the effect of cell size.

Table 6.2 Effect of spherical particle size on the sedimentation distance after 1 h in water at 293 K

Particle radius (nm)	*Distance settled*
0.1	0.08 nm
1	8 nm
10	800 nm or 0.8 μm
100	80 μm
1,000	8 mm

The sedimentation rate is proportional to the square of the diameter of the cell. Thus, any approach that leads to cell aggregation will help increase sedimentation substantially—albeit not as much as the "diameter square" relationship would seem to imply. The relative increase is less than that implied by the "square" relationship if the cell aggregate is porous and its density turns out to be less than that of the individual cells. In one study, the settling of larger-diameter activated sludge flocs was found to be proportional only to (floc diameter)$^{0.55}$.

Table 6.3 Typical sizes of various cell types

Cell type	*Size range (nm)*
Bacterial virus	0.08
Bacteria	0.1–2.5
Yeast cells	2–10
Human red blood cells (erythrocytes)	6–8
Human white blood cells (leukocytes)	8–12
Filamentous fungi	10–40
Human liver cells	30
Plant cells	10–100
Paramecium, didinium (protozoa)	100–200

Bacterial cells are often difficult to separate owing to their small size and low density, and so flocculants such as sodium polyacrylate, polydimethyldiallyl ammonium chloride, and carboxymethyl cellulose are added in some situations to achieve just this benefit. Ultrasound has been used to enhance the sedimentation of monoclonal antibody producing hybridoma cells because the acoustic treatment aggregates the cells. Addition of ferromagnetic particles to a bacterial cell suspension (gluconic acid producing *Acetobacter methanolicus*) followed by an exposure to a magnetic field has been shown to enhance sedimentation as a result of aggregation of cells with the magnetic particles. Two commercially products, Dynabeads and DETACHaBEAD, offer another way to achieve the immunomagnetic separation of cells.

In this approach, sedimentation is enhanced by forming rosetted cells on the antibody-coated Dynabeads. The rosetted cells are washed and concentrated using a magnetic device. The positively isolated cells may then be detached for further study. Because of the dramatic sensitivity of the sedimentation rate to the cell size, clinical researchers have attempted to determine whether the sedimentation rate of erythrocytes can be a useful indicator of patient health, results have been mixed, however. Size differential between cells can thus also be a useful tool when one is trying to separate a specific cell type from a mixture of cells. If one were to take samples at a fixed point below the liquid surface from a container in which sedimentation of otherwise identical but differing cell sizes is under way, the initial samples would contain the largest cells. With time, smaller and smaller cells would be in the sample. This principle is useful in the separation of lymphocytes from erythrocytes where sedimentation through a dense medium in a centrifuge is one approach.

Viscosity

The sedimentation rate is inversely proportional to the viscosity of the liquid medium. This property can also lead to some practical options. Because the viscosity change with dilution is often nonlinear, one may dilute the liquid medium by 25-50% to achieve a greater decrease in viscosity, and hence, better sedimentation rate. Conversely, one may add something that increases the viscosity of the liquid medium to slow down the sedimentation of cells. An example of this is the use of 4% dextran (of average molecular weight 500,000 Da) as a viscosity-increasing agent in the study of antibody-binding rates by FACScans in immunology, where suspended cells must be prevented from settling.

Gravity

Cell separation via centrifugation seeks to increase the sedimentation rates hundreds or thousands of times over simple sedimentation, by operating at a force much greater than the normal gravity. Because normal gravitational sedimentation is limited by the gravitational constant g, the relative efficiency of centrifugal separations has been examined by measuring the sedimentation velocity of disrupted yeast cells with and without a flocculent.

Chemical and Environmental Factors

It was pointed out earlier that polymers and other flocculating agents can aid sedimentation by increasing the effective cell size. In addition, the following factors can affect the behavior of particles in solution, hence can influence sedimentation rates: (i) pH and temperature of the solution; (ii) Presence of salts (e.g., of calcium, magnesium, manganese, sodium, potassium); and (iii) Presence of sugars. Although in most cases the interaction of these parameters is complex, the question is one of how the chemical environment affects the surface charge of the cells and their tendency to form cell aggregates. The idea is that the surface interactions will control the degree of cell aggregation, and hence, the sedimentation rates. Most microbial cells usually have a negative surface charge. The system pH or temperature can affect this charge. Divalent cations, especially calcium, and certain pH ranges can promote cell aggregation, leading to better sedimentation rates. Monovalent cations and sugars can reduce the tendency of cells to aggregate, thus reducing the likely sedimentation rates. The overall effect depends on the specific cell type and the physiologic state of the cells involved.

7

MICROBIAL TRANSFORMATION

The microbial transformation of organic compounds has been studied at the molecular level only during the last century of human history, driven by the development of techniques for determining the structure of organic compounds and cultivating microorganisms in pure culture. As information on these transformation reactions accrued, it became clear that microorganisms were, via their diverse metabolic pathways, principally responsible for the recycling of complex organic compounds on Earth. Almost simultaneously, it was recognized that the enormous diversity of microbial enzymes represents an invaluable resource for biosynthesizing desirable molecules. Historically, applications predate an understanding of the process, for example, the fermentation of sugars to yield ethanol. Although alcoholic beverage production was recorded in early written records, its biocatalytic basis was only proposed in the nineteenth century. Its enzymatic basis was first revealed in 1897 using a microbial cell-free system. Now we appreciate that microorganisms collectively have evolved the greatest enzymatic diversity found on Earth to metabolically degrade disparate organic compounds, and thus obtain chemical energy for adenosine triphosphate (ATP) synthesis. This metabolism, called catabolism or biodegradation, is a major contribution to the Earth's carbon cycle.

Microorganisms are increasingly used in engineered systems to biodegrade hazardous, xenobiotic compounds; this application is called *bioremediation*. In the twentieth century, work on microbial biodegradation of organic molecules and on biocatalysis to yield desirable chemicals has gone hand in hand. The versatile enzymes found in biodegradation pathways are sometimes effective biocatalysts for specialty chemical synthesis. For example, recombinant *Escherichia coli* strains expressing naphthalene dioxygenase, the first enzyme in the biodegradation pathway of naphthalene and other polycyclic aromatic hydrocarbons, are used in an emerging commercial process for the environmentally benign production of the blue jean dye indigo. The process is based on the regiospecificity of naphthalene dioxygenase that dioxygenates indole at carbons 2 and 3 to yield an unstable dihydrodiol that spontaneously eliminates water. The resultant compound, indoxyl, spontaneously dimerizes to yield indigo. In a commercial biocatalytic process, indoleis not supplied exogenously, because of its expense, but is generated in situ via over-expression of the enzymes functioning in the tryptophan biosynthetic pathway. A major impetus for bacterial indigo production is to preclude the formation of hazardous side products that are formed in the chemical process for indigo manufacture. Thus, biocatalysis and pollution prevention by biodegradation are intimately intertwined.

NEW DEVELOPMENTS

New developments in commercial biosynthesis and environmental bioremediation are certain to emanate from a wider availability of information on microbial biocatalysis and biodegradation. Because the number of chemical substances subjected to microbial catabolism is large (10^6-10^7), a comparably large number of distinct enzymes and pathways must occur in nature. The vast majority have not yet been studied, in contrast to the better-characterized enzymes and pathways in common intermediary metabolism. Information on these reactions or pathways is not typically available in biochemistry or microbiology textbooks. The increasing desire for this information has led to a virtual explosion of information published in journals, specialty books, and symposium proceedings over the past 2 years. There are also sources that contain extensive information but that are less readily available, for example, technical reports of government agencies such as the U.S. Environmental Protection Agency (EPA).

The information retrieval problem is exacerbated by the different clientele for this information, including microbiologists, chemists, chemical engineers, and civil engineers. Most of these groups do not routinely survey the same publications and are accustomed to somewhat different presentations of the relevant data. The rapid growth of the World Wide Web (WWW or Web) offers a way to overcome availability, outdatedness, and inconsistency problems because it can provide timely access to large amounts of appropriate, consistently formatted information. The Web builds on the success of the Internet, a collection of thousands of linked computer networks. Since the Internet's beginnings in the early 1970s, many scientists have used electronic mail as a very effective way to convey information. In January 1998, the Internet had approximately 30 million hosts, and its use is still growing rapidly. In 1991, a client/server software system called Gopher was developed to allow individual client computers to access textual information on the computer networks of the Internet. Gopher, developed at the University of Minnesota, appears to be a hierarchical text database that allows the user to "go fer" (look for) Internet resources on various topics and display them on even a modest computer. The World Wide Web, developed shortly thereafter at the European Laboratory for Particle Physics, combines graphics with text and allows for links between related Internet resources. For example, clicking on a highlighted or underlined term transports the user to the designated Internet resource, which can be on a server housed anywhere in the world.

WEB INFORMATION

The Web information of most value to people involved in biocatalysis deals with chemical compounds, microorganisms, metabolic pathways, enzymes, and genes. Users can access each field of information separately, although in some cases there are hypertext links that facilitate finding related information in independent databases. There are commercial databases that cover biocatalysis, for example, focusing on the use of enzymes in organic synthesis. These databases are not freely available and thus are not linked to public Web databases and are not discussed here. The enzyme and metabolism databases focus most directly on the biocatalysts that might be employed in organic synthesis or metabolic engineering.

Most metabolism and enzyme databases, however, deal largely with the reactions common to most living things, often denoted as intermediary metabolism. Emerging applications in biocatalysis are less likely to involve such reactions, being more likely to be derived from metabolism that is carried out by a more limited range of organisms, typically bacteria. As an example, nitrile hydratase from *Rhodococcus* sp. is now used in the commercial synthesis of acrylamide from acrylonitrile. This is not a reaction carried out by animals, or even *E. coli*, but it nonetheless provides for clean catalysis with high product yield and purity.

BIODEGRADATION DATABASE

The University of Minnesota Biocatalysis/Biodegradation Database (UM-BBD) first appeared on the World Wide Web on February 25, 1995, to meet the information needs of this distinct, diverse scientific community. It was designed to provide information on specialized non-intermediary microbial metabolism that is at the heart of scientific efforts in biocatalysis and biodegradation, in an easily accessible format, and to interface readily with Web-based databases of intermediary metabolism, such as the Kyoto Encyclopedia of Genes and Genomes (KEGG). Given that there are 10^6-10^7 organic compounds subject to catabolism, and that their catabolism is largely non-intermediary and unique to selected microorganisms, the UM-BBD has a much larger information domain to cover than intermediary metabolism databases. At the same time, information contained in the UM-BBD has potentially the greatest impact. Hazardous waste treatment is a $200 billion industry worldwide, and bioremediation comprises 5-10% of that total.

There is the potential for an expansion of this market with the development of new methods for metabolic and protein engineering. For example, researchers working on biodegradative pathways have begun to use metabolic engineering to develop new metabolic routes to handle particularly problematic pollutants. One example describes using a knowledge of existing biodegradative enzymes to design new pathways to metabolize polyhalogenated compounds. The UM-BBD could clearly stimulate similar projects by compiling and disseminating the crucial information needed to design such pathways de novo. We briefly describe the present status of the UM-BBD, including its format, use and users, peer acceptance, scientific advisory board, and future plans.

Format

In its first year of operation, the UM-BBD has evolved a standard format for reaction and compound pages. Both can link to many other internal and external information sources. The Entrez Medline link is especially interesting. The link back to the UM-BBD is prominent at the top. Also, when Entrez added a protein structure database, at our request they inserted the "Structure" link that is seen at the top, so that our users would be only a click or two away from viewing and manipulating any available tertiary structure of the enzymes contained in the UM-BBD. The UM-

BBD now includes links to graphics of enzyme reaction mechanisms on selected reaction pages. An index to graphics helps users find them.

The usefulness of a database is a function both of the information it contains and the access users have to this information. Initially, the only way to access UM-BBD information was through browsing lists of pathways, reactions, and compounds. As the UM-BBD grew, browsing became an increasingly less convenient mode of access. The ability to search the database was then added. At present one can search for compounds by full or partial name or synonym (e.g., toluene or methyl benzene), CAS registry number (e.g., 108-88-3), or formula (e.g., C7H8), and for enzymes by full or partial name (e.g., urease) or full or partial EC number (e.g., 3.5.1.5 or 3.5.1). This search option will continue to be developed with input from the UM-BBD user community and may be expanded in the future to include chemical substructure searches.

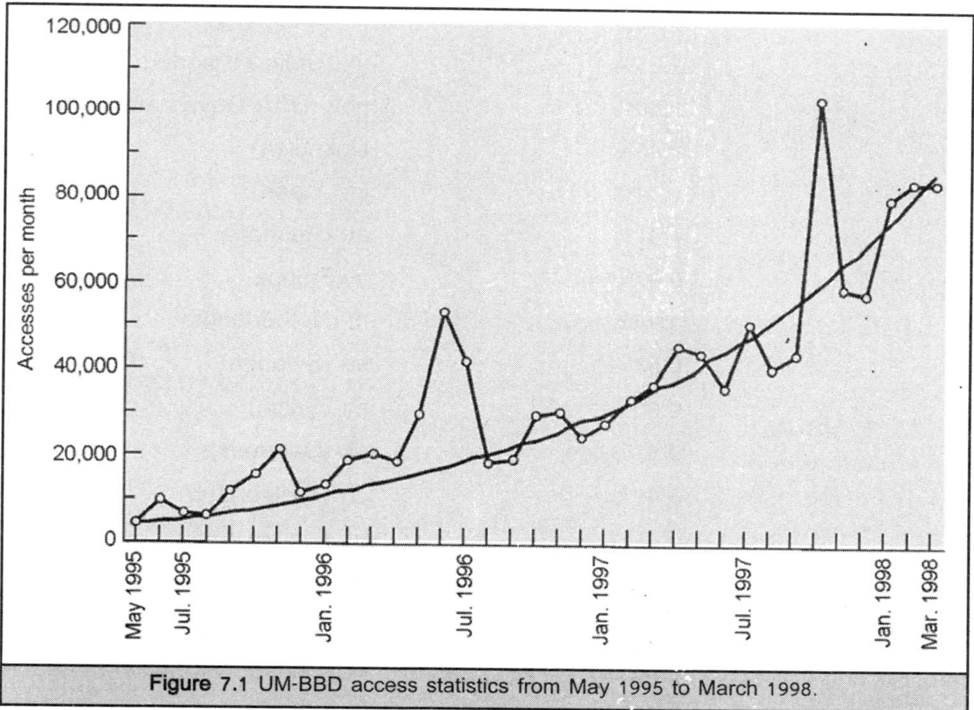

Figure 7.1 UM-BBD access statistics from May 1995 to March 1998.

Use and Users

Two months after the UM-BBD initially appeared on the Web, we began collecting use statistics. Usage has grown continuously; use statistics by Internet domain for 1997 are shown in Table 7.1. Currently, the UM-BBD has approximately 80,000 accesses per month. U.S. educational (30%) and industrial (20%) users together make up half of all present use. UM-BBD users are invited to "sign our guestbook" (i.e., voluntarily complete a seven-item electronic form). Those who sign are given the option to regularly receive e-mail progress reports, and most choose to do so. This e-mail user list presently has about 400 members, who are surveyed on an annual basis. Users have recommended pathways and features to us, and we have acted on some of their suggestions:

A pathway for degradation of a chlorinated aliphatic compound and graphical pathway maps were added in part because of user requests.

Table 7.1 UM-BBD use statistics by domain

#reqs	*%bytes*	*Domain*
187790	29.54%	.edu (USA Educational)
166052	19.66%	.com (USA Commercial)
87740	18.86%	[unresolved numerical addresses]
39050	8.13%	.net (Network)
12560	2.70%	.ca (Canada)
10008	2.40%	.de (Germany)
9527	2.23%	.uk (United Kingdom)
5944	1.56%	.gov (USA Government)
13285	1.45%	[Unknown]
4193	1.14%	.jp (Japan)
4960	1.07%	.au (Australia)
3950	0.97%	.fr (France)
3494	0.94%	.nl (Netherlands)
3718	0.68%	.se (Sweden)
2212	0.60%	.es (Spain)
2352	0.51%	.dk (Denmark)
1954	0.49%	.ch (Switzerland)
1948	0.46%	.us (United States)
2097	0.45%	.it (Italy)
1907	0.42%	.kr (South Korea)
1621	0.36%	.org (Non-Profit Making Organisations)
1473	0.34%	.mx (Mexico)
2210	0.33%	.il (Israel)
1332	0.31%	.be (Belgium)
1224	0.28%	.br (Brazil)
1014	0.26%	.fi (Finland)
1388	0.24%	.tw (Taiwan)
949	0.23%	.cz (Czech Republic)
925	0.23%	.no (Norway)

Table continue...

#reqs	%bytes	Domain
1022	0.22%	.mil (USA Military)
761	0.17%	.at (Austria)
672	0.14%	.sg (Singapore)
507	0.14%	.za (South Africa)
768	0.14%	.nz (New Zealand)
526	0.14%	.gr (Greece)
596	0.13%	.ar (Argentina)
467	0.11%	.pt (Portugal)
422	0.11%	.ru (Russian Federation)
468	0.11%	.my (Malaysia)
390	0.11%	.ie (Ireland)
524	0.11%	.co (Colombia)
406	0.10%	.th (Thailand)
436	0.09%	.pl (Poland)
381	0.09%	.su (Former USSR)
317	0.09%	.id (Indonesia)
284	0.08%	.cl (Chile)
331	0.08%	.hu (Hungary)
259	0.05%	.ee (Estonia)
131	0.05%	.hr (Croatia)
215	0.04%	.si (Slovenia)
112	0.04%	.sk (Slovak Republic)
258	0.04%	.tr (Turkey)
172	0.03%	.hk (Hong Kong)
185	0.03%	.in (India)
125	0.03%	.cn (China)
131	0.02%	.cr (Costa Rica)
78	0.02%	.ro (Romania)
124	0.02%	.lt (Lithuania)
70	0.02%	.bh (Bahrain)
64	0.02%	.pe (Peru)
104	0.01%	.ve (Venezuela)

Table continue...

#reqs	%bytes	Domain
71	0.01%	.uy (Uruguay)
69	0.01%	.eg (Egypt)
77	0.01%	.arpa (Old style Arpanet)
29	0.01%	.jo (Jordan)
57	0.01%	.ae (United Arab Emirates)
63	0.01%	.ph (Philippines)
31		.is (Iceland)
36		.yu (Yugoslavia)
18		.jm (Jamaica)
25		.pa (Panama)
23		.lb (Lebanon)
10		.do (Dominican Republic)
11		.lu (Luxembourg)
16		.pr (Puerto Rico)
11		.bm (Bermuda)
20		.ec (Ecuador)
10		.ua (Ukraine)
5		.zw (Zimbabwe)
5		.az (Azerbaidjan)
4		.md (Moldavia)
8		.gy (Guyana)
3		.am (Armenia)
8		.by (Belarus)
5		.mk (Macedonia)
9		.kw(Kuwait)
6		.bg (Bulgaria)
7		.gb (Great Britain)
9		.mt (Malta)
1		.dm (Dominica)
5		.gt (Guatemala)
4		.dz (Algeria)
9		.et (Ethiopia)

Table continue...

#reqs	%bytes	Domain
6		.cy(Cyprus)
7		.ge (Georgia)
6		.pg (Papua New Guinea)
2		.bo (Bolivia)
2		.pk (Pakistan)
2		.mu (Mauritius)
2		.lv (Latvia)
1		.om (Oman)

Connections

When the UM-BBD was first installed on a server in February 1995, it linked to several Internet resources. Links to others have been added since then. Most of the Web databases the UM-BBD links to have been contacted and collaborate on cross-linkage where possible. For example, the UM-BBD now has reciprocal links to Entrez PubMed, Entrez Nucleotide, Entrez Protein, Chem-Finder, and Ligand, and others are under discussion. Each such interconnection is usually the result of individualized programming.

The UM-BBD links to others in separate text lines, or, especially for reaction information, as links inserted in the text-formatted reaction. Unique among present Web-based metabolic databases such as KEGG, the UM-BBD specializes in a more broadly based set of metabolic reactions. Because 10 million organic compounds are known and most are enzymically biodegraded, the potential range of the UM-BBD is enormous. These reactions are under active study now and will be in the foreseeable future, and the ability to search for current information in remote databases, as opposed to more static links to individual documents, is very important to UM-BBD users.

At present, the UM-BBD provides dynamic search links to EXTOXNET, Entrez/MEDLINE, and Entrez/GenBank. Although more difficult to construct and maintain than links to individual documents (in the first 6 months of 1996, each of these links was reconstructed at least once because the Web sites changed search format), they allow UM-BBD users to obtain the very latest information installed on these servers. Thus, a search is installed in preference to an individual link wherever possible, even if it returns only one document at present.

Tomorrow or next week additional items may appear, and when they do UM-BBD users will have access to them. UM-BBD search links include information on the number of hits that were returned when the search was last tested, the date of that search, and, if possible, the exact records being searched and terms that are being searched for. For example, Entrez/MEDLINE searches may be carried out for MEDLINE titles and abstracts, or if this produces too many hits, they may be restricted to titles alone. Entrez/MEDLINE and GenBank searches require custom formulation to return the best results because of the ambiguities and complexities of chemical and biochemical nomenclature and database search syntax.

Acceptance

As mentioned earlier, the UM-BBD links to other metabolic databases. It also provides users with links to relevant information in several other biological and chemical Internet databases. The five prestigious international Web databases that it links to most closely—ChemFinder, Ligand, Entrez/Medline, Entrez/Protein, and Entrez/GenBank—all link back to it. The UM-BBD has also been selected to appear on many peer-reviewed scientific Web referral lists, including Internet Chemical Resources, the American Society of Microbiology Biology Resources List, and BioTech: Biochemistry, Biophysics, and Molecular Biology.

Advisory Board

During the first year of operation, a world-renowned scientific advisory board was established, including members in three countries from educational institutions, commercial companies, and government agencies: Peter Chapman (EPA); I. C. Gunsalus (EPA); Andreas Kiener; Thomas Leisinger; and Jack T. Trevors. Most recently, Alfred Spormann has joined the UM-BBD advisory board.

Future Developments and Plans

The UM-BBD began with only 4 pathways in February 25, 1995 and grew to 13 pathways by September 15, 1996, 25 pathways by February 20, 1997, and 59 pathways by April 21, 1998. However, with 10 million compounds in the biosphere, it will never, and is not meant to, contain information on the biodegradation of more than a representative fraction of these compounds. It was developed to contain reactions and pathways that span the diversity of organic functional groups and microbial metabolism. The next step in its development is to use this representative information to develop ways to predict microbial catabolic metabolism for compounds that the UM-BBD does not contain.

FUTURE DIRECTION

Biocatalysis will expand in scope as a knowledge of microbial catabolism increases. Most of the more than 10 million organic compounds are thought to be biodegradable, providing a virtually limitless potential for expansion of known microbial enzymatic reaction types. Additionally, information is rapidly emerging from the more than 60 bacterial genome sequencing projects completed or ongoing. This will lead to a new era of microbial metabolism studies, building on the extensive data on intermediary metabolism. These studies will increasingly reveal the vast catabolic capabilities of bacteria from diverse environments and taxonomic classes. Research advances will create a need for more people trained in microbial genomics, metabolic reconstruction, and enzymology. The UM-BBD, and associated courses, can contribute to training the next generation of microbial biocatalysis specialists.

The UM-BBD has been used instructionally at several colleges and universities, including the University of Minnesota, Iowa State University, Pomona College, the University of Waterloo, the University of Basel, and Hebrew University in Rehovot, Israel. A University of Minnesota course on biocatalysis and biodegradation is being offered completely over the Internet, using the UM-BBD as its focal point. The biocatalysis and biodegradation course instructs students on the general principles of microbial environmental biotechnology and microbial catabolic reactions. Using these fundamentals, students peruse the UM-BBD and related World Wide Web databases to research specific microbial transformation reactions. Student assignments are transferred to the instructors for assessment via electronic mail and the Web. This is a venue for advanced college-level instruction that is increasingly becoming available to students worldwide. It reflects a fundamental change in university education: for 6,000 years students were required to come to the scrolls, books, and personal contact with scholars that allowed learning to advance. Now, people can be at remote sites and increasingly access the accumulated knowledge of humanity via electronic means. The UM-BBD is one small part of this profound change in human information transfer and learning.

8

AGGREGATION AND COAGULATION

In colloid science, *flocculation* is defined as an aggregation of solid particles induced by addition of polymers; *coagulation* refers to an aggregation due to van der Waals or electrostatic attraction between particles. *Cell aggregation* was defined by Calleja as the gathering together of cells to form stable, contiguous, multicellular associations under physiological conditions. The criteria of intimate cell–cell contact and of physiological conditions are not met in the case of floc formation by the addition of flocculating agents; therefore, this definition establishes a distinction between aggregation and flocculation. However, polymers formed at a cell surface or excreted into the medium may be responsible for cell–cell interactions of the same nature as those produced by polymers added to the system. Calleja's definition of cell aggregation involves a physical movement, either active or stochastic, which is responsible for the change from a dispersed to an agglomerated state. The criterion of stability may refer to aggregation under dispersive conditions, typically in an agitated suspension (as opposed to formation of a colony on agar), of a centrifugation pellet or a sediment.

The criterion of physical movement emphasizes that a bond is being formed between distinct bodies, as opposed to chains and mycelial structures. However, certain cell aggregates of great significance in biotechnological processes are not simply formed by a gathering of initially dispersed cells. In activated sludges and in cell granules formed in anaerobic digestion, a dynamic situation is obtained in which some newly formed cells remain more or less loosely attached to each other, either by cell-cell or cell-debris interactions, or because a matrix prevents their dispersion, while free cells may be captured by an existing aggregate. From the biotechnologist's point of view, the distinction between coagulation, aggregation, flocculation, and agglomeration is not always clear. The words *aggregate, floc, agglomerate,* and *cluster* are often synonymous. They refer to an assemblage of individual cells, formed by initially free cells or by the lack of cell dispersion and occurring under physiological conditions or after addition of various agents to the medium. Aggregated microbial cells are of great interest in biotechnology. They offer advantages, with respect to fully suspended cells in downstream processing, by facilitating cell-broth separation by sedimentation or filtration. They also allow continuous fermentation processes by keeping the cells in the bioreactor with an up-flow of the broth. Further details concerning cell flocculation can be found in several books. Biofilms are not considered explicitly here; however, as concerns the mechanisms of association, they have many features in common with flocs.

INTERACTIONS BETWEEN COLLOIDAL PARTICLES

The essential features of the interactions between colloidal particles may be understood by considering three different aspects: (i) DLVO theory accounts for van der Waals and electrostatic forces; (ii) when surfaces are close to molecular contact, solvation interactions may play an important role; (iii) when mobile macromolecules are present at the interface, they may exert a particular influence at long distances due to their interactions with the solvent, with the surfaces, and with one another. In the section "Charge Neutralization by Cationic Polymers", a particular mention is made of the possibility to neutralize the surface charge and thereby provoke flocculation by oppositely charged macromolecules.

DLVO Theory

According to DLVO theory, the free enthalpy of interaction, G_t, between two particles approaching each other may be accounted for by the sum of two contributions, corresponding respectively to electrostatic and van der Waals interactions

$$G_t = G_e + G_v \quad \ldots(1)$$

The electrostatic contribution, G_e, is due to the overlap of electrical diffuse double layers of approaching particles. For two spheres of the same nature and same radius, R, suspended in a symmetrical electrolyte solution,

$$G_e = (647\pi kTRNC\kappa^{-2}T^2_d)\exp(-\kappa D) \quad \ldots(2)$$

where k is the Boltzmann constant, T the temperature, N the Avogadro number, C the electrolyte concentration (mol/L), and D the separation between particle surfaces.

$$T_d = \tanh[ze\psi_d/4kT] \quad \ldots(3)$$

where z is the ion charge, e is the electron charge, and ψ_d the electric potential at the inner limit of the diffuse part of the double layer. It differs from the surface potential, $v0$, due to adsorption of ions in intimate contact with the surface, that is, in the Stern layer (chemically adsorbed or retained by localized electrostatic attraction). For computing G_e, the value taken for ψ_d is usually the zeta potential ζ, that is, the electric potential at the plane of shear between the cell surface and the solution. κ, the reciprocal of the Debye length, determines the rate of the variation of the electric potential as a function of the distance to the surface, within the diffuse double layer.

$$\kappa = (1000e^2N/\varepsilon kT)(2I)^{1/2} = 2.32\ 10^9(2I)^{1/2} \text{ (at 25°C)} \quad \ldots(4)$$

where I is the ionic strength of the bulk solution and ε the electric permittivity. At an ionic strength of 10^{-1}, 10^{-3}, and 10^{-5}M, κ^{-1} is close to 1, 10, and 100 nm, respectively. Note that the ionic strength is commonly in the range of 10^{-2} to 10^{-3} M for natural water supplies and near 10^{-1} M for microbial culture media.

When the surface potential is low (<25/z mV), G_e simplifies as follows:

$$G_e \cong 2\pi R\varepsilon\psi^2_d \exp(-\kappa D) = 2\pi R\sigma^2 \exp(-\kappa D)/\kappa^2\varepsilon \quad \ldots(5)$$

where σ is the surface density of charge (Cm^{-2}). This equation is also valid for nonsymmetrical electrolytes.

For two spheres with the same radius, R, the contribution of van der Waals interactions is given by

$$G_V = -\frac{AR}{12D} \quad \ldots(6)$$

where A is the Hamaker constant. This is mainly determined by dispersion forces (surface-surface, surface-liquid) and thus by the polarizability of the surface-constituting molecules. For organic particles in suspension in water, it is reasonable to consider A values in the range of 0.2 to 1.6 kT (kT = 4.1 x 10^{-21} J mol^{-1} at room temperature). Values are close to 5 kT for mica and in the range of 2 to 50 kT for oxides depending on the atomic number of the metal ion.

Figure 8.1 presents the computed G_t vs D curves for spherical particles of 1-μm radius, A = 1 kT, ψ_d = −20 mV, and κ^{-1} = 10 nm. The respective influence of variation of A, ψ_d, and κ^{-1} is also illustrated. The contribution of van der Waals forces is shown in Fig. 8.1b (ψ_d = 0 mV). The figure shows that, even at low values of the diffuse-layer potential, a potential energy barrier appears at a distance of a few nanometers. This barrier may be very high compared with the kinetic energy associated with Brownian motion, equal to 1.5 kT. At sufficiently high ionic strength, the variations of G_e and G_v as a function of the interdistance bring about a secondary minimum that holds the particles together, although not in close contact. The efficiency of electrolytes to provoke coagulation of colloidal particles is usually expressed by the concentration required to achieve a rapid

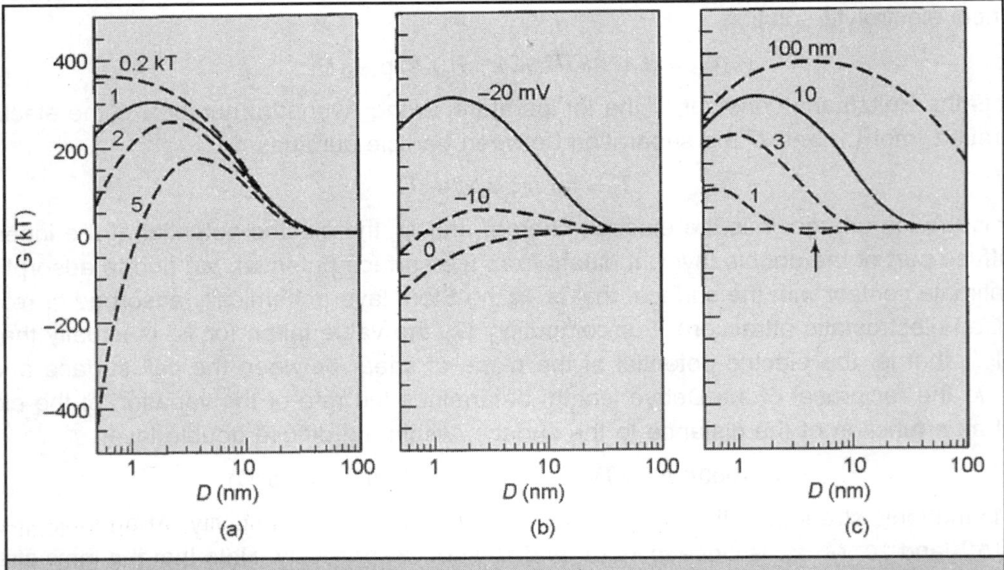

Figure 8.1 Variation of the free enthalpy of interaction as a function of the distance between two identical spherical particles of 1-μm radius, according to DLVO theory.

coagulation, called critical coagulation concentration. According to the valency rule of Schulze–Hardy, this is strongly dependent upon the valency of the ion with a charge opposite to that of the surface (counterion) and the critical coagulation concentration decreases as the counterion valency increases. Consider a negatively charged surface; the decrease of the critical coagulation concentration is explained by a decrease of the potential energy barrier as a result of two effects: (i) if the valency of the cation present in the solution is higher, its adsorption in the Stern layer rises, the potential drop in the Stern layer is larger, and ψ_d is less negative; and (ii) as the valency of the cation increases at a given concentration, the ionic strength rises, the Debye length κ^{-1} decreases, and the potential barrier decreases.

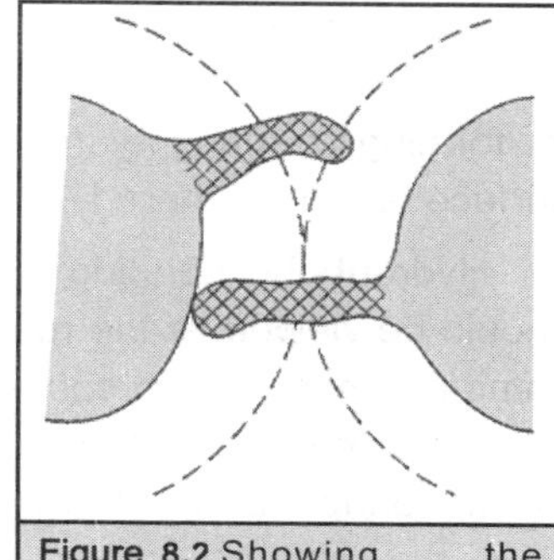

Figure 8.2 Showing the penetration of a surface appendage through a potential energy barrier, the location of which is represented by dashed lines.

The critical coagulation concentration was found to increase in the order $Al^{3+} < Ca^{2+} < Na^{+}$ for *Escherichia coli*. Equations 2 and 6 are restricted to situations where $D \ll R$; within this limit, the interaction energy is proportional to the particle radius. In DLVO theory, the interacting surfaces are considered as homogeneous, smooth, and rigid. This model is very rough for particles of biotechnological interest. Microbial cells may have complex shapes and bear surface appendages such as fimbriae, pili, fibrils, and flagella. As a consequence, the radius of curvature ruling the interactions may be much smaller than the overall cell radius. In particular, it is conceivable that cells become bound via appendages through a potential barrier resulting from average surface properties. It is also possible that certain zones of the surfaces of two particles are in molecular contact, and thus that solvation interactions have a significant influence, while other zones are still separated by an appreciable distance. The higher the ionic strength, the larger the possible impact of this mechanism.

Hydration forces

Between hydrophilic surfaces in water, a short-range repulsive force arises, which tends to prevent the removal of water from the surfaces and is commonly referred to as hydration force. This has been attributed to the structuring of water molecules at the surfaces through H-bonding with hydrophilic surface groups. However, consideration of recent experimental and theoretical results has led to the suggestion of an alternative interpretation, in which the short-range force is an entropic repulsion arising from the confinement of mobile surface groups. The hydration repulsion G_{rh} between two hydrophilic surfaces appears to follow the empirical equation (9)

$$G_{rh} = + G_{rho} \exp(-D/\lambda_{rh}) \qquad \text{...(7)}$$

where the decay length, 2_{rh}, is of the order of 1 nm for 1:1 electrolytes, and G_{rho} (3-30 mJ m^{-2}) depends on hydration of the surface, including hydration of the cations bound to a negatively charged surface, that is, held in the Stern layer.

The effect of electrolytes on the hydration forces may thus determine whether particles will coagulate or not. The effectiveness of cations as coagulants usually increases according to the so-called lyotropic series: $Li^+ < Na < K^+ < Cs^+$ for monovalent cations, $Mg^{2+} < Ca^{2+}$ for divalent cations. The decrease of ion hydration is responsible for a decrease of hydration forces between the surfaces. For some surfaces, it is also accompanied by increased ion binding at the surface, and thus by a larger drop of the electric potential in the Stern layer. A hydrophobic surface cannot bind water molecules via ionic or hydrogen bonds, therefore, an attractive hydrophobic interaction arises between two hydrophobic surfaces in water. This results from the strong tendency of water molecules to associate through hydrogen bonding. To a large extent, hydrophobic bonding is due to the fact that the surface reduces the degrees of freedom offered to adjacent water molecules for forming as many hydrogen bonds as possible; as a consequence, the water structure near the surface is more ordered.

Hydrophobic bonding is thus entropically determined, and the resulting association of surfaces should be viewed as the result of a thermodynamic balance rather than the effect of a force, in the common sense. As a consequence of the entropic character, the effect of hydrophobic bonding increases markedly with temperature. For two surfaces in water, the hydrophobic interaction energy, G_{ah} (ignoring DLVO and other solvation forces), is given by

$$G_{ah} = -2\gamma \exp(-D/\lambda_{ah}) \qquad ...(8)$$

where the decay length λ_{ah} is equal to 1–2 nm and γ is the surface–water interfacial energy (10–50 mJ m^{-2}). The rather long-range nature of the hydrophobic interaction accounts for the rapid coagulation of hydrophobic particles in water and the folding of proteins.

Role of macromolecules

A nonionic polymer dissolved in an ideal solvent (no interaction between polymer segments and the solvent) takes the shape of an "unperturbed" random coil, characterized by the radius of gyration

$$R_g = \frac{ln^{0.5}}{\sqrt{6}} \qquad ...(9)$$

where l is the effective segment length and n is the number of segments. R_g may thus vary from a few to about 100 nm. In a real solvent, the size of the coil may be different from R_g and is given by the Flory radius

$$R_F = \alpha R_g \qquad ...(10)$$

In a poor solvent, the segments attract each other, the coil is more compact, and a is smaller than unity. In a good solvent, the coil swells and its Flory radius is given by

$$R_F \cong ln^{0.6} \qquad ...(11)$$

A poor solvent may become a good solvent by adding certain solutes or changing the temperature. The temperature at which $\alpha = 1$ is known as the theta (θ) temperature, and an ideal solvent is known as theta solvent.

Figures 8.3 and 8.4 illustrate the different situations that may be encountered when two surfaces approach each other while carrying a nonionic polymer. Once surface separation is a few times the radius of gyration, the segments overlap, which leads to a repulsive osmotic force, that is, to the decrease of entropy associated with a closer confinement of the polymer chains between the two surfaces. This is known as steric repulsion and is responsible for steric stabilization of colloids by polymer additives.

As a simple case, consider a low coverage of polymer molecules (no overlap or entanglement of neighboring chains), each being grafted at one end to the surface. This may be relevant to microbial surfaces. In a theta solvent, if the number of chains grafted per unit area is Γ, the repulsive energy per unit area is approximately given by

$$G_s \; 36\Gamma kT \exp(-D/R_g) \qquad \ldots(12)$$

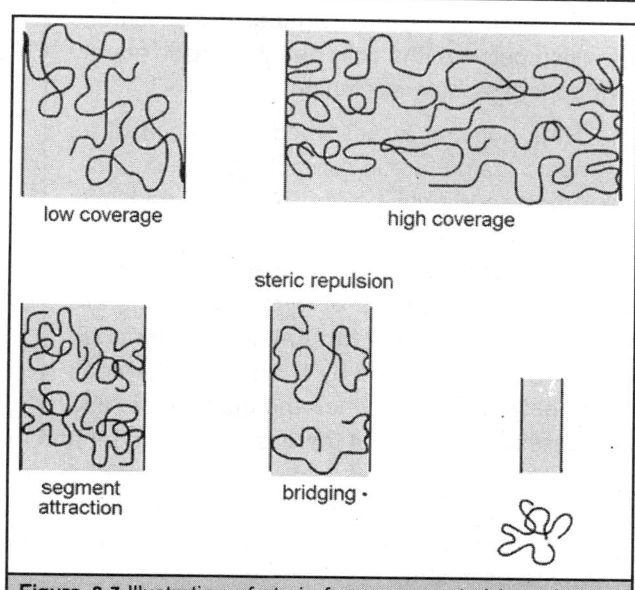

Figure 8.3 Illustration of steric forces generated by polymers between two surfaces.

In a good solvent, the coils swell and the range of repulsion is broader. When the surface density of grafted chains is high, so that they are forced to extend away from the surface much farther than R_g, a polymer brush is obtained. The thickness of the polymer layer, L, no longer varies according to the power 0.5 or 0.6 of the degree of polymerization but is proportional to the polymer length. Once two brush-bearing surfaces are closer than $2L$, a repulsive energy arises. For a good solvent and for $D/2L$ in the range of 0.2 to 0.9, if s is the average distance between the grafting points, the interaction energy per unit area is approximately given by

$$G_s \cong \frac{100L}{\pi s^3} kT \exp(-\pi D/L) \qquad \ldots(13)$$

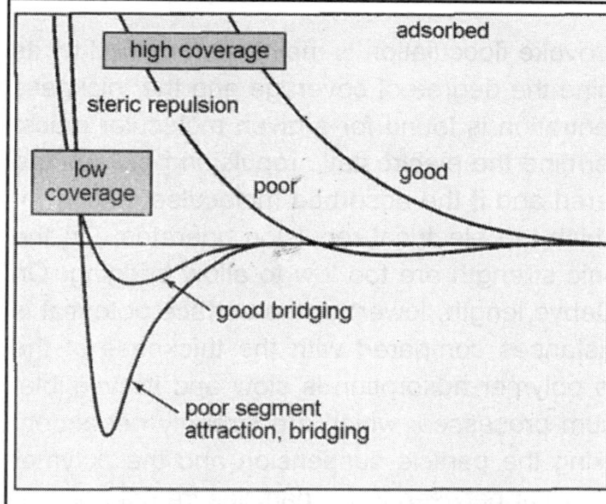

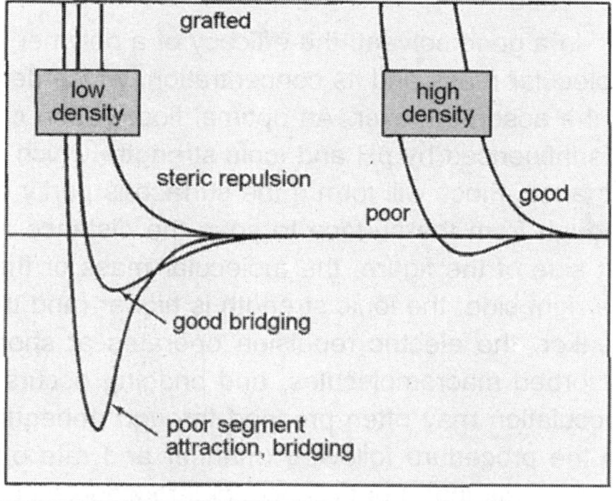

Figure 8.4 Showing the macromolecular interactions; influence of the mode of polymer retention at the surface (adsorption, grafting), polymer concentration at the surface (low, high), and solvent quality (good, poor).

In this case, G_s is the balance of an osmotic repulsion between the coils and the elastic energy that opposes chain stretching.

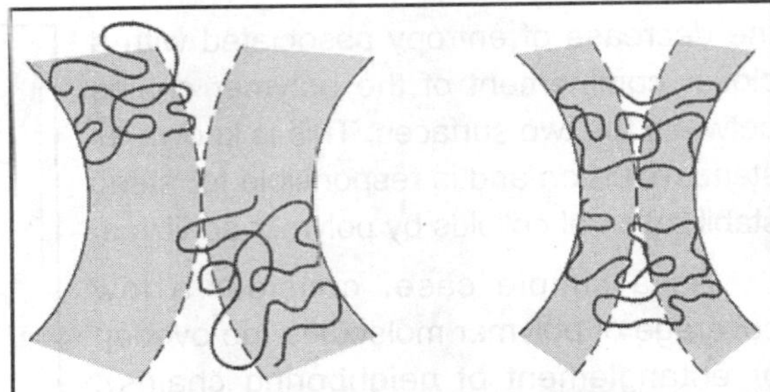

Figure 8.5 Showing the effect of ionic strength or surface potential on polymer bridging; the dashed lines represent the distance at which the electrical repulsion operates.

If the polymer is adsorbed, a fraction of its segments (trains) are in contact with the surface; the other segments belong to loops and tails dangling in the solution. The polymer may adopt a conformation leading to an adsorbed thickness smaller than the radius of gyration. Things are more complicated than for grafted polymers due to the dynamics allowed by a multiplicity of weak bonds between the polymer and the surface, and to a possible variation of the adsorbed amount and the binding sites as the two surfaces approach each other. As a consequence, the variation of the interaction free energy as a function of distance is less sharp. Moreover the force between two surfaces at a given separation distance may take a long time (hours) to reach a stationary value. Attractive forces may also be mediated by adsorbed or grafted polymers.

1. In a poor solvent, segment attraction between molecules bound to opposite surfaces leads to overall attraction between the surfaces until a balance with steric repulsion is reached.
2. When the degree of coverage is weak and the polymer is strongly adsorbed, the macromolecules may form bridges between the two surfaces. Under suitable conditions, bridging forces can extend over longer distances than van der Waals forces.
3. When a polymer does not adsorb, bringing two surfaces at a distance smaller than R_g will push the coils out of the space between the surfaces. The polymer depletion is responsible for a difference of osmotic pressure between the bulk liquid phase and the liquid phase held between the surfaces. This effect is small compared to van der Waals and solvation interaction energies.

In a good solvent, the efficacy of a polymer to provoke flocculation is mainly determined by its molecular mass and its concentration, which determine the degree of coverage and the thickness of the adsorbed layer. An optimal flocculation concentration is found for a given molecular mass. It is influenced by pH and ionic strength, which determine the electrostatic repulsion between the surfaces. Flocs will form if the surface is partly covered and if the adsorbed molecules extend far enough from the surface to span the distance at which the electrical repulsion operates. On the left side of the figure, the molecular mass or the ionic strength are too low to allow bridging. On the right side, the ionic strength is higher (and the Debye length, lower) or the surface potential is weaker, the electric repulsion operates at short distances compared with the thickness of the adsorbed macromolecules, and bridging occurs. As polymer adsorption is slow and irreversible, flocculation may often proceed through nonequilibrium processes, which are strongly dependent on the procedure followed (manner and rate of mixing the particle suspension and the polymer solution; stirring). For instance, heterocoagulation may occur between those particles free of polymer and those that are saturated. In other situations, the surface may be partly covered by macromolecules that have not yet collapsed, thus allowing polymer bridging and producing transient flocs, the lifetime of which may vary from seconds to days, depending on the conditions.

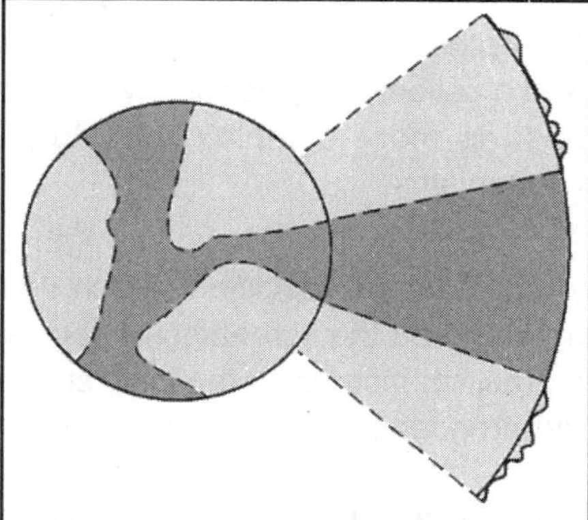

Figure 8.6 Patch model for a cationic polyelectrolyte adsorbed on a negatively charged surface. Left view of the particle surface; right, section through a part of the surface.

Negatively charged cationic polymers

As materials of biotechnological relevance are usually negatively charged, only cationic polyelectrolytes are considered here. Due to electrostatic repulsion between segments, polyelectrolytes display more or less extended conformations according to ionic strength and to their charge density, and thus to the fraction of charged segments τ. For a given molecular mass, the maximum adsorbed amount increases with τ because of the stronger polymer-surface interaction, passes through a maximum, and decreases due to increasing electrostatic repulsion between polymer segments. At low polymer ionicity (small τ), the maximum amount adsorbed (expressed in mass) increases as a function of molecular mass. At high ionicity, the maximum adsorbed amount and the adsorbed-layer thickness are practically independent of the molecular weight; the loop size is small and the polymer chain adopts a flat conformation on the surface.

The flat conformation realizes a local neutralization of the particle surface charge, but reduces the possibility of bridging contacts with other particles. With a cationic polyelectrolyte of high charge density, the influence of polymer molecular mass and of ionic strength can be explained by the electrostatic patch model. The adsorbed polyelectrolyte represents patches of positive charges surrounded by the original particle surface. Particles with this mosaic surface–charge distribution may interact in such a way that positive and negative patches of different particles come into contact, causing a strong attachment. As the surface coverage increases, the particle becomes totally coated by the cationic polymer, electrostatic repulsion arises between particles, and the colloid remains in a dispersed state.

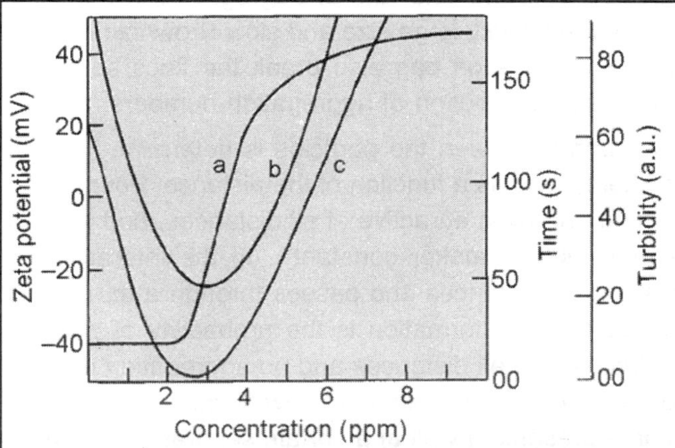

Figure 8.7 Flocculation of Na-montomorillonite by *N,N,N*-trimethyl aminoethyl chloride acrylate. Variation of (a) zeta potential, (b) sedimentation time, (c) relative supernatant turbidity, as a function of the amount of polymer introduced per unit volume. Polymer molecular weight 10^6; τ = 100%; clay concentration 0.05 g L^{-1}; NaCl concentration 1.2 mM.

The optimum flocculation concentration corresponds to a situation where the particle surface is partially covered by patches of cationic polymer. The influence of polymer concentration on flocculation of a clay, evaluated by the relative supernatant turbidity and the sedimentation time. The optimum flocculation concentration coincides with reversal of the zeta potential of the particles. With a cationic polymer of low ionicity, the influence of charge density, concentration, and molecular mass is more complex, due to the combination of the charge-neutralization and bridging mechanisms.

The influence of ionic strength is also complex because it affects the screening of segment/segment lateral repulsion (which decreases chain stiffness and increases loop size and adsorbed amount) and the screening of segment/surface attraction (which reduces the proportion of trains). In practice, increasing the ionic strength reduces the concentration at which the flocs initially appear and broadens the flocculation domain.

Clustering Rate

The rate of clustering/aggregation is determined by the probability of collision between particles, which depends on their concentration, and by the efficiency of a collision to make a bond between the particles. The collision may occur as a result of two types of transport processes: agitation and Brownian motion. If one considers the collision between two identical spherical particles of diameter D^*, the ratio between the probability of collision due to agitation P_o (orthokinetic conditions) and the probability of Brownian collision P_p (perikinetic conditions) is given by

$$\frac{P_o}{P_p} = \frac{\eta D^{*3} G^*}{2kT} \qquad \text{...(14)}$$

where η is the dynamic viscosity, G^* the shear rate, k the Boltzmann constant, and T the temperature. In water at room temperature (η = 1 mPa s), with agitation typical of a magnetic stirrer (G^* = 100 s^{-1}), this ratio is about 100 for particles of 2 μm. The rate of aggregation of microbial cells thus depends strongly on agitation due to their relatively large size and slow Brownian motion, unlike small colloidal particles. On the other hand, agitation can also break the flocs so that a stationary state will be reached, with a characteristic distribution of aggregation numbers.

The chance that a collision gives rise to a bond between the particles is determined by the variation of the free energy of particle-particle interactions as a function of the distance. Four typical situations, may be encountered: (i) the interaction energy is attractive at all distances, and a quick coagulation occurs (low ψ_d, high ionic strength, high Hamaker constant); (ii) the interaction is attractive at short distances but is repulsive at longer distances and passes through a maximum; in this case the probability that a collision leads to bond formation is the probability of passing over the energy barrier; (iii) the interaction is repulsive at all distances and no aggregation occurs; the colloid suspension is said to be stabilized (high ψ_d, low ionic strength, low Hamaker coefficient, steric repulsion); (iv) the interaction energy curve presents a well at a certain separation distance, and the aggregates will be constituted by particles maintained at a certain distance from each other.

Physico-chemical Properties of Particles

The nonspecific interactions that control particle aggregation depend on the physicochemical properties of the surfaces (electrical properties, hydrophobicity, presence of macromolecules) and consequently on their surface chemical composition. Physicochemical methods that are well-established or are being developed to investigate these properties are described in the following sections.

Electrical properties

As discussed earlier, it appears that the electrical properties of particle surfaces strongly influence coagulation by addition of salts, or influence flocculation by polyelectrolytes. The zeta potential, ζ, is usually considered to be close to the potential relevant for DLVO computations. It can be deduced from the electrophoretic mobility, μ. For spherical particles of radius R,

$$\zeta = \frac{\eta}{\varepsilon}\mu B = 12.85\mu B \qquad \ldots(15)$$

at 298 K, if ζ is given in mV and μ in $10^{-8}\ m^2\ V^{-1}\ s^{-1}$. $B = 1$ (Helmholtz–Smoluchowski equation) when $\kappa R > 100$, which is relevant for microbial cells in suspensions of common ionic strength; $B = 3/2$ when $\kappa R < 1$. In the intermediate situations, B is a function of κR.

Equation 15 implies that: (i) the solid plus stagnant layer together behave electrokinetically as a solid particle, (ii) particles are nonconducting, (iii) particle surfaces are smooth and homogeneous, and (iv) the viscosity, dielectric constant, and ion conductivities in the diffuse part of the double layer are equal to their bulk values. When the surface conductance or the bulk (or cell wall) conductance are not negligible, the relationship between electrophoretic mobility and zeta potential is more complex. Therefore it is recommended to report experimentally determined electrophoretic-mobility values instead of ζ potentials.

Surface hydrophobicity

The surface hydrophobicity of microbial cells can be assessed by the water contact angle measured on particle lawns. A greater cell-surface hydrophobicity is of practical concern in separation: It favors cell coagulation and cell accumulation at the liquid–gas interface. The latter effect may be used for increasing cell concentration by flotation. Contact angles of liquids have been used to deduce the surface energy of solids and to predict the stability of solid–solid associations. The model is based on the balance of interfacial free energies involved in the creation of a new solid-solid interface and in the destruction of the interface of each partner with water. This approach is subject to severe limitations. (a) There are different theoretical frames that lead to different values of interfacial energy, indicating that absolute values cannot be considered individually. Nevertheless, classification of different materials according to their surface free energy is the same, whatever the approach used. For polymer materials, this classification is the same as that given by the contact angle of water. (b) Because interfacial energies are deduced from contact angles of liquids, the energy balance does not incorporate the influence of electrostatic interactions, which act at long distances, or the influence of ionic bridges and steric interactions. (c) The system is supposed to be at equilibrium and thus reversible, which is not in accordance with experimental data. (d) A close contact is assumed to be achieved between the surfaces, which is not reasonable in view of the roughness of most surfaces. Other methods of evaluation of microbial-cell hydrophobicity involve adhesion at hydrophobic surfaces or aggregation at high salt concentration.

Surface-chemical composition

The surface-chemical composition of solids, including microbial cells, can be determined by X-ray photoelectron spectroscopy (XPS). The technique involves irradiation of the sample by an X-ray beam, which induces ejection of electrons. The kinetic energy of the emitted electrons is analyzed, and their binding energy determined. Each peak is then characteristic of a given element. Due to inelastic scattering of electrons in the sample, the collected information concerns only the outermost molecular layers of the surface (2-10 nm). As the position of the photoelectron peak is slightly affected by the chemical environment of the element, peak-shape analysis provides further information on chemical functions. The data can be converted into concentrations of model constituents: proteins, polysaccharides, and hydrocarbon-like compounds. For microorganisms, correlations have been obtained between the surface-chemical composition obtained by XPS and surface properties, that is, hydrophobicity and electrical properties. Furthermore, the surface composition has been found to be related to the behavior of microorganisms at interfaces (aggregation, adhesion).

Nanometer-scale surface properties

The physicochemical methods just described have a poor lateral resolution, the information provided concerning only the overall surface properties of particles. Exciting new possibilities are now offered by atomic force microscopy (AFM) to investigate the molecular interactions and properties of surfaces on the nanometer scale. AFM, invented in 1986, consists of measuring the forces acting between a sharp probe and the surface of a sample. It allows the nanoscale mapping of surface topography, as well as the direct measurement of mechanical properties, friction, and interaction forces (electrostatic, van der Waals, solvation, steric interactions). As opposed to more conventional surface analysis techniques, AFM can be operated in aqueous solutions, which makes it possible to study living cells in physiological conditions. A very challenging goal for future research is to directly map the surface properties of living cells in aqueous conditions. The use of functionalized AFM probes, possessing different charges, surface energies, and solvation properties, should make it possible to measure: (i) the local surface hydrophobicity and electrical properties of individual cells, (ii) the surface distribution of solvated macromolecules such as exopolysaccharides, and (iii) steric interactions associated with these macromolecules. These data will provide new insight into the molecular mechanisms of cell aggregation.

IMPLICATION OF SPECIFIC INTERACTIONS IN FLOCCULATION

Flocculation and Flocculence

Flocculation of brewing yeasts, thoroughly reviewed by Stratford, is the most extensively studied of all systems in which cell aggregates are of interest. It is illustrative of the diversity of approaches used to understand flocculation of microbial cells and the interplay between specific and nonspecific interactions. Spontaneous aggregation of yeast cells occurs near the end of fermentation and is of considerable importance in brewing. For instance, early sedimentation in lager beer production will provoke an incomplete or too slow fermentation; failure to sediment will complicate processing and may lead to alteration of taste. In brewing technology, the flocculence of a strain means its ability or tendency to flocculate in industrial conditions. More broadly, flocculence is the ability to flocculate if all environmental conditions are favorable.

The term *flocculability* has been proposed to designate the ability to flocculate in a given medium if hydrodynamic conditions and cell concentration are adequate. Beyond the terminology, it is essential to keep in mind that the occurrence of flocculation depends on (a) the nature of the biological material (strain; physiological state defined by growth conditions and growth phase), (b) the physicochemical environment (temperature, medium composition), and (c) the physical conditions (velocity gradient, cell concentration). Thus, the evaluation of the ability to flocculate must always be related to experimental conditions. In order to test the ability of yeast to flocculate in a given medium, it is important to control the initial state of dispersion and to distinguish the conditions selected for the aggregation process from those allowing evaluation of the final state of dispersion. A screening test may conveniently be performed according to the following procedure. Introduce the washed cells into test tubes containing the desired medium and vigorously stir to ensure dispersion. Agitate gently for a given period to allow flocculation to take place, then let stand undisturbed to allow floc sedimentation.

Finally, measure the absorbance of the supernatant, which reflects the concentration of cells remaining free. The volume of the sediment may also be a way to evaluate the final state of dispersion, as the bed volume of flocculated particles is larger than that of individual particles. Yeast cell behavior in laboratory conditions does not always reproduce the behavior in industrial conditions. Among 45 strains noted as flocculent in brewing, only 20% did flocculate during a culture in laboratory conditions (shaken Erlenmeyer flasks). This was attributed to agitation and aeration rather than to final pH or the requirement of particular wort constituents. Several parameters were used to quantify yeast flocculation: initial rate of flocculation, rate of settling, number of free cells, floc size, and floc strength. It was estimated that the attraction force between flocculent cells is three orders of magnitude larger than the force calculated with the DLVO theory. The dynamic equilibrium between floc formation and floc breaking (i.e., splitting, which determines the average floc size, and erosion, which determines the number of single cells) was modeled. This showed the dependence of floc diameter and number of single cells on bond strength and agitation.

Bottom Fermentation

There are two types of processes for beer brewing. Bottom fermentation, producing lager beer, is carried out at low temperature (near 10°C) by *Saccharomyces carlsbergensis*; at the end of fermentation, the cells flocculate and settle. Top fermentation, typically producing ale, is carried out at higher temperature by *Saccharomyces cerevisiae*; when the cells flocculate, they do not settle but they tend to remain associated with CO_2 bubbles and rise to the top of fermented wort. The surface properties (electrophoretic mobility, water contact angle, and surface composition) of a collection of yeasts were investigated in order to examine their possible relationship with the type of strain, growth phase, flocculence, and occurrence of flocculation.

There is no clear-cut distinction between surface properties of top and bottom strains; however, they differ on a statistical basis. On the average, top strains have more protein at their surface and are more hydrophobic, which may be related to their tendency to remain associated with gas bubbles and move to the top of the wort. Bottom strains contain more phosphate. The difference between the profiles of electrophoretic mobility vs. pH may be explained by modelling the surface charge according to the surface chemical composition, as given by XPS. For the bottom strains, the electrical properties are controlled mainly by phosphate, resulting in a low isoelectric point (pH 2 or below) and an electrophoretic mobility that does not change appreciably above pH 4. For the top strains, the electrical properties are determined mainly by the balance of protonated amino and carboxylate groups in proteins, which give a higher isoelectric point (pH 4) and an electrophoretic mobility changing regularly from pH 2 to 7. Although surface properties of brewing yeasts were statistically related to the top/bottom character, they did not show correlations with the flocculating character. No differences of surface properties were found between strains that did and did not flocculate in laboratory conditions, or between cells from the exponential and stationary growth phases, even for strains where flocculation occurred during the transition from one growth phase to the other.

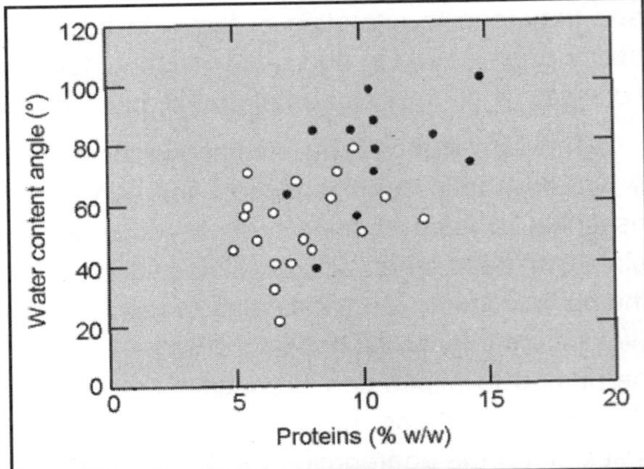

Figure 8.8 Water contact angles of bottom-fermenting strains (○) and top-fermenting strains (●) as a function of the cell-surface concentration (weight %) of proteins as obtained by modeling XPS data.

This is in contrast with reports that ability to flocculate during growth coincides with an increase in cell-surface hydrophobicity correlated with the appearance of fimbriae-like structures. It was also reported that flocculent cells possess a fimbriate or hairy surface, whereas nonflocculent cells are smooth. Certain yeast strains produce aggregates with a particular geometric shape, such as cubes, lozenges, cylinders, and disks, the shape being strongly dependent on agitation. When agitation is stopped, the flocs collapse and, within a few tens of minutes give rise to a cell carpet on the bottom of the flask. This is tentatively attributed to the fact that the cells are coated by polymers that ensure a weak bridging. The viscoelastic properties of the interface would be such that, upon agitation, binding prevails due to chain entanglement, with the hydrodynamic conditions controlling the floc shape; on the other hand, repulsion would prevail in resting conditions.

Mechanism of Aggregation

According to the most widely considered mechanism of flocculation proposed by Miki et al., lectin-like molecules (sugar-binding proteins that need Ca^{2+} to adopt their active conformation) recognize and bind mannans at the surface of adjacent cells. Stratford and Assinder distinguished two flocculation phenotypes, both based on a lectin-mediated mechanism: Flo1 phenotype strains, which included most of the laboratory strains, had *FLO1* genes and possessed mannose-specific lectins; NewFlo phenotype strains, which included most ale-brewing strains, possessed less-specific lectins, binding manno- and glucopyranoses. Masy et al. found the same two phenotypes and called them MS (mannose sensitive) and GMS (glucose mannose sensitive). On the other hand, a different phenotype, called MI (mannose insensitive), representing strains that are not inhibited by sugars, including mannose and glucose, was reported. In the last few years, much progress has been made as a result of advances in the fields of genetics and biochemistry. In a review, 33 genes involved in yeast aggregation have been reported; these include structural genes, regulatory genes, and genes involved in cell wall synthesis and mitochondrial functions. Four classes of strains were distinguished: (a) strains lacking an intact flocculation gene, (b) strains in which intact flocculation genes are constitutively repressed, (c) a strain with flocculation genes that are transcriptionally regulated, (d) strains with genes that encode for constitutive flocculence. For strains in which flocculation was regulated, induction was ascribed to nutrient limitation.

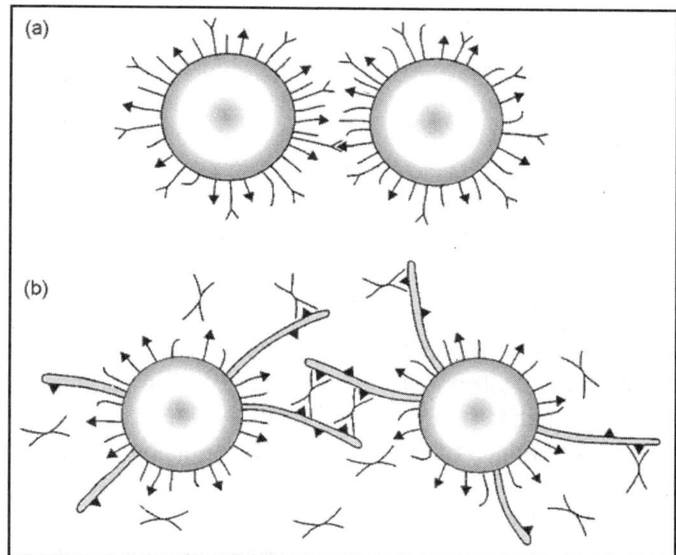

Figure 8.9 Schematic representation of possible cell-cell binding processes involving a lectin-like mechanism. (a) Direct lectin-ligand binding. (b) Implication of surface hairs and agglutinins.

The best-studied dominant flocculation gene, *FLO1*, has been cloned and sequenced. Based on the genetic sequence, it was hypothesized that the Flo1 protein is present in the cell wall, and its N-terminus is exposed to the medium. The location of this protein in the cell wall was confirmed by the use of immunofluorescent probes. It was suggested that the exposed part of the Flo1 protein binds directly with neighboring cell wall mannans. The *FLO8* gene was also cloned; no homology was found with the *FLO1* gene. It was suggested that *FLO8* does not encode a cell wall protein, but is involved in the regulation of the transcription of the *FLO1* gene. Two types of surface proteins involved in flocculation of bottom-fermenting strains were isolated from cell walls.

A mannose-specific calcium-dependent agglutinin was isolated from a nonflocculent strain and from flocculent strains, both in nonflocculating and flocculating growth stages; it was released during the flocculation assay. A protein isolated from cell walls of flocculent cells only, called flocculin, showed homology to the protein coded by the flocculation gene *FLO1*; it was associated with fimbriae-like structures, did not show lectin-like activity, and was thought to act as a lectin ligand. The authors proposed a model in which at least three factors determine the onset of flocculation: (i) release of agglutinin, (ii) appearance or activation of flocculin, and (iii) formation of fimbriae-like structures accompanied by an increase in cell-surface hydrophobicity. For NewFlo phenotype *S. cerevisae* strains, there is evidence that lectins are synthesized continuously from an early stage of growth and rapidly inserted into the cell wall, but remain inactive until they are activated at flocculation onset.

Specific and Nonspecific Interactions

A comparative study of nine brewing strains flocculating in laboratory conditions (three bottom- and six top-fermenting strains) was performed. Flocculation most often occurred only in the stationary growth phase, and the onset of flocculation could not be related to a change of physicochemical properties (chemical composition by XPS, electrophoretic mobility, water contact angle). Laboratory tests showed that flocculation of bottom strains requires the addition of calcium, is inhibited with mannose and sucrose, but not with galactose, and is not affected by ethanol. This is typical lectin-based behavior and the NewFlo/GMS phenotype.

The following flocculation mechanism maybe proposed: (i) there would basically be a repulsion between cell surfaces due to electrostatic or steric interactions (mobile polysaccharides); (ii) in the stationary growth phase, lectins would appear at the cell surface or be made accessible; and (iii) when the sugar concentration in the solution is sufficiently low, these lectins would bind sugar residues of adjacent cells across the repulsion barrier. The onset of flocculation thus seems to be triggered by the conjunction of the appearance of active lectin at the surface and of sugar attenuation, rather than by the calcium concentration. The occurrence of flocculation about 2 pH units above the isoelectric point in laboratory tests, and not at the isoelectric point, indicates a major influence of pH on lectin conformation. The ionic strength has no effect on flocculation, which may be explained either by a minor role of electrostatic interactions or by a compensation of the influences of ionic strength on electrostatic interactions and lectin conformation.

The penetration of the repulsion barrier will be more likely if the lectins or the lectin–ligands stick out of the surface, which is especially the case if they are located on surface hairs. Furthermore, flocculation might be favored by the release of agglutinins. Flocculation of top-fermenting strains is induced by ethanol and other alcohols and does not require addition of calcium. For certain strains, there is no inhibition by mannose, sucrose, or galactose, whereas flocculation of other strains is inhibited by mannose only. The former would belong to the MI (mannose-insensitive)-flocculation phenotype, which has not been frequently reported. For this collection of top strains, there is no evidence of lectin-like interactions.

Other organic solvents (methanol, *n*-propanol isopropanol, *n*-butanol, isopropanol, acetone, 1,2-propylene glycol) have the same effect as ethanol when added at a concentration provoking the same decrease of the surface tension of water. This indicates that the influence of ethanol is related to an adsorption process. The following flocculation mechanism may be proposed. There would exist a repulsion between the cells. As exponential cells, which have similar physicochemical properties as stationary cells, do not flocculate near the isoelectric point, the repulsion would be steric in nature. However, the influence of ionic strength on flocculation suggests that electrostatic interactions play a role as well. In the stationary growth phase, additional macromolecules or hairs would appear at the surface, providing binding sites for specific (lectin-like or other) or nonspecific binding. In the presence of sufficient ethanol concentration and, for the mannose-inhibited strains, sufficiently low sugar concentrations, these macromolecules or hairs would protrude from the surface and manage to bind adjacent cells across the repulsion barrier. The onset of flocculation thus seems to be triggered by the conjunction of a change of cell surface and an increase of ethanol concentration.

RANGE OF DOWNSTREAM PROCESSING

Production of fine chemicals by living cells involves a range of downstream processes, the first of which is usually the separation of cells or cell debris from the broth. This step is largely improved if flocculation can be carried out. Flocculation, as a pretreatment for the continuous centrifugal separation of whole cells, should produce flocs with high mechanical strength. In the production of microalgae, the best practical and economical way of harvesting was found to be chemical flocculation, followed by sedimentation, flotation, or filtration. The following sections review how flocculation can be produced in downstream processes and, to a certain extent, can be understood in the light of physicochemical theories described earlier.

Role of Colloidal Particles

S. cerevisiae forms flocs in the presence of positively charged submicrometer-sized aluminium hydroxide particles. The respective surface charges of yeast and aluminium hydroxide prevent homocoagulation and the process is a heterocoagulation of oppositely charged particles, which can easily be understood in the light of DLVO theory. Positively charged organic-polymer particles [poly-(styrene-divinyl benzene) with quaternary ammonium surface groups; 0.27-μm diameter] were used to obtain flocs of *S. cerevisiae* and *E. coli* whole cells or cell debris. The settling time of the cells was considerably decreased in the presence of the particles, and the centrifugation energy and power required for the removal of *E. coli* cell debris were reduced by one order of magnitude. The optimal polymer particle concentration found for the clarification of suspensions of *E. coli* cells may be related to a partial coverage of the cell surface by the polymer particles, allowing the cells to be bound together by one layer of positively charged particles; complete coverage of every cell would lead to charge reversal and overall repulsion. An optimal polymer-particle concentration was also found for the clarification of *E. coli* cell-debris suspensions. In the recovery of several enzymes by removal of cell debris following this method, complete enzyme activity was recovered in the supernatant, indicating that the polymer particles did not adsorb or inactivate the enzymes.

Role of Salts

Simple electrolytes favor particle coagulation by compressing the electric double layer, as explained in the section “DLVO Theory”. Metal ions of higher charge are more efficient both because they have a higher tendency to adsorb in the Stern layer and because they give a smaller Debye length. Moreover, other factors affect the efficacy of hydrolyzing metal ions, such as Al^{3+} and Fe^{3+}, which are used in water and sludge conditioning. Upon hydrolysis, these metal ions form positively charged polynuclear hydroxo complexes. The latter adsorb specifically and are able to reverse the surface charge. Therefore, depending on the pH and on the concentration of colloidal particles, the Al^{3+} and Fe^{3+} salts may act as coagulants only in a narrow concentration range. The concentration must be high enough to allow compression of the electric double layer or surface neutralization; however, overdosing may reverse the charge of the colloids and restabilize them.

Above a certain concentration of metal ions, colloidal particles may again form flocs by enmeshing in aluminium or ferric hydroxide precipitates. Jiang et al. tested the ability of ferric and aluminium salts to coagulate algae. Hydrolized ferric species produced by oxidation of ferrous sulfate were more efficient than ferric sulfate, aluminium sulfate, or partially neutralized aluminium chloride. Floc size was larger and cell removal higher. Humic substances added to the algae suspension reduced the performances of the coagulants.

Role of Soluble Polymers

Both synthetic and biological polymers have been tested to induce cell flocculation, and results can be understood in the light of the principles presented earlier. Nonionic and anionic synthetic polymers were found to be ineffective for flocculation of *E. coli, S. cerevisiae,* and microalgae, while cationic polymers were good flocculating agents. Increasing the charge density of cationic polyacrylates and polyacrylamides enhanced flocculation. Cationic polyethyleneimine (PEI) was effective to flocculate *Chlorella ellipsoida*. With low molecular mass polymers, no reversal of the cell-surface charge occurred and the amount (mass) required to initiate flocculation decreased considerably as the molecular mass increased from 800 to 2,000 Da. Above 2,000 Da, the amount required to initiate flocculation did not vary much as a function of molecular mass; at high concentrations the polymer reversed the cell- surface charge; when the polymer concentration exceeded an optimal value, the flocs began to redisperse.

Earlier work with silica and *E. coli* showed that an increase of pH from 4 to 6 or 9 increased the PEI concentration required for optimum flocculation, as expected from the variation of polymer ionicity. With *C. ellipsoida,* little effect of pH was observed between pH 4 and 7. The polymer concentration needed for flocculation was one order of magnitude larger for *E. coli,* compared to crystalline silica. With diethyl-aminoethylmetacrylate acrylic acid copolymers, washed cells of *E. coli* flocculated at polymer concentrations at which the isoelectric point was reached, indicating that the mechanism of flocculation was in agreement with the patch model. Downstream processes are large consumers of polyacrylamide derivatives because they are economical and effective flocculating agents. However, neurotoxic and strong carcinogenic activities of acrylamide monomer were reported. Hence, there is a need to produce safe, biodegradable compounds that will minimize environmental and health risks.

Several microorganisms have been reported to produce yeast-cell flocculating agents; they were effective between pH 3 and 5 and lost their effectiveness above pH 5. Cultures of *Rhodococcus erythropolis* were found to flocculate *E. coli, Microcystis aeruginosa,* yeast, activated sludge, kaolin clay, and other particles. Flocculation was considerably improved by adding Ca^{2+}. However, the respective roles of *R. erythropolis* cells, possibly provoking heterocoagulation, and of polymers released into the suspension or at the cell surface, were not clarified. A *Paecilomyces* culture filtrate, containing a polysaccharidic compound, was shown to flocculate several microorganisms, including *Bacillus subtilis* and *E. coli*. Its effectiveness with *E. coli* did not change significantly between pH 4 and 8, and in an ionic strength range of 0.001 M to 0.1 M, except in the presence of CO_3^{2-} and Fe^{3+} ions. However, it decreased steeply below pH 4 and above pH 8, or at ionic strength of 1 M.

Compared to synthetic polymers, chitosan, also commercially available, possesses several advantages: monomers are not toxic, it can be heat-sterilized without degradation, and it is biodegradable. Chitosan, a partially *N*-acetylated poly (glucose-amine), is derived from chitin, the major constituent of crustacean exoskeleton. At acidic pH, chitosan is soluble and exists as well-extended linear chains due to repulsion of charged $-NH_3^+$ groups. At alkaline pH, the positive charges gradually disappear (the neutralization point is about pH 7.9), and chitosan tends to coil and precipitate.

Chitosan-induced flocculation of washed *E. coli* cells at concentrations much lower than those required to reach the isoelectric point, as opposed to the use of diethyl-aminoethylmetacrylate acrylic acid copolymers. This was attributed to the rigidity of chitosan, which undergoes weak deformation during adsorption and is able to form bridges between the cells. The optimal dose for flocculation of microalgae was about twice as low for chitosan as compared with polyacrylamides; this superiority was also attributed to the rigid backbone of chitosan. With washed cells of *B. subtilis* and *E. coli*, the effectiveness of chitosan and cationic polyacrylamides did not vary between pH 4 and 7. Above pH 7, the effectiveness of chitosan dropped due to a decrease of ionicity, whereas the change of polyacrylamide effectiveness was less marked. None of these compounds was effective in flocculating *Zymomonas mobilis*. An optimal chitosan concentration was found near 0.2 g L^{-1} for flocculation of *Euglena gracilis* cultures. Addition of chitosan above this concentration led to a partial cell redispersion due to repulsion of positively charged cells.

With the optimal chitosan concentration, a sharp pH optimum was observed at pH 7.5, the decrease of flocculation at lower and higher pH being attributed to a lower negative charge of *Euglena* cells and to deprotonation of chitosan $-NH_3^+$ groups, respectively. At pH 7.5, flocculation did not decrease appreciably at high chitosan concentration, due to the low ionicity of the polymer at this pH. Using chitosan, cell debris and nucleic acids could be separated selectively from β-D-galactosidase in *E. coli* disintegrates. For a highly charged chitosan, a pH decrease induced a reduction of the optimal flocculant dosage. Highly charged chitosan can achieve a more efficient removal of cell debris than less-charged chitosan. The required flocculation dosage of highly charged chitosan was relatively independent of the charge density, but increased with the molecular mass. At the required flocculation dosage, electrophoretic-mobility measurements showed that the flocs formed were not completely neutralized. It was therefore suggested that the main mechanism of flocculation was a "nonequilibrium" bridging process. This occurs when reconformation following polymer attachment to the surface is slow compared with polymer attachment or particle–particle collision. It is especially pronounced with high molecular mass polymers and leads to trapping of surface-negative sites between the bridges formed by the polycations. Highly charged and high molecular mass chitosan produced large and shear-resistant aggregates suitable for filtration, settling, and centrifugation, whereas highly charged low molecular mass chitosan gave smaller and more shear-sensitive flocs suitable only for centrifugation.

Cell-produced polymers

Ionic strength can dramatically affect the flocculant effectiveness. Bilanovic et al. studied the effect of salinity on microalgae flocculation with cationic polyacrylamides (1×10^7 and 2×10^7 Da) and chitosan (4×10^5 Da) at pH 7. For the two types of polycation, removal of more than 90% was obtained with freshwater cultures of *Chlorella*, whereas removal was very poor for cultures of marine species of the same genera. The efficiency of the cationic polymers to flocculate *Isochrysis galbana* dropped considerably as the ionic strength increased from 0.02 to 0.70 M, without appreciable alteration of the optimal dose. While the rise of ionic strength was expected to increase the screening of polymer–cell surface attraction, it also provoked a shrinkage of the macromolecule, as revealed by a decrease of the intrinsic viscosity. The flocculation is also strongly influenced by the extracellular polymers (ECP) released by the cells in the solution and by the components of the culture media. The optimal dosage of cationic polymers was lower for washed *E. coli* cells than for the initial culture suspension.

The interaction between anionic polymers in the medium and added cationic polymers led to an increase in flocculant requirement. On the other hand, it has been reported that the strongly acidic polymers released in solution by *Lactobacillus delbruekii* and *Lactobacillus fermentum* act as flocculating agents (strong anionic polyelectrolytes) that only need a small amount of cationic substances to be sensitized. Polyacrylamides could flocculate *E. coli* and *B. subtilis* cells in defined media, but were ineffective in complex media because of precipitation. Chitosan did not precipitate below pH 7.5 in a complex medium and allowed flocculation of *E. coli* cells, while requiring twice the dose used in a defined medium; the flocculant efficiency decreased above pH 7 in both media and washed cells. On the other hand, flocculation of *B. subtilis* by chitosan in complex media occurred at pH 4 but dropped as the pH was increased from 4 to 7, and did not change with washed cells. Flocculation of yeast by chitosan in a complex medium occurs simultaneously with the precipitation of chitosan with components of the medium. These can be regarded as competitive processes with respect to the availability of chitosan.

The detrimental effect of precipitation (increase of dosage, contamination of biomass, removal of the fermentation product) can be minimized by a rapid addition and efficient mixing of the flocculant with fermentation media of high biomass concentration. Cells may also become entrapped in the precipitate of polyelectrolyte complexes, with the unexpected benefit that strong flocs are formed. The effectiveness of flocculation is also dependent on the cell-surface constituents but is not simply related to the cell wall type (Gram positive or Gram negative). Polyacrylamide and chitosan were effective in flocculating *E. coli* (Gram negative) and *B. subtilis* (Gram positive) washed cells. Their inefficacy to flocculate washed cells of *Z. mobilis* (Gram negative) was attributed to the presence of nonionic extracellular polymers attached to the cell surface. It was reported that strongly acidic polymers adsorbed on *L. delbruekii* and *L. fermentum* surfaces induced an increase in the flocculant dosage, probably due to a high charge density and, therefore, a high polycation demand.

Flocculating Agents

The combined use of two flocculating agents has often proved to be an effective way to form flocs. Chitosan, polyethyleneimine, and Ca^{2+} were used simultaneously to perform flocculation of *E. coli* cell debris. This was applied in a process using a continuous centrifugal separator and allowed an at least 10-fold improvement in cell debris removal, compared with the same process without flocculation. Sewage sludges were successfully flocculated by the addition of a cationic, followed by an anionic, polymer. The recovery of bacterial extracellular proteinases obtained in *B. subtilis* culture was investigated by adding bentonite (a clay with a negatively charged surface), followed by a cationic aminoamidic polyelectrolyte. The yield of recovery was more than 84% at the laboratory scale and up to 82% at the pilot-plant scale.

Polyelectrolytes used singly, or combined with an oppositely charged polymer, were inefficient in flocculating *Z. mobilis* in complex medium. However, effective flocculation could be performed by adding highly charged polyacrylamide or chitosan, followed by bentonite, providing high broth clarities between pH 5 and 8.5 or between pH 4 and 8, respectively. It was suggested that a combination of electrostatic interactions and bridging capability was responsible for the higher effectiveness of chitosan at low pH. Addition of bentonite after chitosan allowed avoidance of precipitation of the latter at high pH. Flocculation effectiveness was also influenced by the floc collision during settlement.

Control of Floculation

Flocculation can be controlled by genetic factors, as discussed for brewing yeast earlier. Addition of starch to *E. coli* suspension resulted in the formation of bacterial flocs, allowing a very efficient cell-broth separation by settling. Flocculation was dependent on the high expression of a cell-surface maltoporin receptor for starch: floc formation was evident with bacteria containing wild-type maltoporin, but was faster and occurred to a greater extent with strains expressing a high-affinity allele of the starch receptor or with strains grown at high levels of maltose. Because expression of the maltoporin receptor can be genetically controlled, its presence at the cell surface can be made largely independent of the variations in surface properties, thus avoiding the variability of classical flocculation.

The optimal starch concentration was 1-2 mg mL^{-1} for a strain with a high-affinity allele. Decrease of flocculation at higher concentrations was attributed to saturation of binding sites. The optimal starch concentration was 5-10 mg mL^{-1} for a strain with a low-affinity allele. The optimal conditions for floc formation were pH 6-7 and a cell concentration of 2×10^9 cells mL^{-1}, which should be readily achievable in both laboratory and large-scale cultures. It was suggested that the presence of maltodextrin and that the lack of accessible α-1,4-linked chains in the starch preparation could decrease the starch effectiveness. This starch-dependent aggregation is potentially applicable to other microorganisms if they comply with the following requirements: (a) the starch receptors are expressed, (b) the cell surface is not masked by a polysaccharide or a protein layer, and (c) the cell does not produce extracellular amylolytic enzymes. For harvesting bacteria, starch offers several advantages: (i) binding to cells is rapid; (ii) it is nontoxic; (iii) unlike electrically charged flocculating agents, it does not affect the pH, and its binding is effective in the pH range usually found in fermentation conditions; (iv) it is a low-cost material; and (v) it is easily separable from bacteria, if necessary.

DOWN STREAM OPERATION

A biomass proliferating in the form of flocs is a system that is much more complex than flocs produced in a given downstream operation, due to the diversity of interfaces with respect to space and time. Brewing yeast, discussed earlier, illustrates the fact that a floc is the result of the integration of different processes and of their development on the time scale. In this section, even more complex systems of practical relevance are considered. Whenever possible, the presentation points out physicochemical features that provide a guide for understanding and controlling the biosystem behavior.

Continuous Ethanol Fermentation

The ability to flocculate is an important characteristic in the continuous production of ethanol, so that a high concentration of cells can be maintained within the fermentor. This has been used in up-flow tower fermentors, which were successfully implemented at industrial scale. The yeast producing the particular flocs was used in a gaslift laboratory fermentor. A more complete understanding of the phenomena controlling both the shape and size distribution of the flocs appeared to be a prerequisite for a full control of the highly promising fermentation process. Continuous ethanol fermentation has also been performed with *Zymomonas mobilis*. In a laboratory up-flow fermentor (volume of 0.92 L), flocs were obtained by starting-up (10 days) at a low dilution rate (0.065 h^{-1}).

After this period, the dilution rate was incrementally increased to 2.5 h^{-1}, allowing an ethanol productivity as high as 100 $gL^{-1}h^{-1}$. However, no attempt was made to evaluate long-term performance. A simultaneous addition of cationic polyacrylamide and titanium hydrous oxide was used to flocculate *Z. mobilis*. This provided an ethanol productivity of 16.4 g L^{-1} h^{-1} for at least 100 h in a gaslift tower fermentor (volume of 5 L). In similar conditions, productivity of unflocculated cells was evaluated to be about 2 g L^{-1} h^{-1}. Mixed cultures of flocculent strains of *Z. mobilis* and *Saccharomyces* sp. gave higher ethanol production, yield, and volumetric productivity than pure cultures.

Activated Sludge

A common biological treatment of sewage waste is the activated sludge process, which is essentially a way to concentrate the organic matter of the waste. In its simplest form, it is a two-stage operation involving aeration followed by sedimentation. The aeration stage allows proliferation of different organisms and floc formation. An essential requirement for successful operation is the formation of settleable flocs. Among the physical characteristics of the flocs, the size distribution and structure are most important.

Effect of physicochemical factors

The floc behavior can be affected by hydrophobic interactions. Zita and Hermansson isolated, by hydrophobic interaction chromatography, hydrophilic and hydrophobic bacteria from flocs coming from an aeration tank of a wastewater treatment plant. Attachment of these strains to activated sludge liquor containing flocs and wastewater was evaluated. The results showed a correlation between adhesion and surface hydrophobicity up to a threshold, above which bacterial attachment did not increase with surface hydrophobicity. This was attributed to an equilibrium between reversibly attached cells and free-living cells. Urbain et al. showed that a high internal hydrophobicity of the flocs is correlated with a decrease of the sludge volume index (SVI), which reveals an improvement of settleability.

The internal hydrophobicity of the floc was determined by a salt aggregation test performed on a sonicated sludge sample mixed with increasing concentrations of $(NH_4)_2SO_4$ in phosphate buffer. Bacterial strains of different hydrophobicity were separated by subjecting sonicated sludge to an octane–water mixture. The amount of exopolymers released by sonication of a hydrophilic strain cultured in the laboratory was greater than that of a more hydrophobic strain. This observation may suggest that the hydrophilic strains are not simply trapped in a matrix, but are also implicated in the floc organization. No clear relationship has been established between the electrophoretic mobility of sludge particles and their settling properties. A strongly negative electrophoretic mobility of sludge particles was found to induce a high SVI. However, an inverse correlation and the absence of correlation between electrophoretic mobility and SVI were also reported. Floc stability was evaluated by gentle mixing in $CaCl_2$ and KCl solutions of increasing ionic strength. Floc dissociation decreased as the ionic strength rose from 50 μM to 5 mM; it decreased more slowly between 5 and 50 mM. This was explained by DLVO theory, the flocs being stronger as the diffuse double layer was more compressed. The reversibility of the process was tested by reflocculation of degraded activated sludge.

Reflocculation increased between 0 and 5 mM ionic strength and remained constant up to 50 mM. Between 50 and 500 mM, floc dissociation increased and reflocculation decreased. According to the authors, at high ionic strength, the distance between particles is fairly constant and the floc structure appears to be influenced by mechanisms of unknown nature, unlike the mechanisms considered in DLVO theory. The influence of wastewater on floc stability was similar to that described for salt solutions. Polyvalent ions can act by binding ECP and bacteria, or by binding bacteria with one another, through electrostatic interactions. The addition of Ca^{2+} to activated sludge induced a decrease of both the sludge volume and the residual turbidity in the supernatant after 30 min of sedimentation. Removal of Ca^{2+} from activated sludge by ion exchange (H^+, K^+, Na^+, Mg^{2+}) or by a chelating agent (EGTA) induced a sludge deterioration: small particles appeared in the supernatant, causing an increase of the specific resistance to filtration. On the contrary, exchange of Ca^{2+} by Cu^{2+} decreased the specific resistance to filtration and thus resulted in a more stable floc structure. This was attributed to the ability of Cu^{2+} ions to maintain the 3-dimensional structure of the exopolymers.

Effect of biological factors

Bioflocculation in activated sludge also involves extracellular polymers (ECP). Among the ECP, proteins, polysaccharides, lipids, humic acids, and DNA were reported. Humic acids originate from wastewater itself, whereas other compounds arise from cell metabolism or from cell lysis. The influence of ECP on sludge settling (notably on SVI) has been widely studied. However, because there is no unified method of extraction (e.g., centrifugation with or without heating and/or sonication), care must be taken in comparing results from different studies. A general trend found in the literature is that an increasing concentration of ECP extracted from activated sludge is related to an increasing SVI; however, the absence of a relationship between ECP production and SVI has also been reported. ECP were isolated after thermal extraction from sludge and their molecular mass was determined by gel filtration.

A major fraction was found between 30 and 100 kDa, and an increase in its apparent molecular mass was correlated with an increase of the stirred specific volume index (SSVI). This could be due to steric interactions, allowing greater amounts of interstitial water to be retained in the sludge. The presence of growing filamentous microorganisms is an important feature determining floc properties. It was suggested that they could form a backbone to which floc-forming bacteria could attach. This idea is supported by the presence of filamentous microorganisms in numerous activated-sludge samples. However, sludges have poorer settling characteristics when filaments grow out of the flocs. The outgrowth was quantified by the total extended filament length, that is, the sum of the lengths of filaments extending out of the flocs. An increase of SVI or of SSVI was observed as the total extended-filament length increased up to 10 m mg^{-1} of suspended solid. When the total extended-filament length exceeded this value, SVI increased dramatically and bulking phenomena were observed.

Microbial Aggregates

Anaerobic processes, resulting in the production of biogas, are used for the decomposition of organic compounds in wastewater; the main problem encountered is to efficiently retain the biomass in the reactor. Several reactor designs have been considered. Bacteria may be immobilized on a packing material or carrier particles; the ability of bacteria to aggregate can be used to produce granules with good settling properties. This latter is the case of the up-flow anaerobic sludge blanket reactor (UASBR). Bacterial aggregation ensures the conservation of the biomass inside the reactor, but also enhances the interactions between the various microorganisms needed for anaerobic digestion. The formation of bacterial granules in UASBR is governed by the surface properties of the bacteria, which may themselves depend on the operational parameters.

Hydrophobicity is the surface property that has been the most studied for bacteria originating from granules. Contact- angle measurements on cells found in the reactor effluent and on cells originating from granules led to the conclusion that more hydrophobic bacteria remain in the reactor in the aggregated state, whereas more hydrophilic bacteria are isolated and leave the reactor. It was also reported that the more hydrophobic bacteria found in UASBR had a better ability to flocculate. Moreover, the ECP produced by flocculent bacteria were more hydrophobic than those produced by nonflocculent bacteria. Different populations of bacteria found in UASBR are responsible for different steps of waste digestion and are classified as acidogens, acetogens, and methanogens. Most acidogens found in granules are more hydrophilic than the acetogens and methanogens; moreover, granules are formed of layers of different hydrophobicities, each layer being responsible for one step of the degradation process.

The surface charge of the bacteria is also recognized as a parameter governing the flocculation behavior in UASBR. The negative charge is attributed to carboxyl groups located on the ECP; it is indeed reduced when the cells are submitted to shear. Divalent or trivalent metal ions, which can be added to the reactor, may influence granule properties by acting on cell–cell interactions, as described by DLVO theory, but also by affecting the conformation of ECP or by cross- linking the ECP to form a kind of matrix. It was shown that the size of the granules, and thereby their settleability, is significantly increased in the presence of Ca^{2+} (100 mg L^{-1}), which leads to shorter retention times in the reactor. When EDTA, a chelating agent, was added to the medium, the aggregation decreased but was not suppressed. It was thought that part of the Ca^{2+} was involved in irreversible bonds between the cells. The flocculation could also be induced in an anaerobic digestor by the addition of various divalent and trivalent cations, and the tendency to flocculate was related to a reduction in the negative charge of the cell surface. However, divalent and trivalent cations may also play a role through other mechanisms.

Precipitates of calcium or iron may provide the bacteria with nucleation sites for granule formation. An alternative mechanism is by an increase in surface hydrophobicity provoked by complex formation between divalent cations and carboxyl or phosphate groups. Other operational parameters may influence granule formation or the efficiency of the process. When the substrate is rich in protein, highly hydrophilic cells are favored, which leads to the formation of fluffy flocs rather than granules. On the contrary, a substrate rich in volatile fatty acids produces granules with a very hydrophobic surface; this may be detrimental to the process because such granules may be eliminated from the reactor by flotation. Lowering the surface tension of the medium by adding alkyl benzene sulfonate was shown to favor the formation of granules with a more hydrophilic surface, thus preventing their flotation.

SUMMARY

Cell aggregation, which is of considerable importance in bioprocess engineering, can be understood and controlled to a certain extent in the light of established principles of colloid and surface chemistry. A major key is the interplay between electrostatic interactions and interactions involving macromolecules. In this respect, microbial cells represent a complex system that can be regarded as a solid surface, a surface grafted with macromolecules, or a surface with adsorbed macromolecules, depending on the strain and the approach followed. Furthermore, the extracellular polymers may constitute a matrix in which the cells are more or less loosely entrapped. For certain microorganisms, another important mechanism is provided by specific interactions between the cells, or between the cells and foreign substances.

In opposition with this complexity, most of the methods used so far to examine the physicochemical properties of microbial surfaces consider them to be, more or less implicitly, homogeneous and molecularly smooth, although key factors are clearly the spatial distribution of constituents and properties. Atomic-force microscopy is undoubtedly a very promising tool in this respect. In the near future, it may provide direct information, with molecular- scale spatial resolution, as regards chemical composition, physicochemical properties, and interactions of cell surfaces, in conditions relevant to practical interest. This constitutes an exciting field in which a strong synergy may develop between biotechnology and nanotechnology.

9

EXTRACHROMOSOMAL REPLICATIONS

Plasmids are extrachromosomal replicons found in Gram-negative and Gram-positive bacteria as well as in some lower eukaryote organisms. They are present in bacterial cells replicating at a specific number of copies per cell. Their size varies from a few to several hundred kilobase pairs, and bacterial cells can harbor more than one plasmid species. The term *plasmid* originally was used by Lederberg to describe all extrachromosomal hereditary determinants. Currently, the term is restricted to the autonomously replicating extrachromosomal DNA of bacteria. Although plasmids replicate autonomously, they generally rely on some host-encoded factors for their replication. Although not essential for the survival of bacteria, plasmids may encode a wide variety of genetic determinants that permit their bacterial hosts to survive better in an adverse environment or to compete better with other microorganisms occupying the same ecological niche.

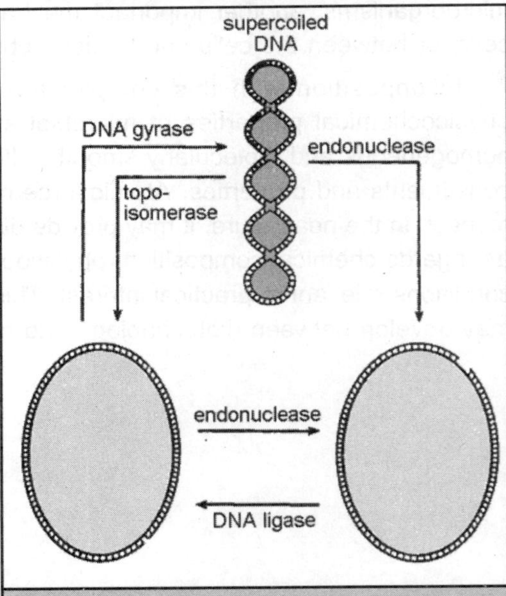

Figure 9.1 The interconversion of supercoiled, relaxed covalently closed circular DNA and open circular DNA.

Plasmids are found in a wide variety of microorganisms, and it is as difficult to generalize about plasmids as it is to generalize about the microorganisms that harbor them. Plasmid DNA is mostly isolated as covalently closed, circular double-stranded DNA molecules, although recently linear plasmids have been described in certain bacterial species. Furthermore, single-stranded DNA plasmids were identified as intermediates during the process of replication in certain Gram-positive bacteria.

Plasmids also include the replicative forms of filamentous coliphages and the prophage state of phages such as P1. The medical importance of plasmids that encode for antibiotic resistance (R plasmids) and those that contribute directly to microbial pathogenicity, such as for instance iron transport in several pathogens or the presence of adhesions, invasins, or antiphagocytic proteins, is well documented, as is the role played by plasmids in bacteria of importance in agriculture and industry. Some plasmid genes encoding antibiotic resistances and other traits are frequently located in transposable elements or integrons. This produces great variation and flexibility in the constitution of plasmids. These extrachromosomal elements are of equal importance, however, for the study of the structure and function of DNA. Plasmids have taken on paramount importance in recombinant DNA technology. In this section we describe the mechanisms of the control of plasmid DNA replication and examine the replication of plasmids ColE1 and R6K, found in Gram-negative organisms, and pT181, found in Gram-positive bacteria. In addition, we also discuss linear plasmids.

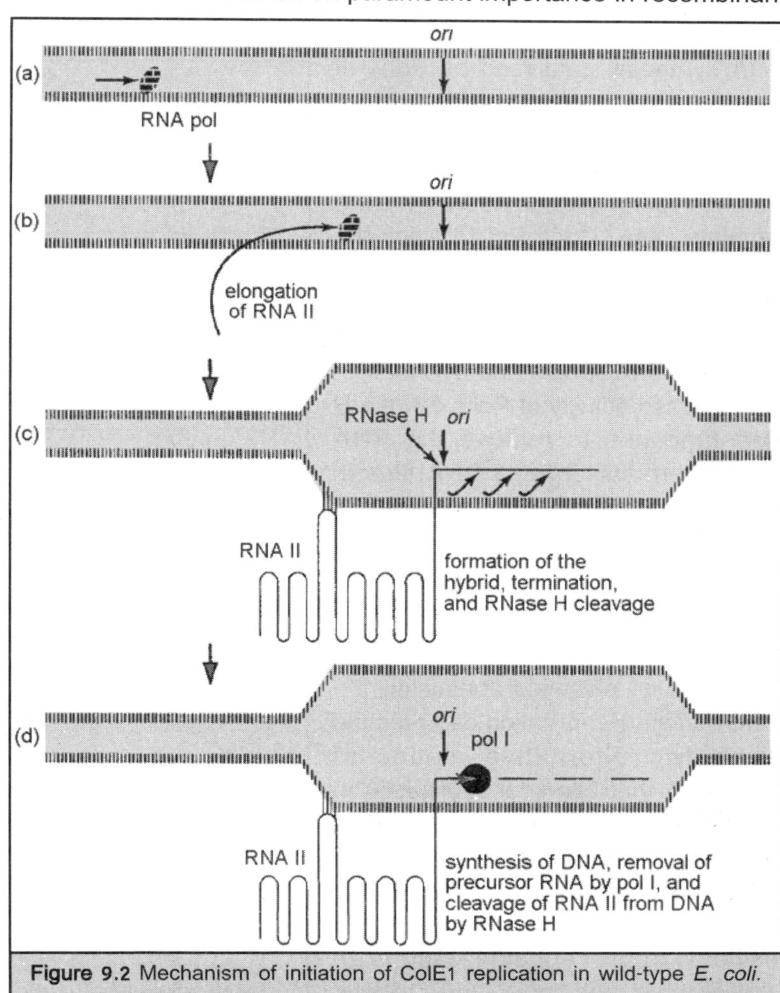

Figure 9.2 Mechanism of initiation of ColE1 replication in wild-type *E. coli.*

COE1-TYPE PLASMIDS

The ColE1-type family encompasses both a large number of naturally occurring plasmids and many of the most commonly used cloning vehicles. The mechanism controlling initiation of replication of these plasmids is among the best understood, perhaps reflecting the importance of this family of plasmids in recombinant DNA applications. Replication of plasmids containing the ColE1 origin is initiated at a unique start-site (*ori*). Unlike in other plasmid families, initiation of replication of ColE1 is not mediated by a plasmid-encoded protein; instead ColE1 requires the host DNA polymerase I (PolI). In addition, initiation of replication of ColE1 requires the presence of a host-encoded RNA polymerase and ribonuclease H (RNase H). Replication of the plasmid is initiated by the synthesis of an RNA molecule by the host RNA polymerase 555 bases upstream of *ori*. This RNA species, called RNA II, extends about 700 nucleotides from its initiation. The 3′ of this RNA molecule subsequently forms a duplex with the DNA of the plasmid at a region close to the origin of replication, which is located at nucleotide 555 of RNA II.

The formation of the duplex structure (known as coupling) begins following the synthesis of a portion of the RNA II molecule. The coupling event is strongly dependent on the formation of a specific secondary structure at the 5′ end of the RNAII molecule that results in an interaction between the RNA portion on the *ori* and an upstream region of this molecule with the template DNA. The duplex RNA II–DNA structure acts as a substrate for the subsequent action of RNase H, which cuts the RNA at the replication origin, leaving a 3′-hydroxyl group that serves as the primer for DNA synthesis catalyzed by Pol I. It has been shown that certain point mutations in RNA II prevent RNA II from adopting the right conformation, which affects hybridization with the template DNA. Once Pol I begins the addition of deoxynucleotides, the remaining RNA hybridized to the DNA strand is digested by RNase H at other sites on the molecule and by the 5′-3′ exonuclease activity of Pol I. RNase H also functions to remove the RNA primer from the growing DNA strand. Replication of ColE1 then proceeds unidirectionally with the initiation of the lagging strand synthesis at specific ColE1 sites.

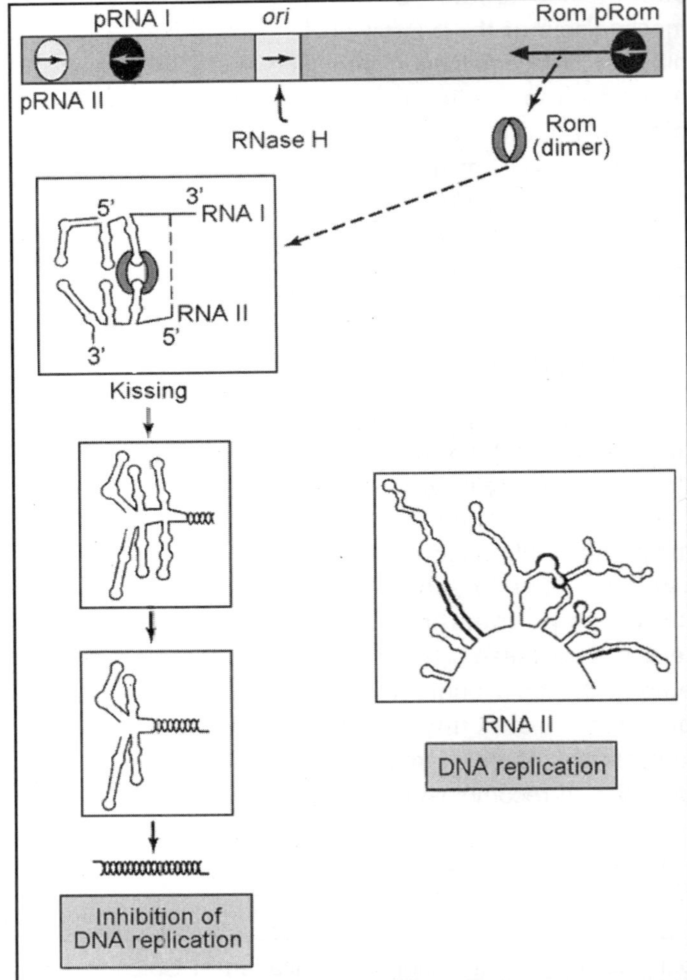

Figure 9.3 Proposed mechanism of the succession of events involved in the regulation of the initiation of ColE1 by RNA I and Rom.

It has been further shown that replication of plasmids containing the ColE1 origin of replication can proceed using two alternative modes of replication that allow for replication in the absence of either RNase H alone or RNase H and Pol I. These two mechanisms appear to not be utilized in wild-type cells, however, and probably represent adaptations to specific host mutations. One of these mechanisms is independent of the presence of Pol I and involves the elongation of RNA II coupled to the template DNA. The lack of RNase H allows the formation of a single-stranded DNA region that can extend to a length that is adequate for assembly of a replisome and initiation of synthesis on the opposite DNA strand (lagging strand). Another initiation mechanism has been demonstrated in RNase H–deficient mutants that is dependent of Pol I. In this case, the extended RNA II species can be recognized by Pol I and used as a primer. However, this mechanism of initiation has proved to be inefficient in in vitro experiments. Regardless of the replication mode, however, RNA II must be produced in the correct structural configuration to act as the primer for DNA replication.

Control of initiation of replication of ColE1 is mediated by the interaction of RNA II with a 108-nucleotide anti- sense RNA transcript, called RNA I, that is encoded in a region that overlaps the coding region for RNA II. RNA I binds to nascent RNA II by complementary base pairing, triggering a conformational change in RNA II that prevents coupling with DNA, thereby inhibiting replication initiation. Within the secondary structures of both RNA species there are several stem-loops. Interaction between both RNA species starts at complementary loops in a reversible process known as "kissing," which results in the generation of an unstable initial complex. After this initial contact takes place, an irreversible process leads to the formation of an RNA duplex between RNA I and RNA II. RNA I achieves its inhibitory effect only if it is present during a short, specific interval in the synthesis of RNA II. This situation occurs most frequently because RNA I transcription occurs as much as five times more often than transcription of RNA II. This results in the formation of a productive primer only in about 1 out of every 20 RNA II molecules that are started.

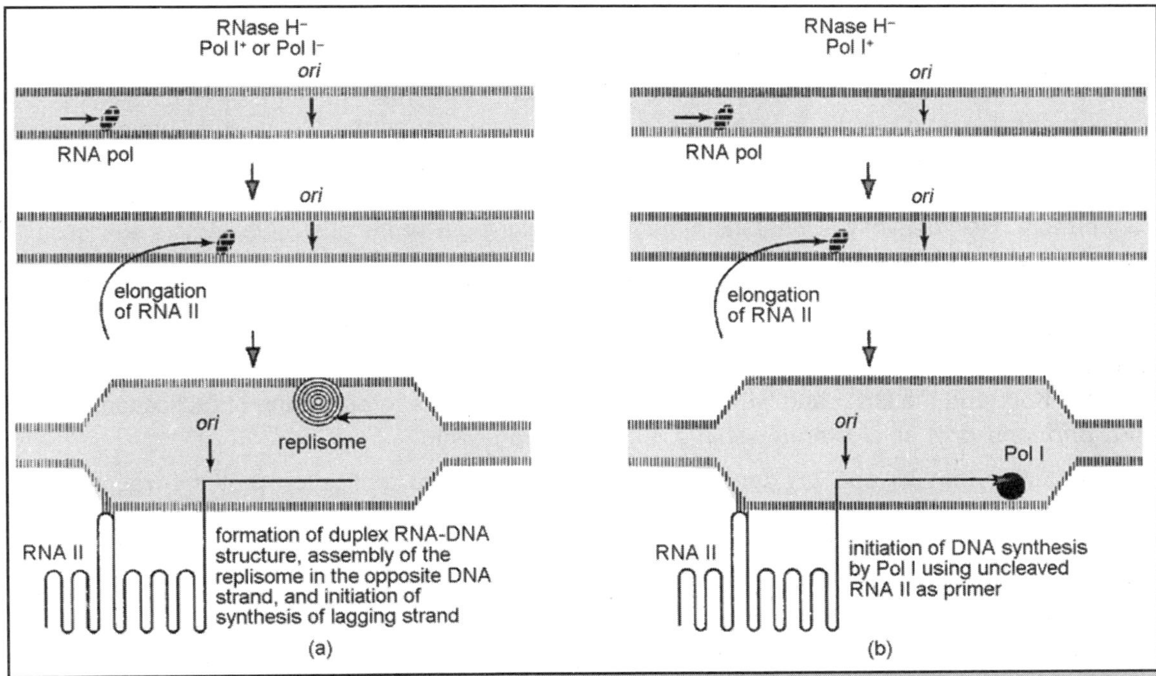

Figure 9.4 Alternative mechanism of initiation of replication of ColE1 in RNase H-deficient mutants. (a) Proposed mechanism of initiation in RNase H-deficient (rnh) and Pol I positive (polA$^+$) or negative (polA) mutants. (b) Mechanism proposed to occur in rnh, polA$^+$ mutants.

The available amount of RNA I in the cell is influenced by polyadenylation and the bacterial RNase E, which cleaves the 5′ end of RNA I, rendering it inactive. A second factor involved in the control of initiation of replication of ColE1 is the plasmid-encoded 63–amino acid protein Rom (RNA one modulator), also known as Rop (Repressor of primer). The gene encoding Rom is located downstream of *ori*. In the early stages of plasmid replication, Rom binds to the stem-loop portions of the unstable initial complex, reducing the dissociation constant of the complex, thus creating a pathway for stable binding of RNA I and RNA II. As a consequence, ColE1 deletion derivatives that lack *rom* present a higher copy number than plasmids that have the complete ColE1 replication region. The presence of Rom, however, is not needed for viability of the plasmid, and in some RNA-regulated replicons is not present. Incompatibility is the failure of two plasmids to be stably inherited in the same cell line. This phenomenon is a consequence of sharing elements of plasmid inheritance functions such as replication or partition. RNAI is the main incompatibility determinant in ColE1-type plasmids. Two plasmids that depend on the same RNA I species for regulation of initiation of replication can not coexist in the same cell. It has been shown that even single nucleotide changes can have profound effects in the incompatibility properties of ColE1-type plasmids.

R6K PLASMID

R6K is a naturally occurring conjugative plasmid that encodes resistance to the antibiotics ampicillin and streptomycin. This plasmid is an example of an extrachromosomal element of intermediate size and copy number because it is about 38 kb in size and has a copy number of 13 to 40 per cell. These features, together with a unique mode of replication, make R6K an attractive system to study the genetic and molecular mechanisms involved in plasmid DNA replication. In addition, this plasmid and its replication components were among the first used to generate gene fusions, transcription enhancement, protein tagging, and site-specific proteolysis; it is currently utilized for high expression and rapid isolation by affinity chromatography of a large variety of proteins. Furthermore, R6K derivatives were instrumental in designing a series of suicide vectors successfully used to generate mutants by allelic exchange or transposition mutagenesis in Gram-positive bacteria. Plasmid R6K is a member of a growing group of replicons encoding an initiator protein that binds to nucleotide sequence repeats, called iterons, located within *ori*. This group of iteron replicons includes the *Escherichia coli oriC* and the plasmids F, pSC101, P1, PMJ101, Rts1, the REPI replicon of pColV-K30, and the RK2- and RP4-related plasmids. R6K has three origins of replication, named *ori*α, *ori*β, and *ori*γ, all clustered within a 4-kb DNA fragment.

In addition, the *pir* and *bis* genes encoding the π and Bis replication proteins, respectively, were found within this replication region. Near this region is located the replication termination site containing the τ sequence that binds the host-encoded ter protein and arrests the movement of the replication fork. The main active origins in vivo are α and β, while *ori*γ tends to remain inactive due to the synthesis of a silencer RNA encoded immediately downstream from the *ori*γ sequence. The silencing activity of this RNA was explained by the formation of a silencer RNA-activator RNA hybrid, driven to completion of the R6K-encoded replication protein, that leads to the inactivation of *ori*γ. However, *ori*γ can replicate autonomously when the other two origins are deleted and the *pir* gene is provided in *cis* or in *trans*. Thus, *ori*γ has served as the simplest model system derived from R6K to study the replication of an iteron-containing DNA molecule. The γ origin is the major initiator binding site and is required in cis for initiation of plasmid replication from the other two sites. Therefore, the γ origin behaves as a prokaryotic enhancer-type element in the sense that DNA-protein interactions at this site induce significant changes in DNA structure that facilitate initiation of DNA replication from distantly located origins.

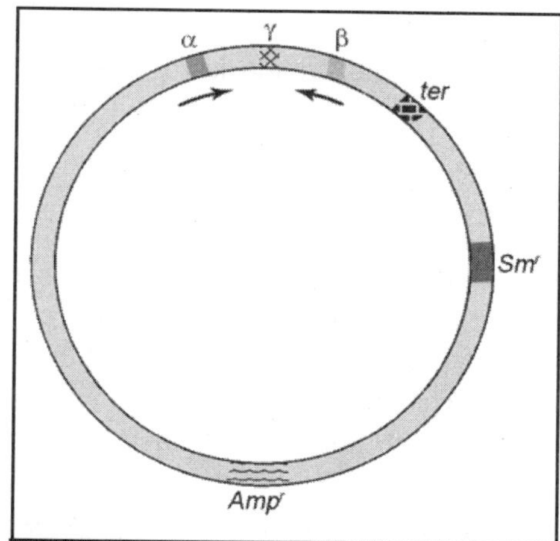

Figure 9.5 Diagram of R6K showing the 4-kb replication region containing the α-, β-, and γ-origins.

The molecular organization of this origin is similar to other plasmid origins, although it contains two functionally distinct segments. The 277-bp core segment, common to all three origins, is essential for replication, and consists of three distinct regions: (i) the AT-rich region, bound by π and the integration host factor (IHF) protein, (ii) seven 22-bp repeats bound by π, and (iii) a multiprotein-binding region interacting with DnaA, IHF, and RNA polymerase. Immediately to the left of the core lies the 106-bp origin enhancer, which includes a DnaA box and a small segment containing the *stb* locus. Even though the enhancer is not absolutely required for replication at low level of wild-type π-protein, it was found that the enhancer region is required for stable maintenance of γ-origin-containing plasmids. Although the *stb* locus has some similarities with the *par* locus of pSC101, the partition systems of these two plasmids differ from each other in several aspects. It has been proposed recently by Wu et al. that a host-encoded protein may be binding to the *stb* repeats. The finding that R6K derivatives carrying the intact three origins can bind in vitro to both inner and outer membrane fractions of *E. coli* appears to support this hypothesis.

An alternative explanation is that *stb* mediates plasmid partition by altering the structure of γ-origin-containing plasmids. The host-encoded protein F is also binds to 10 sites in the γ-origin that overlap all the previously identified binding sites for the R6K-encoded π-protein and the host-encoded DNA binding proteins DnaA, IHF, and RNA polymerase. However, it appears that the Fis protein is required for plasmid replication only when π-copy-up variants and the penicillin-resistant marker are simultaneously used, suggesting that plasmid sequences such as those encoding antibiotic resistance play a role in plasmid DNA replication. The analysis of the functional α-replicon revealed that it is composed of two elements that must be present in *cis* and oriented as in the intact R6K: (i) a 580-bp fragment containing the α-origin, and (ii) the 277-bp core segment located within the γ-origin. These two elements are separated by approximately 3 kbp, which are not required for the γ-origin activity. The 580-bp fragment contains a long 98-bp palindrome, and it was suggested that this sequence serves as the recognition signal for initiation of DNA replication in the α-origin.

DNA homology analysis also revealed the presence of a 23-bp sequence that resembles the seven 22-bp iterons found in the core segment and plays a role in the π-protein-mediated looping between the γ- and α-sequences. In addition, this fragment contains a DnaG binding site that can serve as DnaB loading site by DnaB–DnaG interaction. The minimal β-replicon was defined as a 2-kbp fragment that encompasses the following elements: (i) the γ-core region, (ii) the *pir* gene encoding the π-initiator protein, (iii) the *bis* gene encoding the 17.2-kDa Bis protein, and (iv) the β-origin. The Bis protein is required only by the β-origin, and its synthesis is coupled in cis to the expression of π-protein from an unaltered *pir* gene. The β-origin also contains a half 22-bp iteron and a 98-bp palindrome that has 96% homology with the α-origin hairpin. The presence of these pallindromic sequences at the *ori*α and *ori*β regions was previously assessed by electron microscopy. The half iteron and the long palindromic sequence are required for the π-mediated looping between the γ- and β-sequences and the initiation of DNA replication at the *ori*β, respectively. The π-initiator protein is essential for replication from each R6K origin and, although it can be provided in cis or trans to regulate the activity of the γ- and α-origins, it is required in cis for activation of the β-origin. This protein is a homodimer with a 36-kDa molecular weight for the monomeric form, which is lysine rich and weakly basic. In its native dimeric form, or as a hybrid with β-galactosidase or collagen, it has been shown to bind to the seven iterons in the γ-origin as well as to an eighth iteron and a smaller inverted pair of repeats located in the operator-promoter region of the *pir* gene.

In addition, this initiator protein interacts with the iterons located in the α- and β-iterons; however, these contacts are weak and require the enhancing effect of π already bound to the seven γ-iterons. The π-protein has both positive and negative roles in the replication of R6K. The positive role of π is displayed by its ability to enhance replication of *ori*α and *ori*β. This activity is mediated by the promotion of conformational changes, such as DNA unwinding and bending, in the replication region that cause activation of these two origins by looping out intervening sequences located between the *ori*γ core region and the α- and β-origins. Site-directed mutagenesis showed that π-protein is necessary but not sufficient for activation of *ori*β and probably *ori*α. This observation suggested that the DNA looping process must also be required for the transfer of a multiprotein complex capable of initiating DNA replication. Supporting this hypothesis, it was recently reported that π specifically interacts with the host-encoded helicase DnaB replication protein. This result indicates that DnaB is initially recruited by α, already bound to the γ-origin, and then delivered to the two other origins by the DNA looping induced by the R6K-encoded replicator protein. The π-protein has been shown to possess two negative activities on the replication of R6K. One of them is as an autorepressor at the level of transcription and involves its binding to iteron sequences located within the operator-promoter site of the *pir* gene. These interactions act by either preventing the binding of the RNA polymerase to the promoter or displacing the RNA polymerase from promoter–enzyme complexes.

The other negative role of the π-protein is its negative control of R6K replication. Biochemical and genetic experiments showed that increased intracellular levels of initiator protein will either lower the plasmid copy number or completely prevent its replication. The copy-up phenotype of some of these mutants correlates with base changes within a 100-bp fragment of the *pir* gene and can be suppressed by the presence of the wild-type initiator protein at concentrations lower than that of the mutant protein. This negative regulatory activity of π was explained by molecular models involving either direct π–DNA interactions or association of π with either other π molecules or with host-encoded proteins. One of these models, termed *handcuffing*, is based on the observation that π has the ability to associate two DNA molecules containing γ-origin sequences and to enhance the DNA ligase–catalyzed multimerization of a single DNA fragment carrying this R6K origin of replication. In addition, it was shown that the negative domain of π is located in the N-terminal region of this protein and that the π-mediated inhibition of R6K replication does not require direct binding to DNA. Consequently, the origins located within these DNA–protein complexes are unable to initiate replication, most likely because π-induced DNA structure alterations in the origins or potential π–host protein interactions are prevented.

In summary, all these observations suggest that the dual activity of π and the regulation of R6K replication is the result of a competition between positive γ–α and γ–β interactions and the inhibitory aggregation of γ-containing molecules, all mediated by the π-replication protein. Recently, three novel R6K-encoded proteins, designated DDP1, DDP2, and DDP3, were identified. DDP1 and DDP2 are encoded by two tandem genes located at the 5′ end of the long inverted repeat of the α-origin, while the gene encoding DDP3 was mapped at the 3′ end of the β-origin long inverted repeat. These three proteins are required for the distortion of the DNA structure of the two R6K active origins. Although the distortions caused by these proteins are potentially linked to R6K replication, they are not equivalent to those described in other replication regions previously characterized. It was also suggested that this distortion system serves to synchronize the initiation of replication and establishes the direction of replication from the α- and β-origins by generating a “locked” preinitiation protein–DNA complex.

PT181-TYPE PLASMIDS

The rolling circle type of replication is most often utilized by a number of small, high-copy-number plasmids from Gram-positive bacteria. However, plasmids that replicate using this mechanism have recently been identified in a number of Gram-negative bacteria. Gram-positive plasmids that replicate using rolling circle mechanisms are classified into five families based on sequence comparisons and genetic organization of the replication regions. All plasmids in these families replicate by an asymmetric rolling circle pathway that resembles that of the single-stranded filamentous bacteriophages. One of the best-understood replication mechanisms is that of the pT181-type family. We describe the replication of pT181 as a paradigm of rolling circle replicating plasmids. A more comprehensive description of replication of plasmids in other families can be found elsewhere. This plasmid, derived from *Staphylococcus aureus,* encodes a 38-kDa initiator protein, RepC, that has sequence-specific endonuclease and topoisomerase I–like activities. To initiate replication, RepC nicks one of the pT181 DNA strands (leading strand) at a specific site, generating a free 3′-OH end that is used as primer for subsequent DNA synthesis.

The functional structure of the RepC protein is a homodimer (RepC/RepC) that recognizes and binds to a site (Rep binding site) that encompasses an inverted repeat. A domain of six amino acids has been identified in RepC that is important in the recognition and interaction of the Rep binding site. The binding efficiency of RepC is increased by the presence of *cmp,* a 100-bp cis-acting replication enhancer located about 1 kb from the nicking site. It has been recently demonstrated that an *S. aureus* protein, CBF1, binds *cmp* and increases distortion of the already bent *cmp* locus. Whether this binding is associated with the enhancing activity of *cmp* is still not known. Upon binding of the RepC homodimer to IR III, DNA in this region is induced to bend. This is closely followed by a change in structure of RepC, DNA melting, and formation of a cruciform structure at the IR II region. The melting step is facilitated by the presence of an AT-rich inverted repeat region (IR I) located upstream of IR II. The formation of a cruciform structure may result in an approximation of the nicking site of the leading strand and the active site of RepC. This process involves a tyrosine residue that appears to facilitate the generation of the nick. After the endonuclease attack, the RepC protein remains covalently bound to the 5′ end of the DNA by a phosphotyrosine bond, probably remaining attached throughout replication of the leading strand. This is not true for all plasmids replicating through a rolling circle mechanism, however, for it was recently shown that in a derivative of pMV158 the initiator protein does not remain covalently bound to the DNA after nicking.

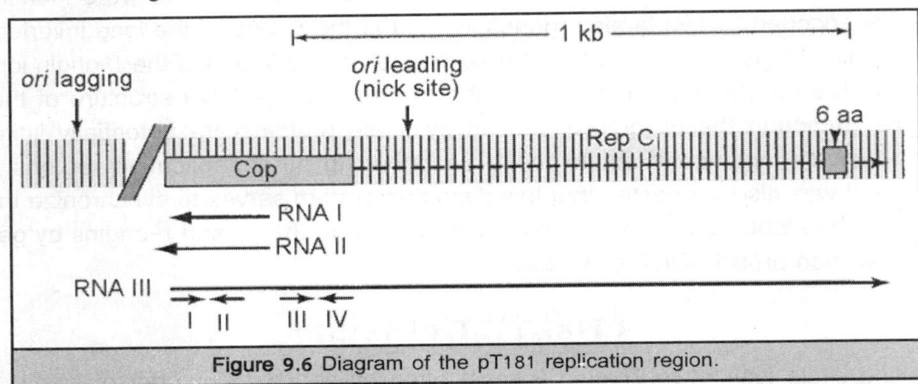

Figure 9.6 Diagram of the pT181 replication region.

After generation of the 3′-OH terminal end by RepC, an initiation complex is formed with the probable participation of DNA polymerase III, the helicase PcrA, and single-strand binding protein. Following (or during) replication, the initiator protein RepC becomes modified (RepC*) by the addition of a 10- to 12-mer oligodeoxynucleotide identical to the sequence located immediately 3′ to the origin of the leading strand. In addition, the two enzymatic activities of RepC are lost during this time. Further analysis has demonstrated that the active initiator homodimer RepC/RepC becomes the inactive heterodimer RepC/RepC* after it has been used for replication of pT181. Replication of the lagging strand starts at a different origin than replication of the leading strand. This origin, also known as SSO (single-strand origin) or *palA,* comprises about a 160-bp palindromic DNA sequence.

Synthesis of the leading strand generates a displaced single-stranded DNA that allows the origin of replication of the lagging strand to adopt the correct conformation to serve as a priming site. Replication of this strand does not require any protein encoded by the plasmid. In addition, some rolling circle replicating plasmids have more than one origin of replication for its lagging strand that is functional in different hosts. Taken together, these facts suggest that lagging strand origins of replication may be important in determining the host range of the plasmid. Termination of synthesis of the leading strand occurs at the nick site by a strand transfer mechanism that is mediated by the replication initiator protein RepC. In pT181, once the nick site has been replicated and extended a few nucleotides beyond this site, one of the subunits of the RepC dimer contacts the growing strand. This interaction initiates a strand transfer reaction, resulting in the formation of a single-stranded monomer (old strand), a double-stranded molecule where one of the strands is newly synthesized, and a dimer in which one of the monomers is attached to the oligonucleotide resulting from the extension of replication beyond the nick site. Although the presence of the nicking site is required, the initiator protein recognition site is not essential for termination. Regulation of initiation of replication of pT181 is achieved through the control of the synthesis of RepC.

The mRNA (RNA III) encoding RepC contains a leader sequence typical of genes that are regulated through attenuation. This leader sequence has four interacting sequences (I–IV). Sequence I, a nine-nucleotide sequence called *preemptor* can form duplexes with sequences II and III. In addition, there are two antisense RNA species (RNA I and RNA II) complementary to the leader sequence of RNA III. Analysis of the secondary structure of the leader portion of RNA III in the presence or absence of antisense RNA indicated that upon interaction of the two complementary RNA species, a stem-loop structure between inverted repeats III and IV is formed followed by an AUUUUUU sequence that acts as a transcription termination signal. On the other hand, if antisense RNA is not present, the interaction between inverted repeats I (the preemptor) and III generates a structure that prevents the formation of the III-IV stem-loop. Under these conditions, transcription of *repC* takes place. Mutations in the preemptor have been shown to lead to an absolute requirement for RepC in trans. In absence of antisense RNA, the stem-loop structure involving inverted repeats III and IV is formed, no RepC synthesis occurs, leading to a plasmid that has an absolute requirement for RepC in trans. The main incompatibility elements in pT181-type plasmids are the

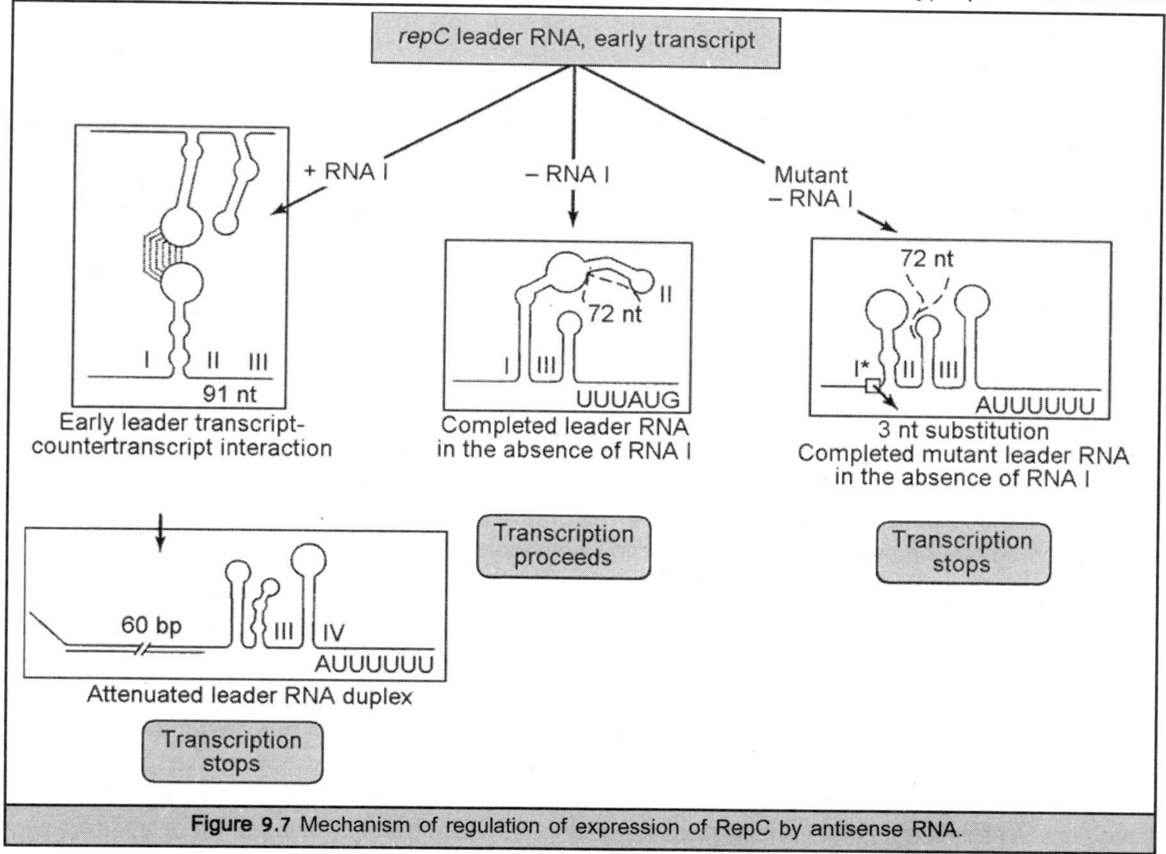

Figure 9.7 Mechanism of regulation of expression of RepC by antisense RNA.

antisense RNA and the origin of replication of the leading strand. The origin of replication of the leading strand directly binds RepC, titrating the protein, which is present at limiting levels in the cytosol due to the controlling action of the antisense RNAs.

LINEAR PLASMIDS IN BACTERIA

A Unique Plasmid Anatomy

Bacterial genomes have long been considered to contain only circular DNA molecules. However, the development of orthogonal field-pulsed agarose gel electrophoresis in the mid-1980s allowed the discovery of the first linear bacterial chromosome in *Borrelia burgdorferi*, the causative agent of Lyme disease. Since then, linear replicons have been found in many bacterial genera including *Agrobacterium, Streptomyces, Thiobacillus, Nocardia, Rhodococcus,* and even *Escherichia*. The first linear plasmid was found in *Streptomyces rochei* in 1979, and others have been detected in the Gram-positive bacteria *Rhodococcus fascians* and *Nocardia opaca* as well as in the Gram-negative *Thiobacillus versutus*. The *Streptomyces* linear plasmids range in size from nine to several hundred kilobases, and all of them contain terminal inverted repeats of different lengths according to the size of the plasmids where they are present. The linear plasmids in *Borrelia* also show a wide range of sizes, typically ranging from 15 to 200 kb and are distributed widely among the members of this species.

The *Borrelia* plasmids are unique among extra-chromosomal elements because some of them carry critical genes such the *guaA* and *guaB* biosynthetic genes, as well as genes encoding the essential major outer surface Osp or Vmp lipoproteins. In addition, it was determined that some of these linear plasmids are present stably in low copy number, about the equivalent of one per chromosome, and can be cured by exposing *Borrelia* cultures to the DNA gyrase inhibitor novobiocinin. These characteristics suggest that the replication and partition processes of linear plasmids are well controlled during cell division. All the facts, together with the observation that the *Borrelia* chromosome has attributes of a linear DNA molecule, has led to the idea that these linear plasmids might be regarded more properly as minichromosomes. The presence of these linear DNA molecules brings an interesting biological problem associated with the protection of the DNA ends from exonuclease degradation, and the complete replication of these plasmids. These problems are solved in eukaryotes by the addition of telomere repeats by reverse transcription of a short RNA incorporated in the telomerase. On the other hand, prokaryotes can protect and replicate completely these linear plasmids using two alternative mechanisms.

The ends of linear *Streptomyces* plasmids have telomeric structures also found in adenoviruses and some prokaryotic phages as well as in almost all eukaryotic plasmids. Sequence analysis of pSL2, a linear plasmid present in *S. rochei*, revealed the presence of an inverted terminal repetition of 614 bp and 11 sites of interrupted homology. These terminal DNA structures, together with specific DNA binding proteins, are thought to be involved in the juxtaposition of the two plasmid termini containing identical or very similar nucleotide sequences. This structure, known as the racket frame-like DNA model also includes a terminal protein covalently attached to the 5′ DNA termini, which is required to protect the DNA from degradation and to complete the replication of the 3′ overhanging ends, as described later. The structure of the 16- and 49-kb linear plasmids of *Borrelia burgdorferi* consists of a double-stranded DNA chain connected at each end by a perfect palindromic AT-rich hairpin loop. In addition, each end contains a conserved 19-bp inverted repeat sequence, a telomeric structural feature found in all linear plasmids analyzed in *Borrelia* to date. These features have similarities to the telomeres of other linear double-stranded replicons, including among them the genomes of vaccinia and other animal poxviruses, the mitochondria of the yeast *Pichia*, and the iridovirus, which causes African swine fever. The fact that the latter and at least one *Borrelia* species share a common tick vector suggests that the *Borrelia* linear plasmids may have originated by horizontal transfer between kingdoms. These telomeric structures indeed play a role in preserving the integrity of these plasmids, and in their replication before cell division.

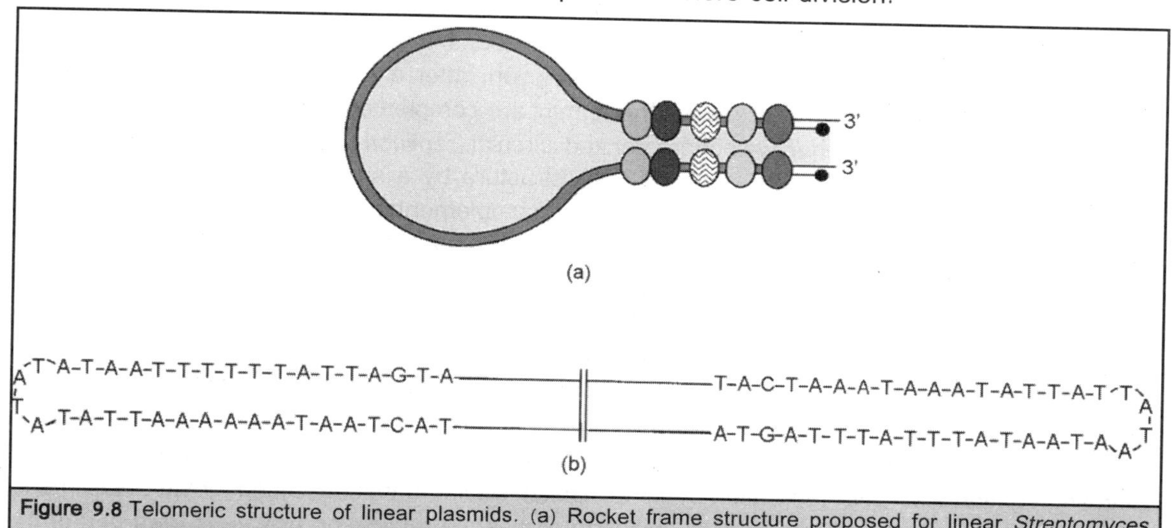

Figure 9.8 Telomeric structure of linear plasmids. (a) Rocket frame structure proposed for linear *Streptomyces* plasmids. (b) Structure and partial nucleotide sequence of the telomeric termini of linear plasmids isolated *Borrelia*.

Replication of Linear Plasmids

It was proposed initially that the linear plasmid of *Streptomyces* replicates by a protein-primed replication mechanism that has been well characterized for adenovirus, phage φ29, and *Bacillus subtilis*. In this mechanism of DNA replication, the telomere is the origin of replication, and it is recognized by specific DNA binding proteins that promote the unwinding of the double helix and serve as the primer for a specific DNA polymerase. However, it was later observed that the *S. clavuligerus* linear plasmid pSCL can replicate as a circular DNA molecule when the telomeres are removed and the ends are ligated. This observation was further confirmed by the detailed analysis of the mechanism of replication of pSLA2, another *Streptomyces* plasmid that contains protein covalently attached to the 5′ end. Ligation of pSLA2 deletion derivatives lacking the telomeric termini to a selectable marker resulted in a high-copy-number circular extrachromosomal element. This observation suggested that this linear plasmid must contain an internal site capable of promoting DNA replication. Two-dimensional agarose gel electrophoresis analysis showed that this plasmid is indeed replicated primarily bidirectionally from an internal origin located near the center of the plasmid, toward the ends, rather than by full-length strand displacement initiated at the telomeres. However, pSLA2 still requires a protein-primed strand-displacing mechanism to complete the replication of a 280-nucleotide segment at both plasmid termini.

It was suggested that the synthesis of the 5′ terminal segment of the lagging strand of this linear plasmid uses as a template the 3′ overhang of the leading strand and is primed by the protein covalently attached to the 5′ end of the mature plasmid. The replication mechanisms of linear plasmids in *Borrelia* are poorly understood; however, our knowledge of these novel replicons encoding essential genes has been extended by a series of very recent publications. The replication of some of the *Borrelia* linear plasmids involves circular intermediates. In addition, it was recently reported that the 180-kb plasmid of *Borrelia hermsii* can exist and replicate either as a linear or a monomeric circular element. According to these observations, it was proposed that the monomeric circular intermediate is formed by a head-to-tail junction, after a nick introduced in the inverted repeat opens the plasmid termini. Because the termini are complementary, the plasmid circularizes and can replicate as any other well-characterized circular replicon. This circular intermediate is then resolved into a linear double-stranded DNA structure by a second nick within the terminal inverted repeats. Because the terminal repeats are complementary, a linear molecule should be reconstituted. More recently, it was found that *B. burgdorferi* sensula to contains atypical large linear plasmids ranging from 92 to 105 kb. These plasmids carry *p27* and *ospAB*, genes that were also detected in other isolates on the 50-kb linear plasmid pAB50.

A more detailed analysis of the larger plasm ids demonstrated that they are formed by tail-to-tail dimerization of pAB50. The presence of such dimers can explain the unusual plasmid variability observed among different isolates and may provide new information regarding the replication mechanism of these linear replicons. It was postulated that these linear dimers are the result of failed segregation after DNA replication by a mechanism similar to that described for vaccinia viruses. In this model proposed by Marconi et al. initiation of plasmid replication proceeds from one of the termini, after the head hairpin loop in this particular case is nicked, allowing the formation of tail-to-tail dimer intermediates. Alternatively, these dimers can arise by DNA replication initiated from an origin located within a linear monomer that results in a circular replication intermediate. In the normal replication process, the circular intermediate is resolved into two linear monomers by independent cleavage events at each telomere junction. Cleavage failure at the tail hairpin by a specific DNA cleavage system results in two monomers linked tail to tail, a possibility that was confirmed experimentally by restriction analysis and Southern blot DNA hybridization experiments. However, this same analysis demonstrated that not all *Borrelia* large plasmids were originated via dimer formation, and alternative DNA replication systems that are still uncharacterized must exist. Some of the proteins involved in the replication of the linear plasmids in *Borrelia* were identified by determining the complete nucleotide sequence of the 16-kb plasmid lp 16.9 isolated from *B. burgdorferi* B31. This study revealed the presence of 15 open reading frames named A to O.

The predicted proteins encoded by OrfM and OrfN showed homology to proteins involved in plasmid and cell replication. OrfM is homologous to MinD, a cytoplasmic membrane protein with ATPase activity required for the correct placement of the division site. OrfM also has homologies with plasmid partition proteins such as ParA or SopA, and the RepB protein encoded by the pheromone-responsive plasmid pAD1 in *Enterococcus faecalis* and involved in the control of the copy number of this extrachromosomal element. The predicted features and the primary sequence of the OrfN hypothetical polypeptide are similar to those of the CopB DNA binding proteins, also known as RepB or RepA2, involved in the control of plasmid copy number in Gram-negative bacteria. This sequence analysis also led to the hypothesis that OrfN may interact with the short repeat sequence located within the promoter region of OrfM and, thus, control the expression of the latter. In summary, sequence analysis of *Borrelia* plasmids showed that the few proteins identified are similar to prokaryotic proteins involved in plasmid replication and maintenance in Gram-positive and Gram-negative bacteria. In addition, these results do not support the previous conclusion that the linear plasmids in *Borrelia* are indeed derived from animal viruses. Furthermore, it seems that circular and linear plasmids of this bacterial genus have a common origin and share DNA replication mechanisms.

10

BACTERIAL BIOFILMS

Biofilms are collections of microorganisms, predominantly bacteria, enmeshed within a three-dimensional gelatinous matrix of extracellular polymers secreted by the microorganisms. In an aqueous environment, a support, termed a substratum, will be immediately biased by dissolved organic molecules and macromolecules that adsorb rapidly from the liquid phase. Bacterial cells present in the fluid contact the substratum by a variety of transport mechanisms. Once at the substratum, the cells can adsorb either reversibly or irreversibly. Provided the cells remain at the surface for a sufficient time, they will secrete extracellular polymers that serve to attach the cells tenaciously to the substratum. Attached cells metabolize prevailing energy and carbon sources (either dissolved within the surrounding fluid, adsorbed to the substratum surface, or existing as a constituent of the substratum itself), reduce terminal electron acceptors, grow, replicate, and produce insoluble extracellular polysaccharides, thus accumulating an initial viable biofilm community. Inert particles, bacterial cells of the same or different species, and higher life forms (e.g., algae, amoeba, protozoa) continue to be recruited from the fluid and incorporated into the biofilm community.

BIOFILM COMMUNITIES

As the *biofilm bacterial communities* mature, the adherent populations will oxidize electron donors and reduce existing terminal electron acceptors in an order of decreasing redox potential. As the biofilm community becomes denser and thicker, the penetration depth of one electron acceptor will overlap that of another; thus, a succession of different microbial populations can be established. Biofilm layers near oxygen-saturated water will harbor aerobic heterotrophic or autotrophic activity, reducing the available oxygen. As the distance from the aerobic interface increases and oxygen is depleted, a sequence of terminal electron acceptors is utilized by specific microbial populations: facultative denitrifiers using nitrate and nitrite; anaerobic sulfate reducers using sulfate; fermentative microbes partially reducing organic carbon compounds; and finally, anaerobic methane producers. Depending upon the balance between mass transfer rates and microbial reaction rates, biofilms can develop stratified ecosystems, on a microscopic scale, similar to those observed in lake or marine sediments. In engineered systems, highly active aerobic activity within a biofilm can completely deplete oxygen concentrations over distances of 10–20 μm, creating anaerobic layers capable of supporting obligate, anaerobic sulfate-reducing bacteria. Consequently, it should come as no surprise to discover anaerobic sulfate reduction and hydrogen sulfide corrosion of the metallic surfaces in heat exchange systems that utilize highly aerated cooling water to condense steam.

As a result of hydrodynamic forces and stresses exerted by replication, there can be a continual erosion of cells and extracellular material from the biofilm back to the bulk fluid. A more random stochastic process known as "sloughing" can also occur where either large sections or the entire biofilm becomes displaced from the substratum and enters the liquid. Biofilms are as versatile as they are ubiquitous. Intentional use of biofilms can serve many benefits, for example, in the water and waste treatment industry, in bioremediation applications, and in industrial biotechnology. Unintentionally formed biofilms can create such detriments as biofouling of heat exchange systems and marine structures; microbial-induced corrosion of metal surfaces or the deterioration of dental surfaces; contamination of house-hold products, food preparations, and pharmaceuticals; and the infection of short- and long-term indwelling biomedical implants and devices. Such detriments can range in severity from being a mere nuisance to being life threatening.

Measurements

Figure 10.1 illustrates, for an arbitrary analytical measure, the typical accumulation of biofilm at a surface as a function of time. Also known are the changes in frictional and heat transfer resistances observed in engineered systems experiencing biofilm formation. Uncontrolled biofilm formation within natural, engineered, and biomedical systems can create numerous detriments as detailed in several definitive reviews. Detriments arise by way of a biofilm's influence on the transport of mass, momentum, and energy. Detriments attributed directly to bacterial biofilms include: (i) material deterioration and corrosion, (ii) increases in both frictional and heat transfer resistances, (iii) attachment to and infection of biomedical implant devices, and (iv) operational problems that plague lab-scale and full-scale bioreactors.

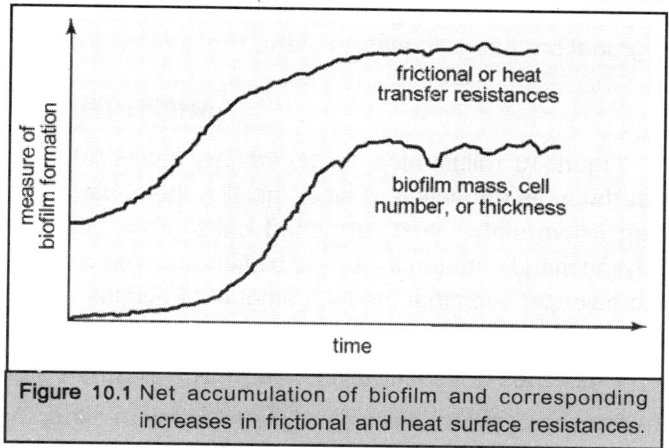

Figure 10.1 Net accumulation of biofilm and corresponding increases in frictional and heat surface resistances.

Applications

Benefits afforded by biofilms in a continuous reactor situation arise chiefly because the cell population is immobilized, and thus the residence time of the cells in the reactor is independent of the fluid-phase residence time. In continuous suspended culture bioreactors (i.e., chemostats), the mean residence time of the system cannot be less than the generation time of the bacterial species, otherwise cells do not have sufficient time to replicate within the reactor and are eventually diluted from the system. Immobilized and biofilm-bound cells remain in a continuous reactor system independent of the fluid phase, thus the mass loading of limiting substrate (or influent pollutant in the case of a wastewater treatment reactor) can be increased well beyond the growth rate limit imposed on suspended cultures.

Consequently, immobilized-cell or biofilm reactors can provide: (i) added volumetric reactivity, (ii) more stable operating performance, (iii) an inherent ease in biomass–fluid separation, and (iv) the prospect of staging different bioconversion processes in sequential reactors. Due to these inherent advantages, the use of biofilm reactors is not confined to just bacterial cells, but also comprises plant and animal cell applications. Bacterial biofilm reactors are employed either in commodities production or in wastewater treatment applications.

Biofilm reactors have been reportedly used to produce acetic acid, L-lysine, gluconic acid, kojic acid, and ethanol, and in the epoxidation of propene. Such biofilm reactors are operated either as packed- or fluidized-bed reactor systems, with cells either attached to an inert support particle or artificially immobilized within a gel matrix. One major application that relies on a microbial culture's ability to form biofilms is wastewater treatment. Biofilm reactor configurations, applied in both pilot- and full-scale wastewater treatment, include packed-bed trickling filters, high-rate plastic media filters, rotating biological contactors, fluidized-bed biofilm reactors, and membrane-immobilized cell reactors. Examples of biofilm reactors employed for wastewater treatment reported in the literature include polychlorinated hydrocarbon degradation, toluene degradation, denitrification, cadmium removal, anaerobic butyrate degradation, nitrification, glyphosphate degradation, anaerobic propionate degradation, phenolic wastewaters, uranium removal, and anaerobic carbon removal.

METHODOLOGY

Figure 10.1 illustrates, for an arbitary analytical measure, the typical accumulation of biofilm at a surface as a function of time. Initially, the substratum is conditioned and cells attach reversibly, then irreversibly. Next, attached cells grow, reproduce, and secrete insoluble extracellular polysaccharidic material. As the biofilm matures, biofilm detachment and growth processes come into balance, such that the total amount of biomass on the surface remains approximately constant in time. Multiple physical, chemical, and biological processes operate in each phase.

Processes governing biofilm formation and persistence include the following:

1. Biasing or preconditioning of the substratum either by macromolecules present in the bulk liquid or, intentionally, as in the case of pre-coating biomedical materials with adhesion molecules (e.g., fibronectin, vitronectin, fibrinogen, von Willebrand's factor).
2. Transport of planktonic cells from the bulk liquid to the substratum.
3. Adsorption of cells at the substratum for a finite time followed by desorption (release) of reversibly adsorbed cells.
4. Irreversible adsorption of bacterial cells at a surface.

5. Transport of substrates into the biofilm, substrate metabolism by the biofilm-bound cells, and transport of products out of the biofilm. These processes are accompanied by cellular growth, replication, and extracellular polymer production.
6. Biofilm removal can be either continuous, as in the case of biofilm detachment brought on by forces associated with either fluid flow or cell replication, or discrete, as in the case of biofilm sloughing.

Initial Events

Processes 1–4 are often grouped in mathematical models as a single process termed deposition.

Pretreatment of the surface

In the case of intentional pretreatment of a surface (i.e., adhesion molecules for biomaterials), there is no rate to be considered; the substratum essentially enters the system with a biased surface chemistry. However, in cases where a clean substratum material is exposed to an aqueous environment, transport of dissolved organic molecules or macromolecules in laminar flow is basically by molecular diffusion; in turbulent flow, transport must also consider convective transport effects. Once at the surface, adsorption of macromolecules occurs almost instantaneously, immediately changing the surface chemistry of the exposed material. Loeb and Neihof and Depalma et al. measured adsorption rates of organic molecules to various solid substrata in seawater, and Bryers reported analogous adsorption rates in a freshwater laboratory system. The net rate of adsorption in these studies can be described by equation 1.

$$R_{\text{adsorption}} = k_a S_i[1 - (S_i/k_s)] \quad \ldots(1)$$

where $R_{\text{adsorption}}$ is the rate of net adsorption ($ML^{-2}t^{-1}$); k_a is the adsorption rate constant (Lt^{-1}); k_S is the surface saturation coefficient (ML^{-2}); and S_i is the real concentration of adsorbed species (ML^{-2}). Although experiments indicate the maximum amount of adsorbed material may not exceed 0.10 μm in thickness, the surface properties resulting from adsorption of an organic film can significantly bias subsequent microbial events.

Conditioning of substratum

The net deposition rate of cells upon a "conditioned" substratum is itself a composite of the individual rates of cell transport to the substratum, cell adsorption, and adherent cell desorption.

Many models exist to describe bacterial transport from the fluid phase to the target surface. Bowen et al. and Beal derive continuity equations describing the transport of suspended particles from a flowing fluid to the surfaces of a surrounding conduit that accounts for both convective and molecular transport mechanisms. From their work, the maximum flux of particles transported to the surface, L distance from the inlet, of a rectangular flow cell, with gap height $2h$, can be written as

$$J(\text{cells-}L^{-2}t^{-1}) = k_{tr}\, X D_x\, X/h \quad \ldots(2)$$

where k_{tr} is the particle transport coefficient $((2/9)K^{0.33}/\Gamma(4/3))$; $K = (1/Pe)(8L/3h)$; Pe is the Peclet number $= 4vh/D_x$; v is the average velocity of the fluid; Γ is the gamma function ($\Gamma(4/3) = 0.89338$);

and D_x is the diffusivity of cells in the liquid. For nonmotile cells, the Brownian diffusion coefficient can be estimated from the Stokes-Einstein equation,

$$D_x X = k_b T / 3\pi\mu d_c \qquad ...(3)$$

where k_b is the Boltzmann constant (1.38 x 10^{-23} J/K); T is absolute temperature; μ is absolute viscosity; d_c is cell diameter. To compensate for cell motility, Jang and Yen propose the following equation,

$$D_x X = v_m d_r / 3(1 - \cos\sigma) \qquad ...(4)$$

where v_m is the velocity of motility, d_r is the free length of a random run, and a is the angle turned by the motile cell.

Once at the substratum interface, cells stick. Observations have indicated that cells can adsorb to a surface for a period of time and then may desorb from the substratum, returning to the bulk liquid. If cells spend sufficient time at the surface, they become permanently bound, initiate within minutes the secretion of extracellular polymer production, and can be removed only by rather aggressive physical or chemical means. One can conceive of this over-all process as much like two reaction steps in series, with the reversible adsorption process followed by an irreversible step pertaining to permanent adsorption,

$$X \leftrightarrow X_{rev} \rightarrow X_{irr} \qquad ...(5)$$

where X is the suspended cells, X_{rev} is the reversibly adsorbed cells, and X_{irr} is the irreversibly adsorbed cells. Assuming no cell replication, the total amount of cells observed at a substratum, X_t, at any one time comprises both reversibly and permanently adsorbed cells, that is, $X_t = X_{rev} + X_{irr}$.

Deposition of cells

Attachment is the deposition of cells from the fluid phase onto an existing biofilm. This process differs from adsorption in that the target surface is now a developed biofilm. The biofilm itself poses a different type of substratum in that biofilm surface morphology is generally more irregular (perhaps filamentous), very porous, compliant, and gelatinous. Experimental results indicate that the increased attachment of cells to a biofilm-covered substratum may depend on far more subtle mechanisms than mere changes in surface morphology. The local chemistry of a biofilm surface will reflect the composition of the bacterial cell and that of the insoluble polymer matrix, both of which could affect subsequent chemical interactions with incoming cells. Banks and Banks and Bryers report that the rates of cell deposition increased when cells were exposed to biofilm surfaces versus clean glass.

Two species were investigated, *Pseudomonas putida* and *Hyphomicrobium ZV620*, as to their deposition rates, under laminar flow conditions, onto both clean glass and biofilm-covered surfaces. *Pseudomonas putida* cells attached to a *P. putida* biofilm at a rate (zero order in attached cell concentration) that was 3.6 times higher than the cell attachment rate to clean glass. *Pseudomonas putida* cells attached to a *Hyphomicrobium* biofilm (with zero-order kinetics in attached cell concentration) that was 2.4 times higher than to clean glass. *Hyphomicrobium* illustrated an attachment rate to a *Hyphomicrobium* biofilm (again a zero dependency on attached cells) that was 3.3 times higher than the *Hyphomicrobium* attachment rate to clean glass. Attachment rate of *Hyphomicrobium* spp. cells to a *P. putida* biofilm was only 1.3 times that to clean glass. The attachment rate of *Hyphomicrobium* cells to a *Hyphomicrobium* biofilm was about half of the rate of *Pseudomonas* cells attaching to a *Pseudomonas* biofilm, even though suspended *Pseudomonas* cell concentrations were three times higher. Results insinuate that species-dependent enhancement of cell attachment may occur by specific rather than nonspecific adhesion mechanisms.

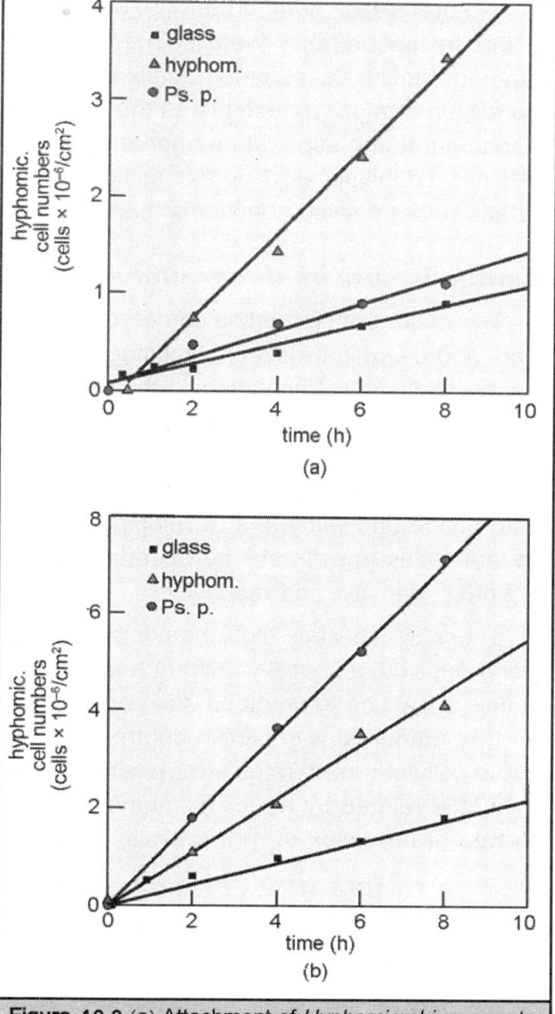

Figure 10.2 (a) Attachment of *Hyphomicrobium* spp. to various surfaces. Suspended cell concentration is 4.5E + 0.7 cells/mL. (b) Attachment of *Pseudomonas putida* to various surfaces. Suspended cell concentration is 1.5E + 0.8 cells/mL.

Physiological Response

Transformation collectively refers to those processes in which molecular rearrangements occur, that is, reactions. Once a cell adsorbs to a surface or becomes embedded within an existing biofilm, it will continue its metabolic processes in response to its immediate environment. Three fundamental rate processes can be identified: (i) alginate or exopolymer secretion, (ii) cellular growth and replication, and (iii) cell death and lysis. Prior to the mid-1990s; it was often easier to measure "per reactor area" performance using observed rates determined from bulk liquid parameters (e.g., electron donor removal rate or electron acceptor uptake rate) than to directly measure growth, replication, and death of cells in a biofilm. This experimental limitation led to the development of a number of unstructured mathematical models that were useful in estimating the flux of growth-rate-limiting substrate into a biofilm of fixed thickness, density, and reactivity, taking into account both external and internal molecular diffusion and biological reaction. It was not until the advent of various molecular methods and advances in computer-aided microscopy that it was possible to determine various biofilm processes directly within the biofilm.

Synthesis of alginate

Reporter gene technology has been used to observe the regulation of the alginate biosynthesis gene, *algC*, in a mucoid strain of *P. aeruginosa* in developing and mature biofilms. In vivo detection of *algC* up-expression in developing biofilms was performed with a fluorogenic substrate for the plasmid-borne *lacZ* gene product (β-galactosidase) using microscopy coupled with image analysis. By this technique, cells were tracked over time and analyzed for *algC* activity. During the initial stages of biofilm development, cells attached to a glass surface for at least 15 min exhibited up-expression of *algC*, detectable as the development of whole-cell fluorescence. However, initial cell attachment to the substratum appeared to be independent of *algC* promoter activity. Furthermore, cells not exhibiting *algC* up-expression were shown to be less capable of remaining at a glass surface under flowing conditions than were cells in which *algC* up-expression was detected.

Metabolic rate of donor and acceptor

The major transformation carried out by cells in the biofilm is the metabolism of both an electron donor and a terminal electron acceptor to produce soluble by-products, extracellular polymers, carbon dioxide, and water. Depending on the microbial population in question and the ambient concentration of electron donor and acceptor, a biofilm can be either aerobic, anoxic (denitrifying), anaerobic (sulfate-reducing bacteria, methane formers), or fermentative. Analysis of biofilm bacterial metabolic rates are frequently complicated by the effects of significant mass transfer resistances in both the liquid phase and within the developing biofilm. Steady-state mathematical models were derived in the late 1970s to estimate the observed flux of growth-rate-limiting substrate into a biofilm of fixed thickness, density, and reactivity.

These steady-state biofilm models are based on assumptions of a constant biofilm concentration (tacitly implying a spatially uniform reactivity) and a constant diffusion path (biofilm thickness). Such models allow one to predict (i) the concentration profile of limiting substrate (and, by stoichiometry, all other nutrients) with biofilm depth, and (ii) the maximum substrate uptake or flux to the biofilm. Biofilm cellular growth rates are typically simulated using a classical microbial growth rate expression, R_g, that is related by a stoichiometric coefficient, Y (in units of cell mass per substrate mass), to the rate of utilization of growth-limiting substrate, R_S ($M_{substrate}\ L^{-2}t^{-1}$), that is,

$$R_g(M_{cells}L^{-2}t^{-1}) = -\ R_S Y \qquad \ldots(6)$$

where the rate of substrate utilization is typically a Monod-like dependency on the concentration of one or more growth-rate-limiting substrates. For the example of a single substrate limitation,

$$R_S = \mu^* SB/(K_S + S) \qquad \ldots(7)$$

where μ^* is the species maximum growth rate constant (t^{-1}); S is the local dissolved substrate concentration ($M_{substrate}L^{-3}$); K_S is the saturation kinetics half-saturation constant ($M_{substrate}L^{-3}$); and B is the local biofilm cell concentration ($M_{cell}L^{-2}$).

Some differences in biofilm kinetics versus typical suspended cell microbial kinetics include the following:

1. The concentration of active bacterial cell mass metabolizing the substrate, B, is a per area concentration and may also be a function of time and space, $B = B(t,z)$. In a mixed culture biofilm, each bacterial species could exhibit a different dependency on time and space.
2. The concentration of substrate S in the biofilm will be affected by mass transfer resistances and biological reaction rates, such that S may also be a function of time and space.

3. Endogenous decay, maintenance energy, or death terms for cells not receiving adequate nutrients are treated as simple first-order functions of the biofilm population, for example,

$$R_{death} = k_d B \qquad \ldots(8)$$

where k_d is the cell death specific rate constant (t^{-1}). Thus the local, instantaneous net production of a species biomass in the biofilm can be modeled as

$$R_g(t, z) = [\mu^* S(t, z)B(t, z)/Y(K_s + S(t, z))] - k_d\, B(t, z) \qquad \ldots(9)$$

Procedure of Removing Biofilms

Biofilm removal processes involve the actual loss of material from the biofilm—either cells or cell–biofilm matrix debris. Desorption is the release of cells from the substratum and occurs during the early stages of substratum colonization. Biofilm removal represents loss of biomass from a developing or mature biofilm that arises by way of two distinctively different mechanisms: detachment and sloughing.

Loss of individual cells

Detachment is the continuous loss of biomass or individual cells from a biofilm. The rate of biofilm detachment has been mathematically modeled in a number of ways. In about the only definitive experimental study on biofilm detachment, Applegate and Bryers reported that the growth conditions of the biofilm strongly affect biofilm removal processes. Carbon-substrate-limited growth conditions fostered biofilms that (i) contained less extracellular polymer per cell and per area, (ii) bound less calcium per unit amount of biofilm, and (iii) exhibited a higher detachment rate than oxygen-substrate-limited biofilms that accumulated more extracellular polymer per cell, bound more calcium per biofilm amount, and developed a rigid compact morphology that was very resistant to detachment prior to predictable sloughing.

Carbon-limited biofilms did not exhibit sloughing for extended periods of cultivation or after prolonged (300 h) of carbon-nutrient deprivation. Perhaps biofilm detachment processes may be controlled, not by fluid hydrodynamics as once imagined, but rather by cell physiology. Boyd and Chakrabarty suggest biofilm detachment could be genetically controlled. Their work showed that the exopolysaccharide alginate of *P. aeruginosa* was important in determining the degree of cell detachment from an agar surface. Nonmucold strain 8822 gave rise to 50-fold more sloughed cells than mucoid strains 8821 and 8830. Alginate anchors the bacteria to the surface, thereby influencing the extent of detachment.

The role of the *P. aeruginosa* alginate lyase in the process of cell detachment was investigated. Increased expression of the alginate lyase in mucoid strain 8830 led to alginate degradation and increased cell detachment. Similar effects were seen both when alginate lyase synthesis was induced at the initial stage of cell inoculation and when it was induced at a later stage of growth. Apparently, high molecular weight alginate polymers are required to efficiently retain the bacteria within the developing biofilm. When expressed from a regulated promoter, the alginate lyase did enhance detachment of cells through the degradation of the alginate. This suggests a possible role for the lyase in the removal process of bacterial growth films.

Catastrophic loss

The catastrophic, apparently random, loss of large pieces or entire sections of biofilm is known as *sloughing*. At the industrial scale, sloughing creates excursions of intolerably high biomass in bioreactor effluents, and periodic "blooms" of pathogenic bacteria in drinking water distribution systems. At the laboratory scale, all too often, random sloughing of a biofilm will terminate an experiment, resulting in very little quantitative data regarding the fundamental causes of sloughing. A number of situation-specific causes of sloughing have been identified: bubble formation in either anaerobic methane-producing biofilms or denitrifying biofilms, and artificially induced sloughing by way of EGTA addition, which chelates the bound calcium ion used in biofilm alginate cross- linking. Applegate and Bryers reports that the growth conditions of the biofilm strongly affected the biofilm removal processes of detachment and sloughing. "Fluffier" biofilms, produced under carbon-limited growth, exhibited high detachment rates but never sloughed, even when subjected to over 300 h of nutrient starvation. Conversely, rigid biofilms cultivated under oxygen limitations showed little detachment, but the onset of catastrophic sloughing was repeatable between replicate experiments.

Development of Bacterial Biofilm

It comprises the following fundamental processes: (i) substratum preconditioning by adsorption of fluid-phase organic molecules; (ii) bacterial cell deposition to the conditioned substratum (cell transport to the surface, cell adhesion to the substratum, cell desorption); (iii) bacterial metabolism (cell substrate conversion; cell growth and replication; extracellular exopolymer production; cell starvation, death, lysis); (iv) biofilm removal (cell and biofilm detachment, biofilm sloughing). Naturally, the relative influence of each process to the overall rate of accumulation is dependent upon the specific system, the prevailing environmental conditions, and biological changes throughout the lifetime of the biofilm. Simple unstructured models of biofilm formation can be written. Biofilm formation is assumed to occur on surfaces exposed to a well-mixed bulk fluid phase, thus tacitly neglecting any concentration gradients in the bulk liquid. Any mass transfer limitations in the bulk liquid will only complicate the already difficult task of determining the kinetics of a heterogeneous reaction.

Suspended biomass in the bulk liquid arises due to either cell growth and replication or to detachment of biofilm material. Planktonic cells leave the liquid phase by either the effluent liquid leaving the reactor volume or by cell deposition onto the reactor surfaces. A single, sterile growth-limiting substrate, *S*, enters the reaction volume, where it is consumed by either the suspended or biofilm-bound cells. Any reasonable measure of both biofilm and suspended cell concentrations is acceptable (cell number, biomass dry weight, biomass organic carbon). However, to complete material balances over the entire system, one most know the stoichiometric relationship between changes in the suspended and attached cell concentrations due to growth and the limiting substrate utilized. Based on these assumptions, material balances for both the suspended cell biomass and single growth-limiting substrate can be written as equations 10 and 11:

Suspended biomass balance:

$$V dX/dt = -FX + (R_g V)_{\text{suspended}} + R_{det} A - R_{dep} A \qquad \ldots(10)$$

Limiting substrate balance:

$$VdS/dt = F(S_{in} - S) - (R_g V/Y)_{suspended} - (R_g A/Y)_{biofilm} \quad \text{...(11)}$$

where X is the suspended cell biomass (ML^{-3}); S is the limiting substrate concentration (ML^{-3}); V is the reaction volume (L^3); $(R_g)_{suspended}$ is the rate of suspended cell growth ($M_{biomass}L^{-3}t^{-1}$); R_{det} is the rate of biofilm detachment ($M_{biomass}L^{-2}t^{-1}$); R_{dep} is the net rate of bacterial cell deposition at a substratum ($M_{biomass}\,L^{-2}t^{-1}$); $(R_g)_{biofilm}$ is the rate of biofilm cell growth ($M_{biomass}L^{-2}t^{-1}$); Y is the yield coefficient ($M_{biomass}/M_{substrate}$); and A is the reaction area (L^2). Rate expressions for the process rates (R_g, R_{det}, R_{dep}) may depend directly on such changing parameters as biofilm amount, limiting substrate, and suspended biomass concentration. Biofilm net accumulation can be described by equation 12:

Biofilm net accumulation:

$$A(dB/dt = R_{dep} + R_{gb} - R_{det}) \quad \text{...(12)}$$

Because rate expressions for both planktonic and biofilm growth are nonlinear saturation kinetic functions of the instantaneous substrate concentration and first-order functions of X and B, respectively, equations 10–12 are coupled, nonlinear ordinary differential equations that require either simultaneous numerical integration (for the transient situation) or simultaneous algebraic solution under steady-state conditions (i.e., dS/dt, dX/dt, and $dB/dt = 0$). Numerous biofilm accumulation studies have been carried out that can be mathematically described by this set of equations. Should the reactor be operated at a hydraulic residence time shorter than the generation time of the cells, then one can assume the term ($R_{gs}V$) is negligible; however, the corresponding term in equation 11, representing suspended cell substrate metabolism ($R_{gs}V/Y_{X/S}$), cannot be disregarded unless in the specific system this term is determined to be small compared to the other terms, or substrate is not supplied.

In a simple bacterial adhesion study, experiments can be carried out without exogenous substrate, thus the R_g terms in equations 10–12 could be ignored. Thus, equation 12 becomes

$$dB/dt = R_{dep} - R_{det} \quad \text{...(13)}$$

and is valid provided the adherent bacterial replication is zero.

PROGRESS IN BIOFILM TECHNOLOGY

This section highlights recent research advances in understanding biofilm phenomena, which include advances in biofilm experimental techniques, new findings regarding mass transport mechanisms in biofilm, and genetic processes within biofilm communities.

Experimental Approaches

All biofilm experimental systems should contain certain critical components: (i) the biofilm study reactor itself (where the surface accumulation of cells and biofilm will be studied) and (ii) the environmental support system that controls the experimental conditions (e.g., temperature, flow velocity, nutrient concentrations, pH) of the bulk fluid. Under quiescent conditions, the rate of transport of dissolved nutrients and particulates (i.e., the bacteria) to the substratum may be the slowest in a series of rate processes. For example, particle transport may limit the rate of cell deposition in quiescent conditions. Thus, it is recommended that only experimental systems that create a gradient-less, well-mixed bulk fluid phase be employed. Otherwise, instead of determining the true kinetics of whatever biofilm process is of interest, the resultant data will erroneously reflect mass transfer limitations.

Detection of biofilms

Detection of biofilms often requires direct sampling and removal of a finite quantity of biofilm from a reactor surface to be used in a number of destructive analytical procedures that may measure (i) overall biofilm amount (cell mass or biofilm thickness), (ii) a component of the biofilm (biofilm carbon, cell number, biofilm nitrogen content), or (iii) a biofilm cell component or cellular activity (ATP, cell protein, dehydrogenase activity, active DNA synthesis, species-specific 16S rRNA). Reactor designs that provide access for destructive sampling involve removable sections of substratum and are quite numerous. The reader is directed to a number of excellent reviews for further details. One drawback to destructive analyses is that the parameter determined represents an average over the entire sample area, which ignores any spatial variations in that parameter that may exist. Almost any parameter determined by a molecular or chemical assay (ATP, DNA, total protein, biofilm polysaccharide content, 16S rRNA, cellular phospholipid signatures, specific enzyme levels, and oligonucleotide probes), cytochemical (acridine orange, DAPI stain, and Hoechst DNA stains) or immunofluorescent staining, or cellular activity (CTC stains, INT stains, and plate counting) techniques that has been applied to planktonic (freely suspended) cells can be used to quantify changes in biofilm on a substratum.

Rodriguez et al. reports on the potential of a fluorescent redox probe, 5-cyano-2,3-ditolyltetrazolium chloride (CTC), for direct observation of actively respiring bacteria. Oxidized CTC is colorless, but upon reduction by electron transport activity, the insoluble CTC-formazan fluoresces at 365 nm upon excitation. Respiring marine bacteria within thin (10–50 μm) biofilms attached to optically nontransparent polysulfone substrata were rapidly and easily enumerated. Respiring cell counts via CTC were also useful in directly assessing the efficacy of various biocidal control agents. The advent of modern molecular biology has provided many useful diagnostic methods with which to analyze biofilm ecology. For example, Amann et al. characterize the population structure of sulfidogenic biofilms, established in anaerobic bioreactors, by selective polymerase chain reaction (PCR) amplification and fluorescent microscopy. 16S rRNA common to the sulfate-reducing bacteria in the biofilm was selectively amplified by PCR and used to design both general and specific fluorescent hybridization probes. Biofilms, (5–10 μm thick) on glass cover slips immersed within laboratory anaerobic biofilm reactors, were fixed in formaldehyde and dehydrated prior to hybridization with the RNA probes. Hybridized biofilm samples were then viewed under epifluorescent microscopy. Rogers and Keevil report on the destructive sampling of a multispecies biofilm, intentionally inoculated with *Legionella pneumophilia*, followed by both immunogold and fluorescein immunolabeling for this species.

Episcopic differential interference contrast microscopy was employed to simultaneously visualize the total biofilm community and the labeled *Legionella* species. This techniques was able to provide observations of variations in the biofilm community from one position to another (parallel to the substratum) but was unable to discern such variations with depth (perpendicular to the substratum). McCarter et al. report a very elegant series of genetic techniques to investigate bacterial responses to various stimuli, one being the adhesion to a surface. Using a transposon (mini-Mu*lux*), the authors were able to incorporate recombinant reporter gene insertions adjacent to target gene promoter sites that encode for light production as a function of target gene expression. Light production is conveniently measured by exposure of X-ray or photographic film, visual examination, chemiluminescence, or photometry.

Such techniques also obviate the need to disrupt biofilms prior to quantification. McCarter et al. employed a *lux* reporter gene adjacent to the chromosomal gene encoding for swarmer cell differentiation that *Vibria parahaemolyticus* experiences upon association with a solid surface. Resultant mutants thus emit light upon growth at a surface but not in suspension. Dagostino et al. also employed transposon mutagenesis to insert into an appropriate recipient bacteria a marker gene that lacks its own promoter. The premise is that if a suitable target gene is "on" at the surface, then expression of the marker will be observed only in the presence of the surface. Dagostino et al. employed *Escherichia coli* C600 (pRK2013: mini-MuTetr *lacZ*) as donor and two marine bacteria as recipients. The authors were able to successfully isolate a transposon-generated mutant in which the *lacZ* gene was not expressed in either liquid or agar, but was expressed when grown on a polystyrene substratum.

Diagnostic methods

The "holy grail" of biofilm research is a diagnostic method that creates as little disturbance to the biofilm and the reactor operation as possible. Regrettably, most parameters of interest in cell adhesion and biofilm accumulation require destructive sampling and analyses. True non-invasive diagnostics of biofilm accumulation, amount, and biofilm-cell reactivity are limited to relatively few sophisticated techniques. The simplest of these techniques is on-line microscopy, which places the biofilm surface within the viewing field of a microscope. This technique requires a reactor dimension compatible with the microscope, and a flat reactor surface on which to focus. Microscopic observation of adherent cells can be quantified and preserved by use of video recorders or image analyzer systems that digitize observed images and save those images on a computer for future numerical interpretations. For about 14 years, confocal scanning laser microscopy (CSLM) has been a tool in the biological sciences. CSLM uses confocal apertures, or pinholes, to create a thin (0.4 μm) depth of field, which eliminates out-of-focus light emitted from the laser-excited fluorophores. Laser light sources provide the intense, highly coherent, collimated light necessary to penetrate thick specimens.

The laser light is used to excite fluorophores—either those intrinsic (e.g., chlorophyll) to the sample, or selected chemical or immunological stains intentionally applied to the sample. Resultant fluorescence that passes through the aperture is detected by a photomultiplier, and a digital image is collected. Using a computer-manipulated stage, optical sections can be collected from a specimen in three dimensions. This allows, for the first time, the ability to digitally locate cells on a substratum or within a biofilm, non-invasively. Numerous fluorochromes are available that can be used to localize and measure intracellular and extracellular conditions in three dimensions within living biofilms. Camper et al. detail image analysis software that can be used to collect images of bacterial adhesion patterns from the CSLM, digitize those images, then superimpose those images onto digitized images of the substratum itself, collected from any one of a number of other analytical instruments (e.g., X-ray photoelectron spectroscopy, atomic force microscopy, time-of-flight secondary ion mass spectroscopy); all images are collected at the same location on the substratum. Thus, one can now correlate the adhesion of a single cell to the local topographical or chemical properties of the substratum directly underneath the cell.

Lawrence et al. illustrates the ability to optically section fully hydrated biofilms, both horizontally and sagitally, producing optical sections of undisturbed biofilms with a spatial thickness of 2 μm. Any fluorescent probe (cellular DNA stain; immunofluorescent stain; pH-, ion-, and redox-sensitive stains; viability redox stains) or any combination of probes can be simultaneously recorded by the CLSM with great clarity and little background interference, due to the confocal exclusion of any fluorescence originating from excited fluorochrome above and below the focal point. CLSM is a critical tool in population analysis of biofilms consisting of mixed strains, mixed species, and mixed cell lines and will allow detailed examination of the relationships between biofilm structure, adaptation, reactivity, and response to external stress. Another noninvasive technique that can be used to collect information on the molecular chemistry of cell adhesion at the substratum–fluid interface is Fourier transform infrared spectroscopy (FT-IR) coupled with attenuated total reflectance (ATR) waveguides integrated within flow cells. Atoms and groups of atoms within a molecule vibrate with a characteristic frequency and will adsorb light at those frequencies. Light that contains infrared frequencies can be focused on a molecule, and the amount of light adsorbed is then measured as a function of frequency. Specific IR adsorptions can thus be assigned to particular bonds, and alterations in these bonds due to changes in local environment can be assessed from the resultant spectral details.

Recent increases in the capability of FT-IRs and the focusing of the IR wave within specific crystal wave guides (germanium or zinc selenide crystals) allows one to establish a standing IR evanescent wave at the surface of the crystal. Fourier transform signal processing and multiple scanning allow aqueous samples to be processed where the IR spectra of molecules directly adjacent to the waveguide can be collected. In biological systems, the IR absorbance of water must be subtracted via computer manipulations to provide the IR spectra of molecules accumulating at the crystal surface. The effective depth of penetration for the evanescent IR wave is a function of crystal material and the wave number but ranges around 0.3–0.7 μm, thus providing spectra that reflect those chemical species directly adjacent to the surface. ATR/FT-IR has been used to study the effects of protein-conditioning films on alginate adsorption to the germanium surface. Alginate adsorbed in greater quantities to protein-coated ATR crystals than to uncoated ones. For the development of biofilms on ATR crystals, the observed spectrum is the integral of all biomolecules on the surface and over the entire surface.

Bremer and Geesey report the chemical changes as detected by FT-IR that occur upon inoculating with a microbial culture, at a germanium ATR crystal situated within a flow cell. Spectral intensities representing various bonds within proteins (amide I at 1,645/cm; amide II at 1,550/cm) and polysaccharide (C-O stretches at 1,058/cm) are seen to accumulate with time, indicating the feasibility of the ATR/FT-IR system to detect biofilm formation. Because the IR wave can only penetrate less than 1 um into the biofilm, reported increases in absorbance intensities most likely represent increases in cell surface coverage of the crystal and not increases in biofilm thickness. A plot of attached cells per area versus the increases in area under the amide I peak was generated to estimate a lower limit of detection by FT-IR of -5×10^5 cells/cm^2. In a medical application, ATR/FT-IR has been used to follow the penetration of an antibiotic (100 μg/mL ciprofloxacin) into the IR wave region underneath a *P. aeruginosa* biofilm. Several reports exist of the use of invasive but nondestructive analysis of solute concentration profiles, by microsensor chemical probes, as a function of spatial dimension within the developing biofilm.

Microsensors (tip size $\leq$ 15 μm) exist that can detect various dissolved solutes including glucose, oxygen, pH, sulfide, and ammonia. DeBeer et al. report using microoxygen sensors (tip size = 15 μm) to estimate oxygen profiles, local oxygen uptake rates, and oxygen diffusion coefficients in a mixed-culture biofilm community cultivated within a laboratory biofilm reactor. Schramm et al. determined microprofiles of oxygen and nitrate in nitrifying biofilms from a trickling filter and correlated these solute profiles with bacterial population profiles for *Nitrosomonas* and *Nitrobacter* spp., as determined using fluorescence in situ hybridization of cells fixed with 16S rRNA oligonucleotide probes.

Challenges and Limitations

Bacterial biofilms are heterogeneous reaction systems wherein limiting substrates, essential nutrients, cellular antagonists, and cellular by-products are exchanged between the surrounding fluid and the site of the reaction (i.e., the cell). In numerous circumstances, the performance of biofilm processes can be limited by both internal and external mass transfer rates (e.g., antibiotic treatment of a biofilm infection, biocide challenge to eliminate biofilm formation in heat exchange systems, biofilm reactor treatment of wastewaters or waste gases, and in situ hazardous waste decontamination). Recent advances in characterizing biofilm architecture have led to the need to reformulate concepts of the mass transport mechanism in biofilms. For about 25 years, estimates of substrate consumption for biofilms of known thickness have been based on the steady-state solution of equation 14,

$$D_{eff}\, d^2C/dx^2 = vR_c \qquad \ldots(14)$$

where C is the solute concentration, R_c is all reactions transforming solute C, v is the stoichiometric coefficient for the transformation, x is the coordinate of molecular transport, and D_{eff} is the effective diffusivity of solute in the biofilm. These various models are based on the assumption that transport of reactants (i.e., substrates) into the biofilm and export of products is by molecular diffusion alone and that bulk transport (convection) does not occur within the biofilm. These models also do not account for (i) any heterogeneities in the biofilm gel structure, (ii) any locally dependent bacterial concentrations, or (iii) any nonlinearity in diffusion coefficients.

Traditionally, either D_{eff} is estimated as the solute diffusivity in pure water, reduced by an arbitrary fraction, or D_{eff} is determined for the particular biofilm system in question from diffusion experiments. For small ions and uncharged small organic compounds, both Libicki et al. and Siegrist and Gujer indicate that transport in thin microbial layers is indeed by molecular diffusion. Both reports indicate that the decrease in observed diffusivity values relative to those in pure water is a function of the microbial aggregate cell-to-polymer composition. Siegrist and Gujer provided the first report to insinuate that molecular diffusion may not be the only mass transport process acting within a biofilm and that convective mass transport may predominate in the upper layers of thick, filamentous biofilms. Results of Drummond and Bryers and Drummond suggest these dense alginate clusters are themselves porous, with a wide distribution of pore sizes ranging from angstroms to tenths of micrometers. Their results have shown the transport of larger solutes (i.e., macromolecules) through the hydrated biofilm polysaccharide gel matrix is controlled by transport mechanisms far more complex than molecular diffusion.

Overall effective diffusion coefficients for various molecules, increasing in molecular weight and differing in molecular size and stereochemistry, were determined using half-cell diffusion chambers. Transient diffusion studies were carried out at constant temperature and pH using bacterial biofilms of increasing thickness. Half-cell diffusion results were considered erroneous due to the "perforated" nature of the biofilms. The existence of water channels in these biofilms was documented and found to significantly affected mass transfer coefficient estimates obtained from half-cell studies. Consequently, a local microtransport technique was developed based on the fluorescence recovery after photobleaching (FRAP) technique of Poitevin and Wahl.

FRAP is a local, instantaneous, and noninvasive method that can be repeatedly carried out over a wide expanse of biofilm. Corrections of the FRAP "tracer" diffusion coefficients must be made in order to estimate "Fickian" diffusivities. Diffusion coefficients determined for large macromolecule solutes (dextrans, proteins, DNA fragments) moving in biofilms did not correlate simply with increasing molecular weight except within the homologous chemical groups. Transport of macromolecules through the biofilm was affected not only by molecular weight but by the size or length of the molecule in solution. Using a computer-actuated laser and motorized microscope stage, FRAP analysis across a large section of biofilm illustrates that local mass transfer coefficients can vary significantly in a biofilm over very short distances. Thus, research reported by Drummond and Bryers and Drummond and independently corroborated by Stoodley et al. has shown biofilms are not architecturally spatially continuous. Rather, biofilms are structurally nonuniform, consisting of dense microbial clusters surrounded by tortuous water channels that perforate the biofilm matrix. Thus, theory and experiments addressing overall mass transport within a real biofilm system must account for both convective transport and molecular diffusion.

Biotechnological Exposure

Any exposure of plasmid-recombinant microorganisms to an open system environment, either inadvertently or intentionally, mandates research into those fundamental organism–plasmid processes that influence plasmid retention, transfer, and expression. Use of plasmid-recombinant bacteria to impart a desired phenotypic ability to an existing ecosystem has been desired for the bioremediation of hazardous wastes in both controlled reactor systems and in waste-contaminated open environments. In open environmental systems, a majority of the microbial activity occurs associated with an interface, within biofilms. In wastewater or waste gas treatment, efficient reactor design often dictates the use of immobilized, biofilm-bound bacterial communities. Consequently, the fate and ability to manifest a plasmid-borne phenotype within a biofilm community requires basic information on the biofilm's genetic processes. A plasmid is deemed unstable if it either undergoes molecular rearrangement (structural instability) or is not absolutely inherited by progeny (segregational instability). Such phenomena can have significant effects on the outcome of cell cultivation processes. It has been established, both experimentally and theoretically, that plasmid maintenance and cloned gene expression can reduce the overall growth rate of the plasmid-bearing cell relative to the plasmid-free cell.

Reduction of copy number and loss of plasmids from populations under continuous suspended culture have been reported in many cases, even in the presence of selective pressure. That immobilization might stabilize a plasmid-bearing population can be shown mathematically. Plasmid persistence in suspended cultures has been observed in cases where the plasmid-bearing cell was at a growth rate disadvantage; this observation was directly attributed to biofilm formation and cell detachment from the biofilm. Improved plasmid stability upon artificial cell immobilization in κ-carrageenan gel beads has been reported by de Taxis du Poet et al., Nasri et al., and Sayadi et al. Huang et al. cultivated biofilm cultures under controlled hydrodynamic conditions in a parallel-plate flow cell reactor using *E. coli* DH5α containing a recombinant plasmid either with (pTKW106) or without (pMJR1750) a segregational stability factor, *hok*/*sok*.

Both plasmid constructs provided for the inducible expression of β-galactosidase. Results in these studies suggest, for these two plasmids, that the expression of the chromosomal gene for synthesis of the polysaccharide matrix of the biofilm commands a higher priority on cellular intermediates than the foreign genes. Consequently, a series of experiments were carried out to indirectly manipulate bacterial polysaccharide production in the biofilm by varying the nutrient carbon-to-nitrogen (C:N) ratio. Plasmid loss probability, total polysaccharide production, and ratios of total polysaccharide to total protein in biofilm cultures of *E. coli* DH5α (pMJR1750) all increased with increasing C:N ratios. Synthesis rates of total RNA, rRNA, and β-galactosidase-specific mRNA in *E. coli* DH5α (pTKW106) all increased after induction by IPTG, but levels of each parameter seen in biofilm cultures were 10–20 times lower than the same values observed in suspended cell cultures.

Intercellular communication

Cell–cell communication involves a chemically unique molecule, produced biologically by one cell, that reacts with a set of specific receptors located at another cell's surface to subsequently alter cellular behavior. Two examples of cell-cell communication in higher life forms would be (i) the thousands of lectins that control embryonic development in plants and animals, and (ii) signaling molecules, such as the pheromones, that control the behavior of adult insects. Cell-cell communication in bacterial systems, both sessile and planktonic, has been recognized for years, but our understanding of the control mechanisms has entered a new era with the advent of various molecular diagnostic tools. Recent research has identified the specific chemical structures of various signaling molecules that manifest certain cellular responses.

The majority of cell–cell communication studies have dealt with suspended bacteria (e.g., quorum sensing), but recent advances suggest that certain structural architecture in a biofilm arises due to molecular signaling. Bacterial swarming, floc formation, and biofilm formation or aggregation are multicellular, cell-concentration-dependent behavior that can be induced in response to the recognition of surfaces of certain viscosity. Cells differentiate into multinucleated, elongated, and hyperflagellated form and orient themselves lengthwise across the surface. For a number of species, the ability to differentiate into swarmer cells is critical to virulence, surface adhesion, and colonization.

Acylated homoserine lactones serve as one class of signal molecule in bacterial communication. These molecules and their derivatives have been linked to control of bioluminescence, Ti plasmid transfer, production of virulence factors, antibiotic resistance, and swarming motility. Higher life forms have evolved mechanisms to interfere with such signaling processes. For example, certain sea-weeds and sea grasses produce halogenated furanones, structurally similar to acylated homoserine lactones, that interfere with the swarming process and have been shown effective as antifoulants and antimicrobials. Givskov et al. illustrated that two different furanones derived from the seaweed *Delisea pulchra* could progressively inhibit and eliminate the swarming behavior of *Serratia liquefaciens* as concentrations were increased from 0 μg/mL to 100 μg/mL (concentrations far too low to affect the growth of the bacteria). Davies et al. (unpublished results) employed homoserine lactone mutants of the classic PAO-1 strain of*P. aeruginosa* to determine if cell–cell communication plays any role in bacterial biofilm polysaccharide matrix formation.

Wild strains and mutants were separately used to inoculate flow cells to observe cell adhesion to surfaces and subsequent formation of biofilms. Cells of the wild-type strain (PAO-1), retaining the ability to secrete homoserine lactone, adhered to surfaces avidly and formed complex biofilms comprising discrete m icrocolonies and well-developed water channels. Cells of the double homoserine lactone mutant (JP2), which lacked the ability to synthesize either the *Pseudomomas* autoinducer, PAI, or the factor U homoserine lac- tone, did adhere to surfaces avidly but failed to form the complexly structured biofilms. These mutant cells did not form detectable amounts of exopolysaccharide matrix material, did not aggregate into discrete microcolonies, and did not develop the water channels characteristic of the wild-type biofilms. Cells of the double homoserine lactone mutant simply formed unstructured masses of cells that were tightly packed together. These unstructured adherent masses of cells could be easily removed from the colonized surfaces by simple washing with surfactants. Chemical analyses proved that these biofilm contained no detectable biofilm exopolysaccharide matrix material. When a cell-free supernatant from a wild-type biofilm (PAO-1), which contained both types of homoserine lactone molecule, was added to the bulk fluid delivered to the flow cell, adherent JP2 mutant cells gradually began to produce matrix material and form complex biofilms; 12 h after the addition of the homoserine lactone molecules from the wild-type supernatant, the JP2 mutant strain biofilm was observed to be identical to the wild-type biofilm strain. Such research could potentially lead to nontoxic control strategies for disrupting detrimental biofilms, which would be economically and environmentally significant.

Traditional Strategies

One traditional approach to eliminate biofilm formation within an engineered system is to prevent cell adhesion. One such anti-cell adhesion approach used in the biomedical materials sector focuses on modification of the surface chemical properties of the substratum by various methods including photochemical coupling of benzephenone derivatives of polyethylene glycol, polyacrylamide, and poly vinyl pyrrolidone; passive adsorption of pluronic surfactants (copolymers of polyethylene oxide and polypropylene oxide) to polystyrene; incorporation of polyethylene oxide into the upper layers of polyethylene terephthalate by solvent swelling; and a series of neutral, anionic, and cationic surfactants adsorbed onto stainless steel or glass. Rather than prevent adhesion, one alternative control approach for biomedical materials is to retard surface microbial activity by incorporating within the substratum a slowly released antibiotic agent. Remedial approaches to eliminate or eradicate fouling biofilms existing in nonmedical engineered systems consist of mechanical cleaning, materials or unit replacement, or chemical biocide challenges. One novel alternative to chemical antagonism of a biofilm has been to eliminate an nutrient essential for microbial growth.

In the petrochemical, refining, and power industries, either oxidants (chlorine, ozone, bromides) or biocides are employed regularly to suppress biofouling of water-cooled heat exchangers; biocides prove mostly ineffective in that they inactivate the bacterial cells but, unlike oxidants, are incapable of eliminating the source of the system inefficiencies, the biofilm matrix. One of the most innovative technologies to control detrimental biofilm formation was reported by Wood et al. Substrata containing layers of copper and cobalt phthalocyanine catalysts were constructed for bacterial adhesion trials..These immobilized catalysts serve to promote the formation of active oxygen species from various oxidizing agents such as peroxide or persulfates. Biofilms on the treated catalytic surfaces were exposed to various concentrations of hydrogen peroxide. Generation of antibacterial oxidant concentrations right at the interface of the substrata aided in the almost complete removal of biofilm.

11

MICROBIAL PRODUCTION OF POLYESTERS

Microbially produced biodegradable plastics can be divided into four classes: polysaccharides, polymers of amino acids, naturally occurring polyesters, and semisynthetic polyesters. All possess linkages that are susceptible to nonspecific, extracellular, microbial enzymes that exist in nature and thus are inherently environmentally degradable. Their structures can be used as the basis for the design of synthetic biodegradable polymers. Unlike many common synthetic plastics, those microbially produced are often at least partially water soluble or tend to be easily hydrolyzed, especially at extremes of temperature and pH. These characteristics aid in their biodegradability but greatly limit their use as plastics. Although their properties as plastics may be greatly improved through chemical modification, care must be taken to maintain environmental degradability. The ideal plastic is durable, biodegradable, recyclable, and produced from renewable resources such as carbohydrates. Poly-3-hydroxyalkanoates (PHAs) are a family of intracellular polyesters produced by a wide variety of microorganisms. They are water insoluble, relatively resistant to aqueous hydrolysis, and readily environmentally degradable. They can be produced from renewable resources, and, although temperature sensitive, they are potentially recyclable.

PHAs have mechanical properties similar to those of some common synthetic plastics. Like conventional thermoplastics, PHAs can be recycled with some loss of molecular weight and mechanical properties. Unlike many conventional polymers, PHAs are completely environmentally degradable and will decompose rapidly to CO_2 and H_2O during composting. Degradation has also been demonstrated in soil burial studies and in the marine environment. PHAs whose polymeric units are predominantly five carbons or fewer are classified as short-chain-length (SCL) PHAs, and all others are referred to as medium-chain-length (MCL) PHAs. Homopolymeric poly-3-hydroxybutyrate (PHB) has properties that are frequently compared to those of polypropylene, except that it is more brittle and its melting temperature is high enough that there may be significant thermal degradation on melt extrusion. Lower-melting, more-flexible PHB can be obtained by annealing or addition of plasticizers. However, copolymers of 3-hydroxybutyrate and 3-hydroxyvalerate (P[HB-co-HV]) and of 3-hydroxybutyrate and 4-hydroxybutyrate (P[3HB-co-4HB]) are more flexible and have lower melting temperatures than PHB, making them more suitable for the manufacture of extruded products.

MCL PHAs have even lower melting points and glass transition temperatures; they are tough, rubberlike thermoelastomers, especially after cross- links between the polymer chains have been formed by heat or chemical induction. PHAs would seem to be ideal biodegradable plastics, but their production is much different from that of conventional plastics. The following sections describe how PHAs may be produced by widely varying methodologies, with details of diverse approaches to PHA fermentation processes and the separation of the polymer from the biomass. Existing commercial processes and future possibilities for PHA production with an emphasis on cost reduction are also discussed.

SYNTHESIS OF POLYHYDROXYALKANOATES

Biochemical Production

PHB can be produced synthetically from racemic β-butyrolactone, and P(HB-co-HV) is made from a similar mixture of β-butyrolactone and β-valerolactone with triethylaluminum and water as catalyst. Although the stereospecificity of the resulting polymer can be controlled to a certain extent by adjusting the ratio of triethylaluminum and water, synthetic PHAs are usually optically inactive and hence only partially biodegradable.

Genetic Synthesis

SCL PHA

PHB can be synthesized in plants in which the bacterial reductase and polymerase genes have been introduced. Although virtually all organisms possess the remaining gene in the SCL PHA pathway, coding for 3-ketothiolase, its activity and specificity vary and it may not be present in all eukaryotic organelles. Production is usually increased by cloning the 3-ketothiolase gene with the other two. Although transgenic plants are easily created, productivity remains problematic. The SCL PHA pathway was first cloned into *Arabidopsis thaliana*, the *Escherichia coli* of plant biotechnology. The yield was low, 0.1% dry weight, and the plants were stunted. However, granules similar to those found in bacteria were produced. The initial cloning strategy targeted the cytoplasm, and granules were even found in the cell nucleus. Subsequent work has targeted the plastid, which is also the site of starch and oil synthesis in plants, resulting in increased yields of up to 14% dry weight in the leaves of *A. thaliana*. These plants grew well but were chlorotic, possibly because of large amounts of PHB in the chloroplasts. Present research centers on oil- and starch-producing plants because these are already adapted to have large amounts of storage material in the plastids, but other strategies are also being assessed.

MCL PHA

Because of its high melting point and brittle nature, the PHB homopolymer might have limited application even if its production cost in transgenic plants was low. Although SCL PHA copolymers such as P(HV-co-HB) have been synthesized in plants, it is not obvious how the metabolic pathways of plants can be manipulated to maximize their production. It may be easier to produce MCL PHA in plants. Unfortunately progress is limited by a lack of understanding of the link between fatty acid synthesis and MCL PHA synthesis. Nevertheless, small quantities of MCL PHA have recently been produced in recombinant *A. thaliana*.

Microbial Synthesis

SCL PHA

Many bacterial species accumulate PHB, but PHA copolymers are more commercially valuable. Most PHB-accumulating bacteria will produce P(HB-co-HV) under the appropriate conditions. When propionic acid is supplied during a fermentation process, some of the acid is converted to propionyl-CoA. A portion of this propionyl-CoA is condensed with acetyl-CoA via β-ketothiolase to produce five-carbon monomers; this is the most common pathway of P(HB-co-HV) production. Because acetate is needed for this synthetic route and some propionic acid is decarboxylated before monomer synthesis, usually less than 50% of the monomers will be HV even if a feed of pure propionic is added (i.e., no glucose or acetate). This yield can be increased in *Ralstonia eutropha* (previously *Alcaligenes eutrophus*) by decreasing the dissolved oxygen concentration in the bioreactor during the accumulation phase. High yields of HV are obtained when pentanoic acid is used because pentanoic acid is metabolized directly to L-(3)-hydroxyvaleryl-CoA, which is subsequently racemized (at least in *R. eutropha*) to D-(3)-hydroxyvaleryl-CoA. Although some acetate is produced leading to HB monomer formation, PHAs with greater than 90% HV content can be obtained.

SCL PHA producers have not been reported to accumulate monomers with side chains longer than C_2 or C_3 but can produce P(3HB-co-4HB) when fed 4-hydroxybutyric acid or 1,4-butanediol. *R. eutrophus* accumulates terpolyesters of 3HB-, 3HV-, and 5HV-monomers when fed a mixture of pentanoic and 5-chloropentanoic acids. Even among SCL-accumulating microorganisms, there may be differences in the PHA synthetic pathway. For example, in *Rhodospirillum rubrum*, L-(+)-3-hydroxybutyrl-CoA is racemized to D-(–)-3-hydroxybutyrl-CoA using enoyl-CoA hydrases that have not been found in *Azotobacter beijerinckii*. In most SCL PHA-accumulating bacteria, high intracellular levels of reduced nicotinamide adenine dinucleotide (NADH) and acetyl-CoA greatly stimulate the rate of PHA synthesis. Thus, if the growth rate is limited by a nutrient other than the source of carbon and energy, PHA is accumulated in the cell. For this reason, PHA production is usually divided into two phases. In the first phase, balanced growth proceeds with relatively little PHA accumulation.

When the concentration of the limiting nutrient in batch culture falls below a critical value, the rate of PHA synthesis increases while synthesis of other cellular components decreases during the second phase. Because co-substrates such as propionic acid are expensive and can be used for the synthesis of biomass other than PHA, these chemicals are only added during the accumulation phase. This limitation probably results in nonhomogeneity of polymer composition in different PHA granules with some (produced during the growth phase) containing almost exclusively PHB. There is still debate about the existence and importance of monomer turnover in PHA granules.

In contrast, carbon and energy limitations increase depolymerase activity, leading to a decrease in cellular PHA content. The control of these biochemical pathways varies greatly among different bacterial species. For example, *Alcaligenes latus* does not require nitrogen or phosphorous limitation; it can accumulate large amounts of PHB during balanced growth. The most important economic considerations are substrate costs, reactor productivity, and separation costs.

SCL PHA can be produced from hydrolyzed starch (glucose), lactose (using recombinant *E. coli* on cheese whey), cane or beet molasses (using *Azotobacter vinelandii* or *Alcaligenes latus* on sucrose), waste wood (fermented, hydrolyzed hemicellulose), methanol, and even mixtures of hydrogen and carbon dioxide (using *R. eutropha*). Addition of the appropriate cosubstrate such as propionic acid can result in P(HB-co-HV) synthesis in any of these processes although the %HV can vary considerably depending on the bacterial strain. High productivity have been reported for several of these substrates. Using *R. eutropha* growing on glucose, 164 g L^{-1} of biomass containing 76% PHB has been produced in a fed-batch process in only 50 h. Densities well in excess of 100 g L^{-1} have been reported for fed-batch processes using *E. coli* strains with multicopy plasmids containing the key SCL PHA synthetic genes as well as for methylotrophs growing on methanol. In recombinant *E. coli* strains, PHB content is frequently reported as being in excess of 90% of the biomass dry weight, making the separation process much simpler.

MCL PHA

The highest productivity results from using octanoic or nonanoic acids, but these are toxic at concentrations of several grams per liter and their low solubility limits the growth rate as the result of poor mass transfer. The salts of these acids are much more soluble, but the final product concentration in fed-batch culture is limited by accumulation of toxic concentrations of sodium ions in the medium. In any event, the present cost of both the salts and the acids prohibits their use in commercial production. Most microorganisms that make MCL PHA produce lipases and grow well on vegetable oils. If the lipase concentration limits the production rate, the oils can be chemically hydrolyzed to liberate their fatty acids. A *Pseudomonas putida* strain has been grown in fed-batch culture from oleic acid (a fatty acid commonly found in plant oils) to a biomass concentration of 92 g L^{-1} containing 45% PHA in 26 h. If of sufficient low density, these PHAs may float to the surface after breakage of the bacterial cell and can thus be easily and cheaply separated from the rest of the biomass.

ISOLATION AND PURIFICATION

All separation processes to date have been developed for microbial rather than plant biomass, but many of these approaches could also be applied to recovery of PHA from plants. As a first step in the recovery from microbial biomass, the cells are separated from the fermentation broth, usually by centrifugation or flocculation. In the latter case, the pH of the fermentation broth is adjusted to 9; it is heated, then the pH is readjusted to 5 and the biomass recovered by centrifugation or filtration. The wet biomass may be freeze- or spray-dried to remove water. Processes to separate the intracellular PHA from the rest of the microbial biomass may be divided into two categories: (i) selective solubilization of PHA in an organic solvent, leaving the majority of the other biomass components in suspension, and (ii) destruction (i.e., enzymatic solubilization) of cellular components other than PHA, leaving the PHA granules in suspension. Physical treatments such as drying, heating, freezing and thawing, repeated centrifugation, and exposure to acids and organic solvents change the amorphous nature of native PHA to a semicrystalline state.

Biochemical Extraction of PHB

The majority of the patented separation processes describe the extraction of PHB from microbial biomass using organic solvents such as chlorinated hydrocarbons (e.g., chloroform or 1,2-dichloroethane), azeotropic mixtures (e.g., 1,1,2-trichloroethane with water), chloroform with either methanol, ethanol, acetone, or hexane, and cyclic carbonates (e.g., hot [120-150°C] ethylene carbonate or 1,2-propylene carbonate) in which the polymer is soluble. With the wider variety of PHAs that can be produced today, the choice of solvents should be carefully considered. In general, solvents that are suitable for PHB should be equally good for any SCL PHA and MCL PHA. The reverse may not be true. For example, while semi-crystalline PHB is insoluble in acetone, MCL PHA will dissolve in it. Cells may be pretreated to make the polymer more accessible to the extraction solvent. This may be done by washing in a hydrophilic solvent in which the polymer is insoluble such as methanol or acetone for SCL PHA and methanol for MCL PHA or by means of physical disruption such as grinding, wet-milling, French press, or freezing and thawing cycles. Pretreatment with acetone has been shown to have the added advantage of removing certain polar lipid impurities and increasing the amount of PHB recovered.

After solvent extraction, the polymer solution is separated from the cellular residue by filtration or centrifugation. PHB is then precipitated from the chlorinated hydrocarbons or azeotropic mixtures with a cold nonsolvent such as methanol. In the case of ethylene carbonate and 1,2-propylene carbonate, most of the PHB precipitates on cooling. Difficulties are encountered at polymer solution concentrations greater than 5% by weight because these concentrations tend to be very viscous. As a result, large amounts of solvent are required to obtain sufficiently dilute solutions to enable easy separation. Even with efficient solvent recovery procedures, some solvent loss is expected. This method is not considered economical as solvent recovery and raw material costs are high. However, solvent extraction is useful when the PHA content of the biomass is low or when high polymer purity is required. A significant portion of the PHB may remain in the biomass after solvent extraction. Although thermal degradation of PHB in air or a nitrogen atmosphere has been well studied, few examples show that PHB degrades in organic solvents.

The molecular weight of PHB decreased in propylene carbonate and ethylene carbonate at 110-140°C, in a chloroform–methanol mixture at room temperature, and in methylene chloride, chloroform, and 1,2-dichloroethane at their boiling points. Although solvent extraction is commercially uneconomical, it is a commonly used laboratory technique for PHA recovery, especially for molecular weight determination. Usually, methanol-pretreated (or ethanol-pretreated), lyophilized biomass is heated in chloroform under reflux conditions for about 1 h, the polymer solution is recovered by filtration or centrifugation, and the polymer is precipitated by cold ethanol or methanol. The polymer may be further purified by redissolving in chloroform and reprecipitating with cold ethanol. It should be noted that about 25% of the original molecular weight (determined by chloroform extraction under ambient conditions) was lost when refluxed in chloroform for 1 h.

Enzymatic methods

Classical enzymatic treatment (involving lysozyme, proteinases, DNAses, etc.) with or without a surfactant step can be used to solubilize the cellular biomass. Biologists have used these methods to prepare native PHB granules to study intracellular enzymatic degradation. On a commercial level, Holmes and Lim described a process in which the fermentation broth may be first heated to 80°C to denature DNA and some proteins. A series of enzymes are then used to obtain a product that is 90-95% pure with 6-7% proteins and some peptidoglycan as major impurities. There may be a final wash with hydrogen peroxide to make a whiter product. This is the reported method of choice for the current production of P(HB-co-HV), which is sold under the trade name Biopol by Monsanto plc. When PHA granules recovered by enzymatic treatment are freeze-dried, they have a hard, crystalline shell and a noncrystalline core.

Chemical methods

Strong oxidizing agents such as sodium hypochlorite have been used to dissolve the cellular biomass under conditions that result in significant loss of PHA molecular weight. However, by optimizing the operating conditions, PHB with an average molecular weight of 600,000 g/mol and 95% purity has been obtained from *R. eutropha* biomass containing 50% PHB with an initial molecular weight of 1,200,000 g/mol. When the biomass is pretreated with a surfactant, a reduced contact time with hypochlorite is needed to obtain PHB with the same purity. This method allows an even higher molecular weight (800,000 g/mol) to be obtained. Never-dried granules obtained by surfactant–hypochlorite treatment also have an external crystalline shell and an amorphous core.

Physical methods

Although there are many methods (e.g., reciprocating pump, ball mill) to physically disrupt microbial cells, few processes that are based on these principles have been proposed for the large-scale recovery of PHAs. Holmes and Jones patented a process in which a biomass suspension was heated to 220°C under nitrogen in an autoclave, then forced through a fire-jet into a tank of cold water. Cell breakage occurred, releasing PHB granules, which were suspended in 50% aqueous acetone to remove lipids and to cause flocculation.

LARGE SCALE PRODUCTION

PHB remained an academic curiosity until the 1950s and 1960s when W.R. Grace and Co. in the United States produced small quantities for commercial evaluation. They obtained patents for production processes and manufactured articles such as sutures and prosthetic devices but eventually abandoned the project because of low PHB yields and difficulties in the separation process. About 10 years later, Imperial Chemical Industries (ICI) began a program in PHA production and product development. They patented the first commercial process to produce P(HB-co-HV).

Latest Methodology

Tredegar Industries has used recombinant *E. coli* to produce more than 100 g L^{-1} of biomass containing at least 80% PHB by dry weight at the pilot-scale level. Mild detergent treatment and washing steps followed by spray-drying result in a white product that is 98% pure. Fermentation and separation processes were developed for the pilot-scale production of PHB using an *Alcaligenes latus* strain. As much as 1 ton per week of PHB has been produced in Austria in a 15,000-L bioreactor. This technology is currently owned by the German company, Biomer.

The Biomer fermentation process is relatively simple with multistaging from the petri dish to a shaker flask to a small fermenter, which is then used to inoculate the production reactor. If sufficient carbon source (molasses) is added during the fed-batch process, as much as 60 g L^{-1} of PHB maybe obtained. The resulting biomass is washed in water. The concentrated cell suspension is then subjected to methylene chloride extraction. The resulting slurry is separated in a decanter. The solvent phase is then injected into hot water in a precipitation vessel where the methylene chloride evaporates and the PHB precipitates. After water removal (drying), the PHB can then be processed into a marketable product. Biopol has been produced by Zeneca from glucose and propionic acid in 35,000-L airlift reactors and in stirred tank reactors up to 200,000 L. Monsanto uses conventional stirred-tank fermenters to produce 1,000 tons of Biopol per year.

Phosphate limitation (after about 60 h of growth) is used to trigger the accumulation phase in a strain of *R. eutropha*. Phosphate limitation was found to give better results than nitrogen limitation. Although solvents (chloroform or methylene chloride extraction after a hot methanol wash) may be used when high purity is required, the P(HB-co-HV) is usually obtained by treatment of the biomass with a series of surfactants, enzymes, and possibly oxidizing agents to dissolve the non-PHA biomass. The product is essentially a latex that may then be washed, spray-dried, compounded with nucleating agents and plasticizers, and extruded in the form of pellets. Several grades of Biopol are available at a cost of about $8 per pound. Based on process development at IPT, Copersucar, the Brazilian ethanol cooperative is now using hydrolyzed sucrose and NH_4^+ limitation to produce SCL PHA. these are purified by solvent extraction. Pilot-scale production capacity is currently 30 to 60 tons per annum. Cost savings from the burning of bagasse to supply energy and from cheap cane sugar allow Copersucar to sell PHA for much less than $8 per pound.

LAST STEPS OF PURIFICATION

Conventional equipment used to process polyethylene can be used for Biopol to form blow-molded, injection-molded, and film products. A key requirement is temperature control. It is recommended to operate at the lowest possible temperature and to keep the residence time at a minimum because the melting point of PHB is about 178°C and its degradation temperature begins at 180°C. The melting point of the copolymers decreases as the HV content increases, which allows HV copolymers to be processed at lower temperatures with less risk of thermal degradation. Plasticized high-HV Biopol can have a melting point as low as 136°C. For most applications, Monsanto suggests processing at a maximum temperature of 170°C and a maximum residence time of 3 min, and that molds and blow- pin temperatures be kept at 60 ± 5°C. Mechanical properties deteriorate rapidly if the average molecular weight falls below 400,000 g mol^{-1}. PHAs are moisture resistant and have gas barrier properties similar to the best coated films, so that they have potential use in products which require a plastic moisture barrier. Possible markets include mulch films, food and agricultural packaging, drink cartons, feminine hygiene products, disposable diapers, and disposable kitchen items such as plastic films that are difficult to separate for recycling. Coatings may be applied with PHA as an amorphous latex, which can be reconstituted from the crystalline form of the polymer or obtained directly from a bacterial suspension by dissolving the other cellular macromolecules (cell wall components, DNA, etc.).

In the latter case, it is important that the granules are not dried before use as they would become crystalline. The latex may be applied as a spray to a material such as paper, dried, and cross-linked using heat or ultraviolet light. The MCL PHAs are promising as coatings because they are less crystalline than the SCL PHAs (in their crystalline state), and the microorganisms that synthesize them can introduce different functional groups on the molecule to make specific polymers. The mechanical properties, morphology, biodegradability, and thermal and crystallization behavior of PHAs blended with nonbiodegradable polymers (such as polyvinyl acetate) and with biodegradable materials (such as wood cellulose fibers and starch) have been reviewed. In general, immiscible blends or blends with other biodegradable polymers are more biodegradable than miscible blends or blends with a nondegradable polymer.

PROSPECTIVE AND APPROACHES

Cost of production is not a major consideration if PHA is to be used exclusively for specialty applications. However, if PHA is to be economically competitive with synthetic plastics, it must be produced on a far larger scale than any other aerobically produced microbial product. An additional problem arises from the fact that PHAs are intracellular products. Thus, the amount that can be produced per cell has physical limitations. In a typical PHA fermentation process, the cost is divided fairly evenly between substrate cost, capital and operating expenditures in the fermentation process, and the cost of separating the PHA from the microbial biomass. Typically the yield of dry biomass from a hexose sugar such as glucose is about 42% for *R. eutropha*.

If the final biomass contains 75% by weight of recoverable PHA, then the yield of PHA from the total amount of substrate supplied is about 32%. Thus, if carbon substrate was available at $500 per metric ton, the cost of the carbon source alone would be $1,560 per metric ton, making the product twice as expensive as bulk synthetic plastics. New methods such as the incorporation of temperature-sensitive plasmids for the production of autolytic enzymes should reduce separation costs. The direct use of PHA-containing bacteria in melt processing (i.e., no separation process apart from water removal) has also been proposed. However, even assuming a negligible cost for the carbon source by using a waste material such as cheese whey permeate, and neglecting the separation costs, studies have shown that PHA produced in bioreactors could not be economically competitive with synthetic bulk plastics in the foreseeable future. The volumetric production rate cannot compete with chemical processes, and this rate is limited by the oxygen transfer rate of present-day bioreactors. New technologies must be developed for commercial PHA production.

In Vitro Synthesis of PHA

In vitro synthesis of PHA granules may be commercially feasible. In *R. eutropha*, the maximum rate of PHA synthesis occurs early in the accumulation phase and deteriorates slowly thereafter. One would therefore expect that the maximum amount of PHA accumulation would depend on enzymatic activity or on NADPH or substrate supply. However, the most detailed study on this process points to a physical limitation where polymer synthesis slows to a virtual stop simply because there is no more space available in the cell. If this is indeed the case, then a solution is to produce the polymer outside of the cells as has been achieved previously.

Economic exploitation of this concept would require a source of the biosynthetic enzymes, substrates, NADP, and a method for NADP reduction. This would eliminate separation costs because the product could be used directly as a latex. It is likely that, in the near future, starch, oilseed, or leguminous plants will be used to produce PHAs. Other plant materials such as cellulose and starch are already used to make plastics or plastic-like materials. Hemicelluloses have been used directly or converted to organic acids for PHA production at a laboratory scale. Thus, any waste from the separation of PHA from plants could be hydrolyzed, fermented to lactic, acetic, and propionic acids, and then fed into bioreactors for the production of specialty PHAs. Polylactides and other polymers of organic acids could be produced in the same plant as could polysaccharides. Plant materials may become a major source of plastic materials in the next century, with PHAs leading the way.

12

BIOPROCESSING PILOT PLANTS

Several diverse roles, objectives, and purposes are associated with a bioprocessing, biochemical, or biopharmaceutical pilot plant. These include, but are not limited to, manufacture of clinical supplies, process development for new products in the company pipeline, process improvement including technical support for existing products on the market, and production of biologically produced materials (which are not intended for the clinic) to support basic research efforts such as screening, assays, and disease mechanism elucidation. Unlike chemical pilot plants, pharmaceutical pilot plants are used not only to scale up laboratory processes but also to produce developmental quantities of chemicals (and biological substances) for safety, toxicological, and clinical studies. Thus, balancing the demands for scale-up and development research with those for clinical production is a key dilemma for pilot plant operation. The timing of clinical production is critical because expensive and time-consuming clinical programs can be delayed, thus postponing product application filings, approvals, and launches.

Although traditionally pilot plants have not manufactured material for sale, accelerated launch dates, deferred manufacturing facility capital commitments, and other areas of project uncertainty have prompted consideration and use of pilot plants for manufacturing in recent years. The types of products manufactured using biologically based methodologies include primary metabolites, secondary metabolites (ranging from pharmaceuticals to biopolymers), constitutive proteins, recombinant proteins, intracellular or extracellular enzymes and biocatalysts, antigens, viruses, polysaccharides, and DNA itself. These products are made using a wide range of cell cultivation and isolation techniques suitable for bacteria, yeast, fungi, actinomycetes, suspension animal cells, anchorage- dependent animal cells, insect cells, and plant cells. Processing equipment varies considerably among these diverse systems. Designing for extensive processing flexibility can be expensive, potentially in excess of $100 million dollars, depending on the specific process requirements, size, and intended use of the pilot plant. Processing requirements can change dramatically as products are deemphasized and reemphasized in development. These decisions on priorities are likely not to result from the robustness of the process but rather from the state of the clinical data or potential market share.

The ability to process multiple products in an acceptable manner typically is desirable to maximize facility use. Although designers may need to predict future types of processes and products, they should be cautious about over-designing the facility in terms of containment and flexibility. The general areas required in a pilot plant might include support labs for bench-scale development and quality control, production suites for scale-up, utility areas, and personnel offices, as well as ample storage areas for raw materials, supplies, product, samples, and documentation. There should be a clear division between controlled and noncontrolled areas, by utilizing signs at a minimum and perhaps by physically restricting access. General design criteria might focus on modularity, simplicity, future growth provisions, redundancy, construction costs, operating costs, and air and water waste containment. Future operating costs are an important consideration for pilot plants because ample funds are necessary for raw materials, staffing, maintenance, and additional capital modifications to permit improvements.

FORMATION OF PILOT PLANTS

Several operational concepts influencing design requirements need to be resolved before initiation of design activities. Agreement should be sought among end users, operations staff, compliance auditors, environmental consultants, and safety officers as well as from appropriate levels of management directly or indirectly influencing and controlling capital expenditures. A written document summarizing consensus on these issues may then become the basis for conceptual design activities. Major decisions influencing pilot plant capabilities include whether the facility will produce bulk drugs (thus falling under the regulation of Center for Drug Evaluation and Research (CDER) and following guidelines from 21CFR 210/211) or biologics (thus falling under the regulation of Center for Biologics Evaluation and Research (CBER) and following guidelines from 21 CFR210/211 and 600).

The level of validation and the validation philosophy are directly related to the end use of the resulting pilot plant products, specifically whether they are used for process development, clinical supplies (presafety/basic research, safety, phase I, phase IIa, IIB, phase III), or material for sale. Validation strategies, as well as other operational and design issues, are also influenced by whether the pilot plant is administratively located within the manufacturing or research division. Individual company preferences and guidelines for similarity between manufacturing and pilot plant areas need to be identified and evaluated for their impact on the proposed pilot plant facility. Typical areas targeted for standardization within a company are vessel designs, computer systems and interfaces, equipment vendors, and construction specifications. Design requirements center around the desired initial and end product stages of processing, specifically the forms of the pilot plant inputs and outputs. Example stages include master/working seed preparation, fermentation broth, initial captured product (e.g., microfiltration retentate, ultrafiltration concentrate, cell paste), crude product (e.g., rich ion exchange fraction), finished isolated or sterile bulk, and filled vials.

Ambient, refrigerated, and frozen storage requirements based on these inputs and outputs then need to be addressed in the areas of seed, raw materials, intermediates, clinical supplies, and process samples. The scale of the pilot plant relative to current or proposed manufacturing scale and relative to laboratory scale needs to be established. Generally, a 1:5 or 1:10 scale-up is common, but smaller ratios can be advantageous in terms of minimizing risk during factory start-up. The appropriate level of biosafety also must be considered, with good large scale practice (GLSP) being sufficient for most recombinant work. Some pilot plant areas may require biosafety level 1–large scale (BL1-LS) and/or biosafety level 2-large scale (BL2-LS), and careful consideration of multiple-use/decontamination issues for BL1- and BL2-containing facilities is necessary. Solvent use areas need to be identified and minimized if possible because cost implications for explosion-proof design can be substantial. Primary motivations to design a multiuse pilot plant are the conservation of capital and improved response time for clinical manufacturing needs for new development products. The strategy for multiuse should be identified and considered throughout the design process.

Common strategies include using totally dedicated equipment and limiting production to one product at one particular production stage, using campaigned equipment in which multiple products are sequentially processed in the same equipment "campaign style," with documented product change-over procedures completed to minimize cross-contamination potential, and conducting concurrent manufacturing in dedicated equipment in which several products are simultaneously produced in segregated areas with segregation achieved via physical separation or closed systems. These strategies result in minimization of cross-contamination potential through engineering controls, standard operating procedures (SOPs), or temporal segregation. Typical pilot plants combine the second and third strategies by providing segregated and dedicated areas for certain types of processing (e.g., virus and nonvirus areas, GLSP versus BL1/BL2 areas, beta-lactam antibiotic processing, heat-resistant spore-former cultivation). Example layouts for product-specific and common areas depending on the strategy selected have been described. Guidelines for the use of open versus closed systems for various processing stages need to be devised and justified based on available and cost-effective technologies as well as quality requirements. In a closed system, the product is not exposed to the immediate environment.

The quality of materials entering a closed system (such as water, steam, or air entering a bioreactor) is controlled by its quality as well as the manner in which these materials are entering the closed system (such as filter sterilization, autoclaving, or steam sterilization of system connections). In an open system, the product is exposed directly to the surrounding environment, thus necessitating the creation of a controlled or sometimes even a closed system around the open system. One example of this strategy is the transfer of seed vial contents in a biosafety cabinet. Other examples might include the supply of higher-quality filtered air to areas for the cool-down of autoclaved materials and assembly of clean equipment. Thus, levels of containment can be constructed until all open systems are contained within controlled or closed systems to meet quality requirements for the specific processing stage. The specific issue of potential exposure of product to spore-forming organisms has been debated extensively. The closed system concept appears to minimize the concern for airborne spores typically found in the environment.

Cultivation guidelines for spore-forming organisms within a multiuse facility appear to have been redefined to evaluate the robustness of the spore type formed. Careful consideration of segregation has been given when cultivating organisms such as *Bacillus* that form difficult-to-kill spores. A separate, self-contained area with no shared equipment, air systems, or entrances minimizes exposure concerns for other products. The relatively heat-sensitive, substantially larger, fungal spores have not been high-lighted as a substantial cross-contamination concern. Thus, segregation for these types of processes is not as critical, although it still might be implemented due to company preferences for separate bulk drug and biologics processing areas.

ISOLATION OF PILOT PLANTS

The initial generation of flow charts for a bioprocessing, fermentation, or isolation pilot plant is best conducted based on model processes that are developed and critiqued by several end users of the equipment and facility. These model processes should be carefully selected as to give adequate and appropriate representation to current and reasonable future needs of the pilot plant. As they form a preliminary but sound basis for equipment and utility sizing, they should include unit operations as well as expected cycle times. In some cases, simulation or actual laboratory- or pilot-scale testing of critical equipment under typical processing conditions or with expected process streams may be warranted to further define design specifications. Example model processes have been published for vaccines and intracellular and extracellular products.

Resulting flow charts can focus on raw materials for fermentation or isolation, seed development, fermentation, recovery, crude product isolation, final product isolation, product finishing, buffers, utilities, and waste (including chemical waste, biological waste, and any necessary offgas treatment). The requirements for these model processes, coupled with an estimate of the required facility output, can be used to construct a preliminary architectural layout that divides the facility into sections (termed modules, cubes, or suites). These sections are surrounded by controlled corridors and accessed via airlocks. Based on this initial layout, flow patterns are developed to permit the logical, usually unidirectional, flow of personnel, raw materials, product, waste, and clean or used equipment throughout the facility. Flows are designed primarily to minimize cross-contamination from multiple products and from clean and used equipment while maintaining access and flexibility. This separation often is accomplished by using clean and return corridors located on either side of the processing area as well as by temporal segregation using appropriate SOPs. Evaluation of these flow patterns should be conducted carefully and should incorporate end users as well as quality auditors.

In some cases, companies may solicit outside good manufacturing practice (GMP) reviews of the project at this time from external consultants, a second design firm, or even FDA representatives. Gathering and organizing up-to-date accurate information and comments about the impact on model process needs are critical factors in forming a successful basis of design. Example layouts for process flows for biotech facilities have been published. An extensive reference (including (piping and instrument diagrams (P&IDs), layouts, and flow diagrams) for the design of biopharmaceutical plant equipment, utilities, heating, ventilation, and air-conditioning (HVAC), and waste treatment also has been published. Although it was based on acceptable practices in 1991, the strategies and concepts for design presented are still quite relevant. Adequate space around the equipment needs to be reserved for operations and maintenance access as well as for related portable equipment. The movement, segregation, storage, and cleaning of portable equipment such as vessels, skids, hoses, and carts should be fully examined. Various reports of successful and problematic aspects of project design, construction, start-up, and validation have been published. Cost control issues and the benefits of formulating a master schedule need to be reviewed in terms of their potential validation impact.

Construction concerns and "lessons learned" from multiple biotech installations also have been summarized. One case study of a laboratory facility highlights the need for teamwork among client, design firm, and construction manager. This trio can be extended to include the validation contractor in an effort to minimize "finger pointing" and "blame storming" when unforeseen problems are uncovered. Regardless of who may actually be perceived as being "at fault," the company owning the facility ultimately is adversely affected by substantial start-up and validation delays. A comprehensive overview of biochemical pilot plant design and operational guidelines appeared in the mid-1980s.

Skid Mounted Designs

The major decision regarding process equipment is that of similarity versus diversity as it applies to specific process steps as well as multiple products. The extent of commonality required in equipment specifications is an important operational factor to be considered because equipment specifications are based on inputs from model processes. For a pilot plant to accept a manufacturing interface role, at least a portion of the equipment design and selection should match that of the factory which the pilot plant is intended to support. For example, if the factory intends to use continuously sterilized medium for large-scale fermentation, then there should be provisions in the pilot plant to perform continuous sterilization as products get closer to the transfer to production. Alternatively, depending on individual company preferences for the pilot plant role, later clinical batches (phase II and beyond) might be manufactured in the factory to facilitate eventual process transfer and ensure consistency of equipment. Skid-mounted equipment has been often selected in recent installations because it puts the design burden entirely on one vendor, who specifies the appropriate quality, supply rate, and pressure of the required utilities.

Skid-mounted equipment can be fixed or portable, which permits different types of processing in the same space as necessary. Consistency in skid designs with respect to components and spare parts may be a drawback as vendors might not be willing to build skids to individual company specifications without added costs. This situation can be particularly problematic if the skid manufacturer has arranged a low-cost deal with a vendor (e.g., diaphragm valves) or programmed software on a programmable logic controller (PLC) that may not be the vendor of choice for the rest of the plant. There also may be incompatibilities in piping, utility requirements, or software when the skid is connected if miscommunications occurred between the skid vendor and the facility designer. The traditional alternative to skid-mounted designs is to purchase the equipment components separately and request the facility designer to develop the system.

One advantage to this approach is that total flexibility in equipment selection now rests with the pilot plant designers, who may be quite confident in their ability to design the equipment based on their extensive prior experience. Drawbacks of this approach include placement of the design burden on the facility as well as added design costs for one-of-a-kind systems. Specific references are available for flow charts and detailed designs for fermentation and harvest equipment as well as for filtration units, chromatographic equipment, and centrifuges. A list of typical process, utility, and support equipment that might be required in a bioprocessing pilot plant has been compiled. For a bioprocessing facility, certain aspects of equipment and piping specifications should be uniform throughout the installation where possible.

These specifications might include requirements for orbital welding of lines, documented weld inspections, passivation, pressure testing, self-draining (sloped) piping, minimal lengths for dead legs and pockets, restricted use of flanged connections, level of polish on product contact surfaces, stainless steel quality (typically 316L for product contact surfaces), and use of FDA-approved polymers in specific applications (gasket, o-rings, valve seats, distribution piping). Proper storage of fabricated piping awaiting installation as well as prominent identification of valves and lines also must be considered. Valve specifications can be important to determine early in the design estimation phase. Valve capital cost contribution is high because large numbers are required and any installed valves (either diaphragm or ball) need subsequent maintenance. Consideration should be given as to whether live steam, set up as a steam block, should be used to continually purge the back side of product contact valves and whether live steam should be used for tracing in between double o-rings on a port or manway.

While such designs traditionally have increased sterility assurance for lengthy secondary metabolite fermentations, they may create detrimental localized hot spots during animal cell cultivation. Neither the equipment itself nor its disposable or replaceable parts should release any extractable substances (after the initial post-installation rinsing and cleaning steps) into the product. Transfer piping strategy is another major design decision. One common strategy is for several transfer lines to meet at a transfer panel or process manifold, typically located at a high point between two or more areas, and be connected using spool or "jumper" pieces to move material in the desired fashion. Steam entry is at the high point and condensate drainage at the low point for steam-in-place (SIP) during sterile transfers.

Clean-in-place (CIP) solutions are passed along the direction of the transfer. Advantages to this approach include reduced piping distribution costs and space requirements, which can multiply rapidly when multiple tanks and multiple types of process transfers are required. Disadvantages center on operational restrictions concerning the elapsed time between successive transfers through a common manifold system as well as cross-contamination associated with transfers of nonsterile media/buffer, sterile media/buffer, inoculum, harvested broth, CIP solutions, and water through common lines. Both sterilization and cleaning sequences need to be defined and documented during the equipment and transfer piping specifications because retrofits after installation can be time consuming and expensive. P&IDs should be traced through these sequences before approval by both operational and validation personnel.

Requirements for validation testing and in-use monitoring of equipment operation also need to be identified during the equipment specification phase. Examples include sizing condensate lines to permit adequate drainage of condensate when spore strips are present, specification of additional validation ports for thermocouples on vessels and condensate lines, installation of product contact utility sampling stations (especially at worst-case locations), and installation of temperature sensors on critical lines to monitor sterilization and proper trap operation.

Table 12.1 Transfer lines and utility piping possibly required in a bioprocessing pilot plant

Product contact	*Nonproduct contact*
Utilities	Utilities
Clean water	Chilled/cooling water and/or glycol
Clean air	Instrument/plant air
Clean steam	Plant steam
CIP solutions	Contained sewer
Other compressed gasses (CO_2, O_2, N_2, NH_3)	Chemical (non contained sewer)
Process	Sanitary water
Sterilized medium or buffer	Vacuum system
Sterilized nutrient/acid/base/antifoam feeding	Natural gas
Nonsterilized medium or buffer	
Inoculum	
Harvest	

Applications

Product contact utilities might include water, clean steam, compressed air and other gases, and CIP systems. Non-product contact utilities generally include chilled water, instrument air, plant steam, and plant air, although in some cases plant steam may be appropriate for product contact. Multiple utility use points are located within and among suites for both product and nonproduct utilities; these are organized into utility stations for each equipment skid. To minimize piping, it might seem convenient to use product contact utilities in nonproduct contact applications such as compressed clean air on a vessel jacket, relying on a check valve for back-flow prevention. This approach can risk disaster because the operation and quality of utilities can affect processing throughout the entire facility. Cross-contamination for product contact utilities among suites is rare because of positive pressures of utilities piping into each module.

Table 12.2 Example plant steam and clean steam achievable quality comparisons

Quality attribute	*Clean steam*	*Plant steam*
Bioburden	<0.1 cfu/mL	<0.1 cfu/mL
Total organic carbon	<0.5 ppm	<2.0 ppm
Endotoxin	<0.25 EU/mL	<0.5 EU/mL
pH	5–7	5–8
Conductivity	Meets USP 23 Suppl. 5 standards	N/A

Peak utility loads should be based on carefully investigated assumptions that are transferred from design directly to validation testing. Although it may be possible to run several large fermentors simultaneously, each at maximum airflow rate, it may be less likely that several large fermentors are cooled simultaneously after completion of sterilization. Thus, air compressors might be sized assuming maximum load while process chillers might be sized on the basis of an operationally realistic fractional load. Assumptions about processing cycle times and tank volumes are important particularly when sizing CIP systems and liquid waste decon systems. Shorter (20-h) *Escherichia coli* fermentations might require a smaller tank volume [1,000 L] but are run more frequently, compared with longer (400-h) fungal fermentations, which might require a larger tank volume (20,000 L) but are run less often.

Water system design

The most critical product contact utility is often water because of the sheer volume needed during typical processing steps. It is well known that the quality of water needs to be appropriate for the intended application, but water system design decisions often are conservative. In many installations, purified water is used for upstream processes while WFI is reserved for downstream isolation and for cultures sensitive to endotoxin such as animal cells. Water-for-injection (WFI) is often favored for the entire facility for uniformity. Specifications are set for parameters such as bioburden (typically 0.1 cfu/mL) and endotoxin (typically 0.25 enzyme units, EU/mL) as well as USP chemical tests that include conductivity, pH, and total organic carbon (TOC).

WFI systems containing stills generally are designed as a hot 80°C loop with point-of-use coolers that can provide water at a user-selected temperature. These coolers can form deadlegs and thus require substantial flushing before use. It is also possible to design a cold or ambient WFI loop with periodic (typically daily) heating to 80°C or ozonization for sanitization purposes. In some applications, reverse osmosis (RO) systems are used, but care of the membranes is critical to consistent operation. USP purified water is an attractive alternative to WFI for various upstream applications. Specifications include bioburden (typically 100 cfu/mL) and USP chemical tests. Although such systems can be operated hot or cold, recent success with ambient systems sanitized with ozone and ambient systems sanitized chemically has been reported.

Table 12.3 Comparison of example USP purified water, deionized water, and process water achievable quality attributes

Quality attribute	*USP purified water*	*Deionized water*	*Process (city) water*
Bioburden	<100 cfu/mL	<500 cfu/mL	<500 cfu/mL
Coliforms	Absent	Absent	Absent
Total organic carbon	<0.5 ppm	<2.5 ppm	<5 ppm
Endotoxin	<0.25 EU/mL (for information only)	N/A	N/A
pH	5–7	5–8	5–8
Conductivity/resistivity	Meets USP 23 Suppl. 5 standards	>1 Mohm	N/A

Other noncompendial grades of water may be required. In these cases, quality attributes typically are established by the facility to address the needs of specific processing steps. Examples might be deionized water with bioburden (typically <500 cfu/mL) and resistivity (>1 Mohm) specifications or process water (municipal water passing through a break tank). General design elements for water systems include the desired flow rate at each use point and the desired number of simultaneous use points in operation. Use-point positioning should consider the height of the take-off valve, automation of take-off valves, and sink or sewer location for flushing, as well as cleaning, sanitization, and acceptable noncoiled storage for associated use-point hoses. These values are combined to obtain the total "take-off" flow rate; the system and piping are then sized such that velocity of the remaining flow rate is sufficient to minimize bioburden and biofilm growth. Suggested methods for obtaining this residual velocity have been outlined.

In addition, water systems should be designed with specific sanitization procedures identified at the outset. Steam, ozone, or chemical sanitization methods are most common, although steam sanitization generally has not been recommended for plastic loop systems because of concerns about distribution pipe sagging. Plastic systems utilizing polyvinylidene fluoride (PVDF) instead of stainless steel are becoming more common and effective at achieving even WFI conditions. A direct comparison of plastic piping with stainless steel with respect to ion leaching, smoothness, and cfu counts has been made. Clean steam can be generated from a purified water feed using plant steam for supplying heat of vaporization. Increased heat exchanger fouling may occur if a controlled feed water supply is not used. Clean steam distribution piping is designed with regulators to reduce the pressure from 5 atm down to about 2–2.5 atm as needed for typical SIP operations.

Sanitary pressure gauges, placed near or on equipment skids, can assist in determining dynamic pressures during SIP cycles. Steam sampling stations generally take the form of a removable sanitary trap to which a sample cooling device (typically a heat exchanger) is attached. Consequently, a source of cooling water is necessary, either piped from local lines or transported by gravity via a portable holding tank about 20 L in volume. Traps should be placed in header branches and equipment supply lines such that condensate buildup does not occur when skid equipment is isolated for repair. Periodic cleaning of the clean steam system assists in cleaning fouled heat exchange surfaces and in minimization of rouging. Compressed air in product contact generally is used for fermentation sparger air supply and pressurized transfers. Oil-free compressors are favored, with intakes positioned away from other building effluents. Heating up to and holding at 200°F (93°C), typically accomplished by design of the compression ratio and residence time in a receiver, is desirable to minimize airborne biological contaminants such as bacteriophage, although there is some evidence that submicrometer (0.1 μm) filtration also can be effective.

Humidity can then be reduced from the compressed air using refrigeration and desiccant dryers. Moisture removal is desirable to a dew point of –40°C. Air is then microfiltered (0.2 μm) at the point of use to remove microbes and particulates. Compressed air may be sampled for any or all of the following depending on the application: identity, hydrocarbon level, microbial content, and particulates. Other gases may be required for the facility such as oxygen enrichment for microbial cells, carbon dioxide for pH control of bicarbonate-buffered media, nitrogen and oxygen supplementation for animal or insect cells, and appropriate gas blending and control systems. Planning should consider storage of bottled gas cylinders outside the facility or in segregated maintenance areas to minimize the transport of unclean cylinders in controlled locations.

CIP skids can be portable and moved into place in the suite at a designated utility location or one or more fixed main skids can be located permanently in the utility area. CIP systems are composed of supply and return pumps, one or more cleaning agent dilution tanks, provisions for heating cleaning agents, and a PLC for sequence control and associated data acquisition. Because multiple tankage and pumps are required, these skids often are located in utility areas where acid- or caustic-based cleaning agents arriving in drums can be suitably diluted. More than one CIP skid may be necessary to provide product segregation and minimize cross-contamination. Multiple skid requirements depend on design decisions concerning the level of containment and product segregation as well as capacity.

Effect of colling, heating and humidity

The type of cooling/heating system desired, while not in product contact, has widespread design implications. Recirculating temperature control loops containing high surface area/unit volume heating and cooling heat exchangers, an expansion tank, and a circulation pump are common. These systems rely on indirectly heating or cooling of jacket fluid so their response can be slow, particularly if heat exchangers were not sized according to user expectations. The involvement of heat exchangers means that practical application may be limited to vessels a few thousand liters in size. Jacket fluids can be water, glycol, or novel fluids with wide temperature ranges. Operation of the circulation pump and internal (as well as external) integrity of the heat exchangers are also critical.

Cooling heat exchangers can use either chilled water or glycol, whereas heating heat exchangers generally use plant steam. Dual parallel filter housings, containing 10 to 50-μm elements, are advantageous to reduce heat exchanger fouling. Sediment buildup on filters can be monitored using measured pressure differentials across the filters, and the assembly can be designed so that one filter may be replaced while maintaining flow to the second filter. Direct application of heating or cooling fluids to process jackets is the traditional alternative. Consecutive application of chilled water and steam on the jacket is acceptable if the design evacuates the jacket using higher- pressure compressed air such that banging is minimized. Use of glycol and steam on the same jacket can have environmental implications from glycol losses. Regardless of the indirect or direct mode of application, jacket design can greatly affect heat transfer effectiveness. Dimpled or straight jackets are preferred for smaller vessels, with external half-pipe cooling coils used in larger vessels.

The largest manufacturing scale vessels utilize internal cooling coils but these present additional internals for cleaning and increased sterility risks, particularly at welds, because cooling water can be at a higher pressure than the fermentor contents. Filtered dried instrument air, typically at a pressure near 100 psi, is required for solenoids. Pressure specifications for control valve operation vary considerably among vendors. Plastic hosing, although convenient and cost effective, should be evaluated carefully for instrument air lines, especially on hot equipment where heat-induced leaks can develop and in solenoid cabinets where crimping can slice the tubing. A backup instrument air compressor can minimize operational consequences from supply pressure dips. A separate compressed air source, at a higher pressure than chilled/cooling water and jacket steam, is desirable for evacuation of vessel jackets to minimize risk of back-flow into the instrument air system.

Natural gas usage, if required, might best be addressed using bottled gas for safety reasons, which discourage distribution piping throughout the facility. Favored for flaming of openings in many microbiological procedures, the use of Bunsen burners in a biosafety cabinet or laminar flow hood should be examined carefully with company safety department representatives. Care should be taken to minimize the potential for reaching the lower explosive limit near the motor of the cabinet or hood. Compatibility of plastic laboratory ware with flames also should be ensured.

HVAC

HVAC is a key and costly component of the facility. Its design and operation needs to be understood by operating, maintenance, and quality assurance personnel. Requirements for air cleanliness, as well as the ability to achieve them, are determined by HEPA filtration, airflow volume, and pressurization room design as well as the type of operation being conducted. HEPA filters are often used for inlet air but are only used for exit air when required by biosafety containment. The specific classification of processing areas depends on processing step quality concerns. Classification level, representing the maximum number of particles permitted per cubic foot of air sampled, of an area increases from fermentation (class 10,000-100,000; 20-40 room changes per hour) to purification (class 1,000-10,000; up to 100 room changes per hour). Within a room there are cleaner areas surrounding open transfers (e.g., biosafety cabinets or laminar flow hoods) that are class 100 (540 changes per hour). For viable particles, the range is from class 100,000, which permits 2.5 cfu/ft^3, down to class 100, which permits 0.1 cfu/ft^3 with intervening levels generally determined by individual application.

Sufficient air velocities also are necessary to dilute particulate contaminants. In some cases, separate air handlers are considered for critical sterile areas, inoculum preparation areas, high particle or dust areas, and segregated product areas. In other instances, airflow is designed in a once-through pathway, which may increase heating and cooling utility costs substantially. Inlet air enters from the top of the room, flows downward, and exits from ducts near the floor. Any recirculation, which may be desirable to minimize capital costs for heating and cooling equipment as well as operational costs, should be only from the room where the air originated, limited to areas that do not generate substantial amounts of particles, and pass through inlet HEPA filters again. This strategy is beneficial to the life span of the HEPA filters and the ease of maintaining constant temperature and humidity. Exit ducts should be placed near equipment within a processing area that generates the most particulates or aerosols so as not to spread them throughout the area. Specific examples include autoclaves, glass washers, and cell-disrupting homogenizers.

Room pressurizations should be higher for cleaner areas and lower for areas that generate particles. Higher pressurization levels should be assigned to air-locks between adjacent areas for which it is desired to minimize cross-contamination. Relative pressurizations between rooms sharing a common wall should be considered to supplement or minimize reliance on completely sealing rooms during their construction. The use of door sweeps and gasketted ceiling tiles to achieve and maintain pressurization should be evaluated relative to floor cleanability and ceiling maintenance access, respectively. Pressure differentials might be as high as 0.03 in of water between classified areas and as high as 0.05 in between classified and nonclassified areas.

Lower pressure differentials might be acceptable depending on the specific product so long as the direction of airflow remains consistent. Interlocks, control, monitoring, and alarm building automation systems may be designed to reduce pressure losses when doors are opened and to alert users to unsatisfactory conditions. Facility temperature is generally maintained on the cool side because gowning tends to make workers warm. Inaccurate assumptions concerning the impact of radiant heating may cause temperatures to be higher than desired on certain days of operation. Humidity should be comfortable to avoid drying of workers' respiratory membranes during extended hours.

Care should be taken to specify tolerance limits that have quality implications relevant to the processing steps expected to be conducted in that area to avoid unnecessary facility costs. While specific example guidelines are 23 ± 3°C for temperature and 40-60% relative humidity, substantially larger ranges may be acceptable for many processing applications. An extensive P&ID design of HVAC for clean spaces is available as a published reference. In addition, consideration should be given to Federal Standard 209E, Airborne Particulate Cleanliness Classes in Clean Rooms and Clean Zones, as well as European Economic Community (EEC) regulations that establish parallel categories of Grade D (100,000), Grade C (10,000), and Grades A + B (different designs of class 100).

Sanitary Operation

Similar issues exist for product, environmental, and personnel protection. Often the resulting design specifications overlap with those items necessary for sterile or sanitary operation. This concept is addressed in the literature in which solutions to prevent microbial transport for the purposes of containment were compared with those to maintain sterility for fermentors. A unified approach is recommended with the goal of minimizing unnecessarily stringent specifications that can substantially augment capital and operating costs. Environmental and personnel protection issues should be evaluated for each new process undertaken by the facility, along with product protection quality concerns.

Samples of broth, and other product, intermediate, and waste streams, should be submitted for aquatic and human toxicity testing in advance of processing in the pilot plant when appropriate. Preliminary laboratory data then might be available to evaluate potential hazards in conjunction with published documentation concerning the relationships of the questionable compounds, raw materials, or organism with known hazards.

Closed system

The major concept for product protection is the closed system in which all material entering or leaving the process must be controlled in a specified and documented manner. This requirement extends to product contact utilities such as purified water, clean steam, compressed air, and controlled/monitored room air so that when equipment is opened for cleaning the quality of the surrounding environment is consistently at a known and documented level. The equipment is cleaned to an acceptable quality level. After sterilization, material enters the closed system via a sterilizing filter, steamed connection to an autoclaved bottle assembly, or sterilized transfer line. Offgas passes through a 0.2 μm steam-sterilized vent filter, often contained in a low-pressure steam-jacketed housing preceded by a coalescing filter, condenser, or vent heater to avoid pluggage from moisture. Double mechanical seals are utilized with a barrier fluid of a controlled quality (typically condensed clean steam) at a higher pressure than the vessel. A closed system also can be used to surround an open operation such as the use of isolator technology (i.e., glove box) as an alternative to clean rooms or biosafety cabinets.

Environmental protection

Environmental protection concerns focus on emissions in air, liquid, and solid waste. State air permits typically are required for new vessels and need to be altered if additional air treatment devices such as vent filters, incinerators, or scrubbers are added. Hydrophobic vent filters can provide emergency foam control and prevent external product release. These filters can be monitored for pluggage over cultivation time using pressure differentials with plugged filters able to be replaced during a run only if a bypass line has been installed. Liquid emissions include accidental spills/ sewerings, which can be minimized by locking vessel bottom valves, blocking open sewer connections, collecting sample line flush waste, and providing clear written instructions to personnel regarding batch disposition using batch sheets and appropriate signs. Dikes, troughs, and vaults can be used to contain large spills or divert inadvertently sewered material. Sample waste can be minimized using low-volume sampling devices.

Thermal or chemical (less common) batch or continuous kill tanks need to be considered for biological hazards. For chemical hazards, other appropriate waste destruction and minimization systems may include hot caustic hydrolysis, reverse osmosis, and evaporative condensers. Batch leakage into cooling water utility systems can be minimized by pressurizing utility systems higher than the batch, but this might introduce contamination if jacket cooling coils develop leaks. Alternatively, the utility system return headers can be isolated from general use until their contents are treated and/or tested. This solution might be attractive for cooling water as well as condensate, especially when it is returned to river sources. Care must be taken to avoid sequestering large quantities of potential waste for long periods of time. Solid waste can be controlled procedurally using clearly labeled containers.

Biocontainment protection

Personnel protection is not as great a concern now that most recombinant organisms can be grown under typical operating conditions of GLSP. For other pilot plant processes requiring personnel protection, either the culture is rated at or above BL1-LS or a compound produced by the culture adversely affects human heath (e.g., a genotoxin, mutagen, teratogen, or carcinogen). Environmental monitoring of the operating area is necessary for the compound or culture of interest using air or swab sampling techniques. For optimal recovery, swabs should be soaked in a solvent known to dissolve the compound of interest. Screening and monitoring of personnel potentially in contact with the process (operator, mechanics, janitors, engineers, visitors) must be arranged and notification given to those personnel at unacceptable levels of risk relative to their individual health histories.

A key element of personnel protection is personnel protective equipment (PPE). This equipment can include gowns, uniforms, safety glasses, lab coats, and gloves at a minimum but can be extended to include face shields and respirators as required. Spill response equipment, procedures, and training are also necessary. Reduction to practice techniques are presented in the literature for general biocontainment issues and biocontainment regulations and requirements. Specific examples are available for the design of waste inactivation systems, contained facility design and operation, design details for cleanliness and containment for floors, walls, ceilings, and penetrations, and contained sampling systems.

Laboratory Techniques

The main purpose of a pilot plant is to collect data in as many ways as possible. Thus, a large fraction of the pilot plant capital allocation might be devoted to instrumentation. Traditional instruments for on-line monitoring, particularly for fermentation, include pH, dissolved oxygen (DO), temperature, agitator speed, agitator power draw, back-pressure, foam detection, batch level or weight, airflow rate, and vent gas analysis via mass spectrometry. Redundant instrumentation is desired for critical parameters such as pH and DO. Instrumentation for isolation monitoring includes monitoring of conductivity, pH, ultraviolet wavelength intensity, and refractive index detection. Useful utility instrumentation includes on-line total organic carbon (TOC) and resistivity/conductivity meters.

Local readouts of transmitter values can be useful, particularly for pressure and possibly temperature, although the potential for differences between field and control room values may have to be addressed. Many novel sensors have been developed including dissolved carbon dioxide probes, redox sensors, glucose sensors, near-infrared detectors, fiber optic biosensors, fluorometry of intracellular NADPH, and cell density probes. Another novel measurement device example incorporates fermentor side streams to measure viscosity on-line. A description of useful sensors may be found in the literature. Characteristics of on-line and in situ devices include cleanability, minimal drift, minimal interferences, versatility, sterilizability, ease of calibration, and several other factors. Utilization of any sensor should be evaluated for its effect on bioreactor sterility as well as product quality.

Control System

The need for flexibility and change within the pilot plant (as compared with production) needs to be apparent in control schemes, event sequencing, data collection and presentation, and alarming. Increased amounts of automation restrict abilities of the end user to implement changes, but do reduce the number of manual operations and thus possibly headcount. Specifically the control system needs to readily permit set-point and alarm changes. Control systems can be distributed or centralized, with distributed control systems (DCS) favored because there are individual controllers/computers for each unit, often connected to a central computer via a high-speed network. Thus, the loss of one centralized computer is not catastrophic. While there can be "islands of automation" within a facility, connectivity in a centralized location for data monitoring, set-point/alarm changes, and data archiving might be desirable.

Substantial advantages exist for localized controls on skids when many field-based inputs are required based on an operator's evaluation of field conditions. A local readout of monitoring data versus a centralized "control room" monitoring area (or even both) is dependent on the degree of operator supervision versus intervention required for the processing. This decision is related directly to the flexibility of operators to go in and out of the field in terms of gowning procedures and ease of access. The ability to remotely input set-point or alarm changes can be useful when containment or product protection levels are high. Multiple areas for setting parameter values need to be considered carefully from the aspect of operational control to avoid duplication of desired changes and to minimize batch record log omissions. Verbal communications between the field and the control room using radios or intercoms also need to be considered. Distributed control systems are favored for process management. System architecture requirements vary depending on the processing requirements but examples for a non-PC based system have been published.

Distributed control systems also can be PC-based using commercial PC control/data acquisition software. In either case, some type of process-level controller is required such as a PLC, loop controller (LC), personal computer–based controller) (PCBC). The ability for those without computer expertise to easily modify and understand control sequences using newly developed commercial control software is attractive but requires effective change control and documentation procedures. The ability to readily configure the data acquisition system to download and archive data is a consideration important for later operation and process development. There are several types of computer inputs and outputs, which include: (i) digital inputs in the form of an on–off signal used for foam probe, pressure, flow, or temperature switches; (ii) digital outputs in the form of an on–off signal used for the flow of antifoam, open–close valves, or pump/motor start/stops; (iii) analog inputs in the form of an amplified output in the range of 4-20 mA used for temperature, pressure, pH, dissolved oxygen, flow, and power draw; and (iv) analog outputs in the form of a proportional signal that controls the valve opening via an I/P conversion, typically used for back pressure, cooling, and airflow valves or for signals such as motor speed.

The type of control most frequently utilized is the proportional integral differential (PID) control loop. It is tuned using three parameters to minimize oscillations, which can change depending on the specific process (e.g., for dissolved oxygen cascade control of a fermentor cultivating faster-growing *E. coli* versus slower-growing yeast cells) or over the course of the same process (e.g., for glucose and ethanol consumption phases of a baker's yeast fermentation). These three parameters are the size of the error relative to the proportional band with smaller bands resulting in more sensitive systems, integral of the error over time with longer error durations resulting in greater corrections, and rate of error change with time with larger increases in correction with time calculated the faster the error is growing with time. Typically the third parameter is at a minimal value near zero for the optimally tuned loop.

A conscious decision needs to be made concerning the desired level of automation for CIP, SIP, and vessel-to-vessel transfers. Higher levels of automation may ensure reproducibility of operation, reduce reliance on operational staff, and simplify SOPs. Raising the level of automation increases capital, validation, and maintenance costs while reducing flexibility. Operating personnel who routinely observe automated processing steps may be less familiar with the details of automated versus manual equipment operational sequences.

A domino effect may exist in that automation of one valve may lead to automation of several related valves, severely increasing design complexity and cost. Limit switches often are installed that confirm to the control system that the desired automatic valve has opened or shut as requested. Automatic pulsed diaphragm valves have been successful for nutrient feeding as well as for sterilization of transfer lines in automatic plants. One prominent example of a highly automated but apparently reliable fermentation facility is operated for the manufacture of cephamycins as well as for pilot use. Typical diagrams for monitoring and control of a fermentor and for monitoring and control of an ultrafiltration unit have been published.

Collection and Management of Data

The collection and management of data during pilot plant operations are critical, with the collection interval dependent on the nature of the processing step. For fermentations, intervals might be every 15-20 min during secondary metabolite or animal cell cultivations but might shrink to 5-10 min for faster-growing bacterial or yeast cultures, especially when a key process addition or change is planned to be made on the basis of off-gas or dissolved oxygen data. Typically, 5-to 10-min intervals are appropriate for microfiltration or centrifugation operations. Smaller data collection intervals may create substantial data archiving demands that may be alleviated by only saving those data substantially different from previous values, based on some user-defined value. Continuous data collection in the form of chromatograms is typically required for chromatographic operations.

In-process trending of data is attractive to permit timely evaluation of process status and troubleshooting of processing problems. Automatic recording of alarms and changes to the control system, preferably sorted according to the equipment and/or batch on which they occurred, also can help trace back processing problems and actions taken. On-line data might be archived on a common server by some unique batch delimiter, according to the established facility SOP for archiving data. One formidable challenge is to incorporate off-line data in the same or a closely related reference database; this off-line data can get fragmented among the notebooks and files of individual researchers.

Novel customized systems for bar-coding samples taken from the field and creating a database for results can be helpful and can also create possibilities for the use of robotic systems for sample analysis. Remote access is an attractive option for monitoring processes from off-site locations. Remote alarming can be triggered to a satellite pager, thus removing the necessity for round-the-clock coverage. Remote control of operations can be administratively challenging to regulate and document as well as a potential safety issue. Security issues must always be considered for any established remote capability.

Inventory System

Warehousing needs for the facility need to be determined based on expected production capacity. Common items that might be stored include ambient, cold, or frozen solid or liquid hazardous and nonhazardous raw materials (controlled temperature and/or relative humidity storage), disposables, equipment, and spare parts as well as sample and product storage. Computerized inventory systems, organized by process or equipment, can be useful for most of these items with reminders to reorder when supplies decline to a critical level. Marker equipment also might be purchased for long-lead replacement parts for which redundancy was not desired during the original facility design. Sizing of these storage areas should consider typical lot volumes of prepared culture media and solids to minimize repetitive release testing for multiple lots. For a pilot plant, in which several diverse processes may be run, additional process-specific storage often is required. The relative amount of storage necessary in the building should be compared with other on-site and off-site options.

Vendor manufacturing schedules often require purchasing larger quantities of rare raw materials to avoid future shortages. In this instance, if on-site storage is not available, vendor storage or a third-party warehouse alternative may be required. Warehouse access may need to be restricted depending upon the nature of its contents. There should be segregation for released, unreleased, and quarantined materials either by physical separation or by designated labels. While the entire facility should have a comprehensive pest monitoring program, as well as an SOP, there may be a need to directly address pest control, particularly for flours and meals used in certain secondary metabolite fermentations. These food ingredients when purchased for pilot-scale fermentation purposes may not pass through fumigation unless specifically requested on the purchase order.

Typically, larger customers in the food industry have invested capital for on-site fumigators, but commercial nutrient vendors may not have this type of equipment because of their high rate of inventory turnover. Staging areas should be available for collecting raw materials and supplies for a specific batch. Some method of inventory assessment and control should be available; one convenient method is to use computerized bar-coding to check inventory as well as to confirm use of specified materials when charging occurs. In addition, controlled sampling areas (to provide raw material protection, environmental, and/or personnel protection) for sampling raw materials may be necessary.

Redundancy in Pilot Plant

A critical design decision centers on the level of backup or redundancy required in the pilot plant. While typically none of the material created in the pilot plant is "for sale," mechanical problems can delay processing and clinical material production, which may interrupt clinical supplies and compromise the success of ongoing studies. Even if no clinical material is at risk, process development efforts are hampered when experiments are frequently interrupted for mechanical support-type equipment failures. Scenarios that might be protected against include loss of clean steam lubrication to agitator seals, loss of positive air pressure on a sterile tank, and loss of electrical power to a controller.

Corresponding equipment needs might include redundant clean air compressors, clean steam generators, and purified or WFI water generation systems, with the peak plant capacity needs split between two systems rather than relying on a single system sized for peak usage. Such redundancy usually is achievable because utility costs are low compared to equipment costs. An uninterrupted power source (UPS) system sized to include the most critical equipment (typically controllers and data acquisition devices) might also be installed. Backup power for large electrical loads such as motors might not be economical; however, consideration might be given to installing connections for a gas-powered generator if future evaluations altered the economics.

Modular Mobile Validation

Because future processing needs may change, the pilot plant must adapt to maintain its usefulness to the company. Large initially under-utilized spaces might be outfitted with utility headers to minimize processing disruption during future installation. Oversizing of utilities for anticipated future loads as well as unforeseen procedural modifications warranted on the basis of validation testing results such as additional rinsing cycles assures subsequent flexibility. Selection of portable equipment permits exchanges for new processes, provided there are procedures developed for storage and start-up of this equipment. One aggressive approach to increase versatility is the availability of a modular mobile validated pilot plant, composed of a system of interconnected and integrated modules. Modules are selected according to processing needs and can be assembled rapidly to form a processing plant capable of undergoing validation. Production modules include fermentation, recovery, and purification; the utility module includes clean water and clean steam; the personnel support module includes lockers and changing areas; and the HVAC module houses the air handling and filtration equipment. Additional modules can then be added or deleted as required.

PRACTICAL APPROACH

Each pilot plant facility must decide for itself whether it is required to operate in a continuous mode of GMP compliance or whether it can selectively apply GMP practices to certain batches [but not to other concurrently run batches] or during certain specific processing periods. A conservative solution might be to operate continually in GMP compliance, thus insuring facility readiness and preventing omission of a required procedure for an actual batch destined for the clinic. Dual GMP and non-GMP operation serves to create two standards and may cause confusion among operating and maintenance staff. If dual operating modes are considered, then a clear distinction between GMP and non-GMP operation must be established, documented, and maintained.

Preventative Maintenance

Preventive maintenance (PM), performed during an annual facility shutdown, is probably the most important factor influencing smooth operation throughout the upcoming year. The scheduling of an annual facility shutdown is desirable for several reasons including the ease of lockout and tagging of hazardous energy sources, the time economies of concentrating several maintenance workers in a single area, and the ability to collect and document completion of a large fraction of maintenance work at one time. This yearly PM binder can then serve as the model for maintenance personnel for the following year's shutdown. Individual PM work orders, spread over the course of a year, result in substantial time and effort expended in scheduling, preparing, and documenting the work done. Interspersing of annual PM work among batch processing inevitably results in PM delays due to equipment and/or manpower unavailability. Guidelines for planning and executing an annual plant shutdown are available. All PM work orders should be listed on a single spreadsheet to be reviewed before the facility shutdown and augmented with any modifications or one-time replacement work orders that are requested. As PM work is completed, documentation might be signed by the tradesperson as applicable and reviewed by mechanical and operational staff for completeness. Problems identified with equipment or instrumentation should be highlighted as they occur and then entered into change control work order systems so that they can be tracked for their potential effect on validation.

Before the annual shutdown, transfer line manifolds and utility supply lines, which typically cannot be shut down entirely for maintenance throughout the year, might be leak tested to identify faulty valves in need of replacement. After the annual shutdown, equipment start-up procedures should be documented, preferably through the use of equipment start-up batch sheets. These procedures might include equipment cleaning, testing of instrument calibrations such as level, temperature, and pressure, and vessel integrity testing for leaky valves. Instrument calibration accuracy should be evaluated for random or systematic errors, with random error within accepted tolerances being the expected outcome.

Some general PM procedures have been outlined that can apply to several areas in a pilot plant, both product and nonproduct contact. Sanitary fittings, which tend to loosen during repeated heating and cooling cycles, should be checked, with gaskets replaced and clamps tightened as required. Any flanges in product contact should be tightened to a specified torque. PM inspections of vessel internals should be conducted with lock washers on nuts that may loosen due from repeated heating and cooling. A PM should be developed for inspection of vessel cooling coil leaks that can result in inadvertent cooling water contact with product. For agitated systems, there should be a regular examination of foot bearings, motor/drive vibration, oil analysis, and shaft run-out. Double mechanical seals, the failure of which can lead to condensate buildup or batch contamination, should be tested regularly and replaced preventatively based on a minimum life expectancy.

An outline of specific PM procedures for laboratory fermentors can be applied to larger-scale fermentors. An example of documentation for the maintenance of horizontal laminar flow hoods also is available that can be readily adapted to other HEPA filter units. PMs should be conducted on all utility systems, which may include rebuild of air compressor heads, descaling of heat exchange surfaces, and replacing of air dryer desiccant, among other items recommended by the equipment manufacturer. Finally, novel predictive maintenance evaluation techniques should be explored and implemented as appropriate to identify imminent equipment failures. It is most appropriate to perform any annual validation work (e.g., fermenter SIP, autoclave load pattern, controlled temperature unit) at the conclusion of the annual PM shutdown. In this manner, proper equipment operation can be documented and the effectiveness of PM procedures can be evaluated. Often inadvertent mistakes or omissions during maintenance can be identified before they have an adverse, and possibly widespread, quality impact.

Review of outstanding work

A system of tracking outstanding maintenance by equipment or by process should be designed. Each piece of equipment should have a clearly distinguished tag name. Work orders need to be initiated, evaluated for their effect on validation, marked as complete when mechanical work is done, and finalized when process personnel have checked the job. A computerized database can readily track maintenance and identify repetitive maintenance problems. Such a system permits easy review of outstanding work orders for each piece of equipment by the operation group before it is readied for service. All maintenance work orders need to be tracked within one system; care should be taken not to bypass the operations area when maintenance needs are identified by those directly controlling the mechanical staff. Interim maintenance work, such as calibrations and subannual PM work, should be documented and filed in a manner equivalent to the yearly PM results. PM schedules should be reviewed based on the frequency and results of interim maintenance work. Unusually common equipment failures can be identified through maintenance database searches so that appropriate longer-term repair efforts or replacement can be justified and undertaken.

Specific calibrations and maintenance

Additional concerns are associated with calibrations because important processing decisions may be based on the accuracy of the instrument readings. In some cases, specific instruments are designated as "critical" for a processing step; in other cases, desired specification ranges for the processing step are targeted based on certain identified instruments. These two methods provide assurance that product quality is not compromised if any of the other instruments fail in the equipment being used.

Although some facilities distinguish between calibration procedures for critical and noncritical instrumentation, it may be prudent for a pilot plant to calibrate all instruments in a reliable and documented manner because their accuracy may be important for useful process development data. Calibrations should be performed using National Institute of Standards and Technology (NIST) traceable instruments and should include full loop calibration to the process monitor as appropriate. A specific calibration and maintenance SOP for the instrument is most desirable, but a general SOP referencing vendor manual might also be acceptable, particularly in the interim period while instrument calibration SOPs are being developed.

The as-found and as-left conditions should be documented along with the serial number of the standard used. Establishment of an acceptable tolerance for each instrument ensures consistency. When an instrument is found to be outside the tolerance specified, the impact of this calibration error on previously executed batches may need to be evaluated. Thus, a confirmation of the error might be warranted before correcting the calibration. Records should be evaluated for the amount of instrument drift, which may indicate the need to either recalibrate on a more frequent basis or else replace or upgrade aging equipment. An example calibration sheet is available.

Documentation of changes

Any validated facility needs to devise procedures for tracking changes on equipment as well as on processes and computers. Incomplete documentation of changes can obviate extensive time and money spent on validation testing. Although the exact contents may change, such a procedure starts during the initial design phase as equipment specifications are altered, continues through the facility start-up and qualification phases, then moves directly into the facility operational phase. Adherence to change control procedures may be difficult during facility start-up as many unexpected items are discovered to be incorrect and management pressure to have the facility operational may mount. Depending on the philosophy of the individual facility, the scope of this change control procedure may include product and nonproduct contact equipment to varying degrees.

The procedure identifies the individual or individuals that need to approve changes and determine before or in concert with the change implementation what additional validation testing is required. Often, processing personnel, being most familiar with the equipment being modified and the reason for its modification, can provide valuable leadership in this area and initiate contact with quality and validation personnel as required for input. Generally replacements in kind are accepted without additional testing unless the item being replaced was specifically tested or calibrated during the initial validation effort. Example items and associated testing might include a replacement or rebuilt motor or associated drive (measure motor amperage draw at various agitator speed settings) or replacement instrumentation (calibrate transmitter). Replacement items in this category need to be tagged appropriately in the field with the same tag number as the original item and serial numbers updated in any centralized maintenance data bank as well as recorded on installation documentation.

Modifications or obvious changes in equipment or instrumentation need to be evaluated for their potential impact on the validated state of the equipment. This impact is particularly important for SIP and CIP cycles. Such modifications may be minor, such as a change in gasket or o-ring material of construction, or major, such as a change in piping. An example procedure is available that can ensure proposed changes are reviewed by operations, maintenance, and quality for their potential impact. Specific attention should be given to alterations needed in P&IDs, SOPs, PMs, batch sheets, or training programs. Equipment modifications, necessary to accommodate a new process, need to be identified early and their impact evaluated. It is critical to recognize the need to balance the request for change with an evaluation of its validation impact; changes cannot be discouraged in a multipurpose pilot plant but need to be controlled and managed in a logical manner.

Process Development Staff

Personnel needs are best defined based on the production process, but in a pilot plant setting production processes change rapidly. A decision needs to be made early in the design process concerning the level of off-shift activity reasonably expected in the facility because this can severely impact items such as utility capacity and operation. Round-the-clock coverage, while demanding for personnel involved, may be necessary a large fraction of the days to ensure success and timeliness of production. The use of unionized personnel, technicians, or process development staff should be incorporated into the educational and manual work-load expectations for the operating staff. A core operations group, responsible for the facility, can ensure continuity in areas such as SOP updates, maintenance work order tracking, batch sheet content, and equipment operation. Such a group might be supervised by an area manager responsible for coordination and scheduling who might interface with those involved directly in process development. Processing groups might then rotate through the pilot plant as their projects demanded, accompanied by a member of the operations group. Alternatively, a dedicated pilot production process development staff might be instituted, responsible for translation of laboratory processes into GMP batches. A shift report or log should be established and continually updated as to the mechanical and operational status of individual pieces of equipment and utilities.

Experimental Support

To provide adequate and accurate sample analyses, laboratory support needs should be identified and addressed through either on-site or contract laboratories. Targets for testing might include final product, in-process samples, raw materials, sterility, culture purity, environmental quality of controlled environments, and utility quality. While the goal of these laboratories might be to work toward using validated methods and approved SOPs for all assays, this goal might not be attainable for project- specific assays until the later stages of a successful development project. Initially, in the early project stages, only notebook procedures might be available to guide laboratory testing. Information directly impacting quality should be communicated directly to operating personnel and appropriate investigations conducted both in the field and in the laboratory for failing results. A microbiological laboratory is critical for fermentation- based operations to prepare seed cultures and to test new culture seed vials for culture purity and productivity in lab-scale processes.

An examination of up-to-date sterility/culture purity results before vessel-to-vessel transfers should be undertaken as close to possible to the time of transfer to minimize adverse impacts of contaminated cultures. In a pilot plant setting, tension may exist between the time required to adequately test new cultures and the development pressure to use the latest mutant or transformant in a pilot-scale cultivation to evaluate its capabilities. Reasonable procedures need to be established in advance for the introduction of new processes or new cultures into the pilot plant. Laboratory-prepared nutrients, trace ingredients, and sterilized additions need to be evaluated for their quality impact, and the procedures used for their preparation should be documented where possible.

Planning and Standardization

SOPs are essential for the planning and standardization of various procedures conducted by operational staff and maintenance personnel that need to be completed before, during, and after pilot plant batches. Typical categories include SOPs for equipment operation, facility and equipment cleaning, sampling, sterilization/sanitization, maintenance, and documentation procedures for tasks such as product change-over, change control, and batch sheets. When referencing SOPs as part of the batch record, working versions of SOPs can be helpful to track progress in the field. An SOP on the requirements for writing, issuing, and inactivating an SOP is also useful.

SOPs focus operation, guide operations staff, document the order and scope of activities, standardize approaches, define regulatory needs and constraints, and increase the speed of new personnel training. They should be written by a person familiar with the facility or equipment operation that is the subject of the SOP, reviewed by another person who might be asked to follow the SOP, and approved by a quality group. Each SOP should have an effective date of issue, be available to those who need it, and be archived when a future revision is created. The need for structure and consistency within an SOP must be balanced by the need for flexibility in a pilot plant when unexpected process requirements arise. Procedures must be defined for what to do when an SOP is altered intentionally to accommodate a new process or a surprise development in an existing process. Consideration to SOPs should be given during the equipment design phase so that tasks can be performed in a logical clockwise or counter-clockwise fashion. SOPs should be clear so that the expected outcome is assured.

Specific examples of details often omitted include statements regarding settings for pressure switches and acceptable calibration error tolerances, insertion diagrams, part numbers, and manual references to aid in following the procedure, and reliance solely on valve numbers, not valve purposes, to describe the procedure. While SOPs should be reviewed periodically every few years to determine if they need clarification, care should be taken to minimize the number of similar versions of the same SOP issued to the field. Operational clarification memos, issued to all operational staff and signed off when read, might be used to provide additional SOP clarification without reissuing the SOP during the interim period between SOP reviews. All major SOP changes should be documented, evaluated for their effect on validation, and communicated to staff in a documented manner.

Preventative Maintenance

Because concerns regarding product exposure to personnel are similar to concerns for personnel exposure to product, GMP guidelines overlap with safety regulations in many respects such as training, SOPs, preventative maintenance including calibrations, and validation/change control. Safety captions can be required for certain raw materials based on review of material safety data sheet (MSDS) documentation. Broths, as well as isolated compounds, might be submitted for toxicity testing as soon as they are cultivated at a larger scale. Safety is a component of installation qualification as pressure vessel documentation and utility connections are reviewed. Hazard operability (HAZOP) reviews for pressure vessels (e.g., fermentors, autoclaves) and utilities (e.g., clean steam generators) often are required by company policy and summaries included within equipment qualification reports. HAZOP action items need to be addressed promptly and modifications to equipment or procedure made to minimize risk.

Equipment design needs to be reviewed for the ease of lockout and removal of hazardous energy sources (e.g., bleed valves for the evacuation of process and utility lines, field- mounted pressure gauges to confirm depressurization) as well as the individual isolation of skids/units for maintenance. In addition, specific SOPs for preparing equipment for maintenance might be required. Regular safety inspections with documented action item lists communicated to all affected personnel (i.e., operating staff, technicians, engineers, management) can identify and correct nonoptimal working conditions and cluttered processing areas.

Exposure of personnel to excessive sound, heat, or dust should also be considered during facility design as well as monitored during facility operation. Although personal protective equipment such as ear plugs, heat-resistant gloves, and respirators is available, when possible engineering controls such as sound dampeners, insulation, and dust collectors may be better solutions. Safety during validation testing should always be considered, particularly when groups outside of operations are involved in setting up or operating equipment and when sources of hazardous energy are involved. Whether during validation testing or operational use, fittings of any sort should be tightly secured and should not be tightened when equipment is under pressure or hot, specifically during SIP and CIP procedures. Summaries of generally accepted laboratory and pilot plant safety procedures to consider are available. According to ASME code, over-pressurization protection using relief devices is required for vessels more than 6 in diameter. The background for this regulation as well as details of pressure relief valves and rupture disks is available. Sanitary rupture disks are most desirable for product contact; once blown, processing must stop until they are replaced. Sometimes a rupture disk is installed, followed by a pop safety valve rated at a slightly higher pressure with a pressure gauge between to indicate rupture of the disk. If this arrangement is used, then the cleanliness of the area between the rupture disk and pop safety valve must be demonstrated for the batch to not be adversely impacted if the disk ruptures but the pop safety valve does not release. Removal of the entire rupture disk assembly (not the removal of the disk from the assembly) for cleaning should be checked with company policy but may be acceptable in most cases.

Another often-overlooked matter is that the outlets of safety relief devices (which might be unexpectedly spewing steam, hot water, or broth after relief) should be oriented such that personnel safety is considered for those who may be standing near them. One method is to pipe outlets to outside containment troughs, but it may not be desirable to have a potential connection to the outside. A second alternative is to create a designated area within the pilot plant processing area. While outlets to the floor beside the equipment should be minimized, if they are found to be necessary they should be carefully situated outside the normal path of personnel.

Ongoing

Initial and ongoing training programs need to be developed and documented for each individual associated with the facility, based on the facility needs. A training SOP should be developed, which might include a training manual containing useful items such as key SOPs, facility layouts, equipment diagrams, and policies. Initial training might include items from vendor-supplied training sessions to validated load pattern arrangements. Often this training falls at a peak work-load period during the facility start-up and validation when attention can be diverted into seemingly more urgent areas. In other cases, more individuals require training than can reasonably be accommodated in the area surrounding the equipment. When permitted, such training might be recorded for later review by the staff when they have fully assumed responsibility for operation. Another alternative is for those just trained to provide training to others, which may be suitable for certain operations. For best results, training should occur with written documentation available to those being trained for reference. Ongoing training should be done on a regular basis; once every 1 or 2 months seems most appropriate for an active pilot plant facility.

Training should include supervisors, operators, engineers, and maintenance personnel as well as validation and GMP staff. Training might include a process review that highlights critical steps and equipment, SOP reviews, identification of personnel safety and environmental containment issues, and review of PM and ongoing maintenance needs, as well as what operational personnel should look for with respect to proper equipment function. Training sessions are one method to communicate recent SOP changes as well as to explain why certain procedures are necessary. These might also be used for periodic SOP reviews to obtain feedback from those using the SOP frequently. Attendance should be required and sessions recorded when possible; those new to the facility should review prior training sessions for the past few years as applicable.

Process Validation

The recent guideline issued by the FDA makes pilot plants eligible for product licensure should company needs require a product launch from a pilot plant facility. When manufacturing clinical materials, pilot plants are evaluated under the guidelines for investigational new drug products, among other regulations which clearly state that GMPs apply for drug products approved for clinical trials in humans and animals. Individual company philosophies vary on the suitability or desirability of their pilot plants for this task because using a pilot facility for manufacturing reduces its availability for the production of clinical material. The level of validation effort required is best considered during the initial facility design.

Validation can be done retrospectively, although the collection of the required documentation may be more challenging and modifications necessary for testing may be costly. The major steps for validation ensure that installed equipment and utilities operate in a manner acceptable for the process to be run in that equipment. Because multiple processes, many of which are not necessarily identified at the time of validation, may be conducted in a pilot plant, operational ranges need to be set to accommodate a wide range of reasonably expected processes. For example, an incubator might be validated for temperatures between 25 and 37°C to accommodate common bacterial, yeast, and fungal cultures, but not necessarily at 45°C to accommodate thermophiles. The selection of worst-case scenarios can be difficult for a pilot plant if the associated worst-case conditions are unduly burdensome for normal operation.

Specifically, the cleaning procedures required after a 3- week fungal production cultivation are significantly more complex than those required after a 2-day fungal seed cultivation of the same culture. Similarly, the sterilization cycle hold time for a concentrated sugar solution is substantially longer than that required for a heat-sensitive water-like growth medium. Approaches to resolving these multiproduct validation issues can be complicated by desires to follow manufacturing validation guidelines, which most likely were established based on a single product.

Often validation, qualification, verification, commissioning, and start-up are used with specific definitions in mind for individual facilities and organizations. Generally, validation implies the highest rigor of direct reproducible testing of a specific procedure. Qualification encompasses a wider range of documentation collection and testing of capabilities. Verification might imply a single test against an identified standard. Commissioning can refer to documented troubleshooting of newly installed equipment, whereas start-up might refer to initial equipment checkouts directly after installation. The major validation phases are divided to focus on qualification followed by validation. Installation qualification (IQ) focuses of the verification of proper installation, appropriate utility connections, and adherence to manufacturer and purchase order specifications.

Operational qualification (OQ) includes testing of equipment function and an acceptance criteria for performance. Performance qualification (PQ) centers on process-specific testing for SIP, CIP, and autoclave sterilizing and decontamination load patterns. For computers, a system life cycle methodology incorporating specifications, program design, coding, testing, start-up, operation, and maintenance has been adopted. Process validation centers on the reproducibility of the process in the manufacturing area. Parameters influencing this reproducibility might be best studied in the pilot plant to obtain acceptable operating ranges for those processing parameters, such as cultivation temperature, pH, and dissolved oxygen, found to influence product quality.

The set of criteria for proceeding to the next processing step, specifically volumetric or specific productivity, yield, and purity, must be validated in a documented fashion. The critical processing steps important to process quality should be identified and validated. Other noncritical process steps may be studied for their impact on manufacturing productivity. Process validation may require multiple identical pieces of equipment, such as fermentors, at the laboratory and process scale to define acceptable processing ranges in a parallel rather than sequential manner.

Robust System

Documentation associated with a new facility can be extensive initially and can increase substantially during the time that the facility operates. While centralizing all documentation in a single area can be attractive, duplicate copies of some items often are needed in other locations as well as for archival backup purposes. For example, equipment manuals and facility drawings may be needed near the actual location of the equipment, in the maintenance area, and in the centralized documentation area. Validation documentation, consisting of IQ, OQ, PQ, and load pattern and SIP/CIP reports, is generated both during the initial validation as well as during annual retesting and ongoing environmental monitoring programs. Storage of change control documentation might be accomplished through notebooks and file folders for individual pieces of equipment or by individual suites. SOPs might be archived in a centralized location but also posted near equipment for ready reference. Access and sign-out procedures for the control of documentation need to be developed. A robust system for batch records needs to be created, based on individual batch sheets for processing steps or for individual pieces of equipment or some appropriate combination of the two approaches. Issued batch sheets should be tracked by a unique sequential number with the product, processing step, and/or equipment number clearly identified.

There is a need to track the use of equipment by product and batch, particularly for multipurpose facilities. While this might be done through individual equipment use logs, it is also necessary to include such tracking within the batch record itself. Newly authored or completed batch sheets should be reviewed by the operations area and may be required to be approved by a quality group before execution. Batch sheets should have clear written instructions as well as a signature, date, and time, as required, for each step. Handwritten instructions, likely to be added during the processing of pilot-scale batches, may need to be reviewed by representatives from operation or quality groups. Brief justifications for alterations in batch sheet instructions also should be considered. An written explanation documenting that the material from the current step is acceptable for continued processing in a subsequent step is recommended.

Care should be taken to use common templates wherever possible to simplify successive batch sheet differences and minimize the review effort. Executed batch sheets should be reviewed for completeness by the operations area, by those in charge of the batch, and by a quality group as required. A tracking system should be instituted to monitor the review of completed batch sheets as well as to ensure that all issued batch numbers are accounted for in the batch record filing cabinets. Photocopying, microfilming, or scanning of completed batch sheets serves to provide a backup in case of loss or damage. Paperless batch records, as well as electronic signatures, have also been considered to minimize documentation volume, and guidelines for implementation and validation have been published.

13

PHYSICALLY CONFINED CELLS

Immobilized cells are defined as "cells physically confined or localized in a certain defined region of space with retention of their catalytic activities, which can be used repeatedly and continuously". This definition is applicable not only to cells, but also to single enzymes, multienzyme conjugating systems, and cellular organelles, that is, to all types of biocatalysts. Bioprocesses using biocatalysts have increasingly received worldwide attention because of their benefits of low environmental pollution and economical utilization of natural resources and energy. They have also contributed to a wide variety of fields. The use of immobilized biocatalysts increases the efficiency of such bioprocesses because they catalyze biochemical reactions under more stabilized states than their free counterparts; moreover, they can be used repeatedly or continuously.

In particular, immobilized cells can catalyze complicated reactions and have many useful characteristics that other biocatalysts do not have, as described later. The European Federation of Biotechnology, Working Party on Applied Biocatalysis (formerly, the Working Party on Immobilized Biocatalysts) proposed that immobilized cells should be classified as viable or nonviable, where viable includes growing, nongrowing, and respiring. According to this proposal, terms such as treated, resting, growing, and living should not be used. However, in general, these terms are used for classifying the condition of immobilized cells. In the definition of "treated," cells are subjected to chemical or physical treatment before or after immobilization, and resting cells do not show growth during utilization, although they have the ability to grow.

WORK PROGRESS

In 1916, Nelson and Griffin fortuitously found that yeast invertase adsorbed on activated charcoal retained its catalytic ability in the hydrolysis of sucrose. This is said to be the first demonstration of immobilization of a biocatalyst. Although several reports were published on the immobilization of physiologically active proteins by covalent binding on various supports after this finding, immobilized biocatalysts were not used in practice until Grubhofer and Schleith immobilized several enzymes (carboxypeptidase, diastase, pepsin, and ribonuclease) on diazotized polyaminostyrene resin by covalent binding in 1953. Thereafter, many studies on enzyme immobilization have been carried out, and studies on cell immobilization became to be popular in 1970s. The first industrial application of immobilized biocatalysts was carried out by Chibata and coworkers of Tanabe Seiyaku Co., Japan, in 1969. Fungal aminoacylase immobilized on DEAE-Sephadex through ionic binding was used for stereoselective hydrolysis of *N*-acyl-D,L-amino acids to produce L-amino acids and *N*-acyl-D-amino acids.

The first industrial application of immobilized cells was also performed by Chibata and his coworkers in 1973, producing L-aspartate from ammonium fumarate by polyacrylamide-gel-entrapped *Escherichia coli* cells containing a high level of aspartase activity. Several other processes, such as production of L-alanine, L-malic acid, 6-aminopenicillanic acid, milk having a low lactose content, and high-fructose syrup also have been industrialized with the use of immobilized enzymes and immobilized cells, and a vast amount of literature on immobilization is available.

APPLICATIONS

In biological systems, such as microbial cells, plant cells, and animal cells, various chemical reactions are carried out. These chemical reactions are catalyzed by enzymes that are water-soluble globular proteins. Compared to chemical catalysts, enzymes as catalysts have the following superior characteristics.

1. Enzyme-catalyzed reactions take place under mild conditions, such as ordinary temperature, atmospheric pressure, and pH value around neutrality.
2. Enzymes have strict substrate specificity, stereospecificity, regiospecificity, and reaction specificity.

These facts suggest that energy-saving, resource- saving, and low-pollution processes could be designed by using biocatalysts because such processes could be carried out at relative low temperatures and at ambient pressure with little by-product formation. However, biocatalysts are, in general, liable to be deactivated under conditions such as high temperature, extremes of pH, presence of organic solvents, or even conditions suitable for their catalysis. The repeated or continuous use of biocatalysts that are required in order to construct efficient processes is another problem faced when free biocatalysts are used. Immobilization is one of the most useful methods for eliminating some of the disadvantages described above.

In general, immobilization makes biocatalysts stable and easy to use repeatedly in a long-term series of batchwise reactions, or continuously in flow systems. Therefore, the immobilization of biocatalysts is attracting attention worldwide. At present, immobilized biocatalysts have contributed to a wide variety of fields including (i) the production of useful compounds by stereospecific and/or regiospecific reactions, (ii) the production of energy by biological processes, (iii) the selective degradation or removal of pollutants to solve environmental problems, (iv) analyses of various compounds with high sensitivity and high specificity, and (v) utilization in new types of drugs such as slow-releasing medications, in making artificial organs, and so on. These processes require the immobilization not only of single enzymes that catalyze simple reactions, but also of multienzyme conjugating systems that mediate more complex reactions. These systems often require the regeneration system of ATP and/or coenzymes. Because cells contain metabolic systems that mediate such complicated reactions, immobilization of cells serves as immobilization of such multienzyme conjugating systems. Furthermore, in the case of immobilization of cells, the procedure of extracting enzymes from the cells is no longer required. This avoids the inactivation of enzymes during these time-consuming and expensive procedures, making it possible to use the enzymes in a stable state. However, immobilized cells have some disadvantages that must be overcome to realize their practical applications.

1. By-products are easily synthesized because the cells contain the various metabolic systems and the many enzymes that catalyze undesirable reactions. To defeat this problem, selection of a proper strain, mutation, or the proper treatment of cells in order to lower the levels of enzymes participating in side reactions and increase the activities of desirable enzymes, or genetic improvement of a cell line is essential.
2. Cell walls and cell membranes of intact cells often hamper the permeation of substrates, products, and other reaction components into and out of cells. In this case, appropriate treatment of cells before or after immobilization is required to destroy the barriers to permeability.

If immobilized cells are kept in a growing state by a proper supply of suitable nutrients, and if their biological functions are useful, as in the case of conventional fermentation procedures, these immobilized cells (immobilized growing cells) seem very advantageous. Immobilized growing cells serve as self-proliferating and self-regenerating biocatalysts, and catalytic activities of immobilized growing cells are expected to be maintained during long-term repeated or continuous use of cells. However, immobilized growing cells also have some disadvantages as biocatalysts, and ways to overcome these disadvantages must be studied.

1. Various nutrients and energy sources are demanded by immobilized growing cells to maintain the cells growing state(s), and yield may be lowered by the consumption of substrates and/or products as nutrients and energy sources.
2. Contamination of product is liable to occur by cells leaked from carriers.

In spite of these faults, cell immobilization is a key technology used to construct an efficient bioprocess, as immobilized cells have the attractive characteristics described earlier.

PROCEDURES

For the application of immobilized cells, screening or breeding of cells having the desired activity and characteristics is very important. In addition, selection of the appropriate immobilization method is necessary. There is, however, no ideal universal immobilization method fitted to various types of cells at present, although a vast amount of literature on immobilization is now available. The immobilization methods can be classified into four categories: carrier-binding, cross-linking, entrapment, and combined methods, in which combined methods are a combination of former three. Because each method has its own merits and faults, selection should be carried out in consideration of the purpose of immobilization, characteristics of cells, type of reaction, type of reactor, and so on.

Carrier Materials

The carrier-binding method is based on the binding of biocatalysts to water-insoluble carriers through covalent bonds, ionic bonds, physical adsorption, or biospecific binding. Carrier materials should have adequate function to bind biocatalysts, have sufficient mechanical strength; physical, chemical, and biological stability; and nontoxicity. Ease of shaping is also important for applying immobilized biocatalysts to various types of reactors. Furthermore, the economic feasibility should be considered. As carriers, various types of insoluble materials, such as water-insoluble polysaccharides (cellulose, dextran, agarose derivatives, etc.), proteins (gelatin, albumin, etc.), synthetic polymers (polystyrene derivatives, ion exchange resins, polyurethane, etc.), and inorganic materials (brick, sand, glass, ceramics, magnetites, etc.) have been developed and utilized as they are, or after proper modification or activation.

The development of new carrier materials having desired characteristics is also important. Any reactive components of the cell surface, for example, amino, carboxyl, thiol, hydroxyl, imidazole, or phenol groups of proteins or saccharides, can be used for covalent bonding between cells and a modified or activated carrier. The covalent-binding method is one of the most frequently used techniques for the immobilization of enzymes because of the following advantages: (i) enzymes do not leak or detach from carriers during utilization because of the tight binding, (ii) enzymes can easily contact with substrates because enzymes are localized on the surface of carriers, and (iii) an increase in stability is often observed because of the strong interaction between enzymes and carriers. However, this method is not as widely used for cells because of the toxicity of the agents needed for formation of covalent bonds, and the difficulty in finding the conditions for immobilization.

In the limited number of reports on cell immobilization by this method, inorganic and organic carriers have been used as follows: Navarro and Durand immobilized *Saccharomyces carlsbergensis* cells by covalent binding to porous silica beads that had been treated with aminopropyltriethoxysilane and activated with glutaraldehyde. *Serratia marcescens, Saccharomyces cerevisiae*, and *Saccharomyces amurcea* cells were also immobilized by covalent binding to borosilicate glass and zirconia ceramics with an isocyanate coupling agent.

In the case of organic carriers, *Micrococcus luteus* cells were immobilized to the carboxyl group of agarose beads in a two-step process that avoided exposure of the cells to carbodiimide, and the uronic acid–forming activity of these immobilized cells was retained, even though cell viability was lost. A disadvantage of this method is that leakage of the enzymes may occur due to autolysis during the reaction. Recently, Klein and Ziehr have summarized the immobilization of cells by adsorption to carriers. One of the first studies on immobilized cells is on the adsorption of *Escherichia coli* and *Azotobacter agile* cells to Dowex-1 for the oxidation of succinic acid.

This adsorption method is very simple and mild, and can be applied to the immobilization of various types of cells. However, adsorption of cells to carriers is sensitive to environmental factors such as pH, ionic strength, concentration of reactants, and temperature because the adsorption phenomenon is based primarily on van der Waals forces, ionic or hydrogen bonds, or hydrophobic interactions between the cell surface and the carrier. Therefore, cells are liable to leak from carriers. Biospecific binding with lectins introduced on carriers was also used for immobilization of cells.

Lectins that are glycoproteins bind tightly to specific carbohydrate residues on the cell surface. Concanavalin A, which can be obtained from jack beans, is one of the most useful lectins.

Cross-linking Method

The cross-linking method utilizes bi- or multifunctional reagents, but carriers are not used. Cells are linked with each other by the reagents, and the cross-linked cells form pellets. Glutaraldehyde is one of the most popular cross-linking reagents, and combines amino groups through a Shiff base linkage. Diisocyanates, such as toluene diisocyanate and hexamethylene diisocyanate, represent another type of cross-linking reagent, forming a urea linkage with amino groups. *E. coli* having high aspertase activity was immobilized by this method with glutaraldehyde or toluene diisocyanate. A 34% retention of aspartase activity was obtained in the case of glutaraldehyde, but the immobilized cells were inactivated in the presence of toluene diisocyanate. So far there have been very few studies on this method, but *Bacillus coagulans* immobilized by this method is commercially available for immobilized glucose isomerase and is widely used for isomerization of glucose in the industrial field.

Entrapment Method

Entrapment methods can be classified as follows: lattice type, microcapsule type, liposome type, and membrane type. In lattice type entrapment, biocatalysts are entrapped in gel matrices of various types of polymers. Microcapsule type is entrapment in microcapsules of semipermeable synthetic polymers. Liposome type is entrapment within a liquid membrane prepared from amphiphatic compounds. In membrane type, biocatalysts are separated from the environment as spent reaction solution by ultrafiltration membranes, microfiltration membranes, or hollow fibers. Entrapment methods are very popular for immobilization of cells. Microbial cells, as well as plant and animal cells, can be immobilized by this method. However, this method also has disadvantages, as follows:

1. High molecular weight substrates can hardly access the cells entrapped.
2. Renewal of carriers is hardly possible.

Membrane-type entrapment can avoid these disadvantages, although it has no effects on stability of cells. The lattice-type method is most widely applied for preparing immobilized cells, and there is a great deal of literature on this method. Several representative techniques for the lattice-type method are described next.

INDUSTRIAL APPLICATION

Polyacrylamide Gel Method

Bernfeld and Wan reported the entrapment of several enzymes in polyacrylamide gels in 1963, and this method was applied to the cells of the lichen *Umbilicaria pustuluta* by Mosbach and Mosbach in 1966 as reviewed by Chibata et al. Various types of biocatalysts have so far been immobilized by this method. This method was used for not only the first industrial application of immobilized microbial cells as described earlier, but also for the first application of immobilized growing cells, that is, the production of L-glutamic acid by *Corynebacterium glutamicum*. The procedure for the formation of gel to entrap cells is identical to that employed for the preparation of gel commonly used for electrophoresis. For the immobilization of cells by this method, acrylamide, and *N,N'*-methylenebisacrylamide (BIS) as the cross-linking reagent, are mixed with a suspension of cells and polymerized in the presence of an initiator, such as potassium persulfate, and a simulator, such as 3-dimethylaminopropionitrile (DMAPN) or *N,N,N',N'*-tetramethylenediamine.

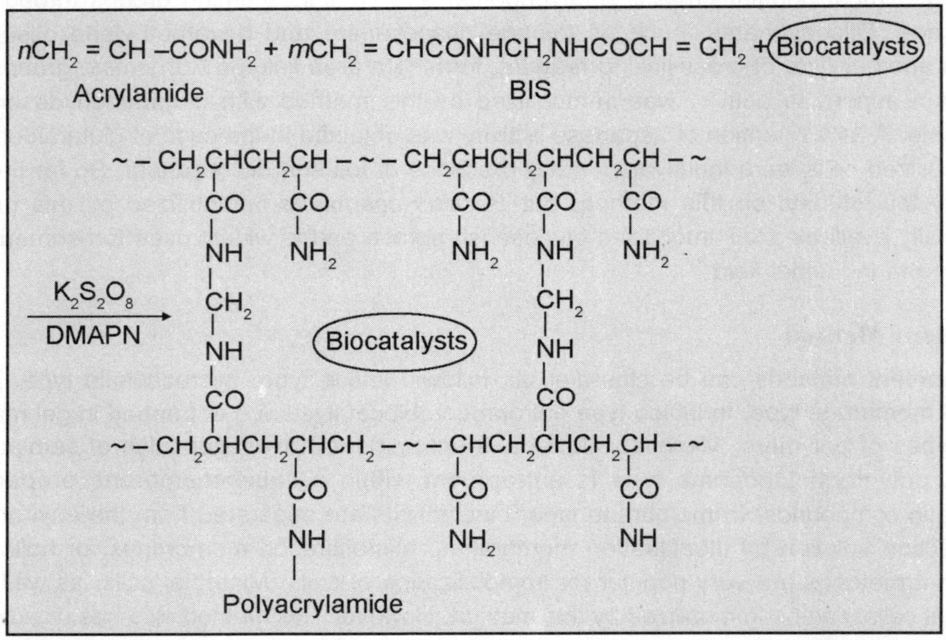

$$n\mathrm{CH_2=CH{-}CONH_2} + m\mathrm{CH_2=CHCONHCH_2NHCOCH=CH_2} + \text{Biocatalysts}$$

Acrylamide BIS

$\xrightarrow[\text{DMAPN}]{K_2S_2O_8}$

$\sim-\mathrm{CH_2CHCH_2CH}-\sim-\mathrm{CH_2CHCH_2CHCH_2CH}-\sim$

Biocatalysts

$\sim-\mathrm{CH_2CHCH_2CHCH_2}-\sim-\mathrm{CHCH_2CHCH_2CH}-\sim$

Polyacrylamide

Figure 13.1 Immobilization with polyacrylamide.

Although this method can be used for various purposes, a major disadvantage of this method is the toxicity of the acrylamide monomer, the cross-linking reagent, the initiator, and the stimulator. In some cases, cells are damaged by free radicals during polymerization. Several analogues or derivatives of acrylamide are also useful for this method.

Alginate Gel Method

Several natural polysaccharides, such as alginate, κ-carrageenan, and agar, are excellent gel materials and are widely used for the entrapment of various biocatalysts. Alginate, which is extracted from seaweed, is a linear copolymer of D-mannuronic and L-guluronic acids and can be gelled by multivalent ions such as calcium and aluminum. For immobilization of cells, sodium alginate, which is soluble in water, is mixed with a suspension of cells and then dropped into a calcium chloride solution to form water-insoluble beads of calcium alginate gels. However, calcium alginate gels are apt to disrupt by mechanical stress and are gradually solubilized in the presence of calcium-ion-trapping reagents, such as phosphate ion and citrate. Leakage of cells is then liable to occur, but free cells can be recovered from the immobilized preparation by this solubilization, which allows the characterization of the cells in gels (e.g., viability). Several multivalent metal ions, such as aluminum ion or strontium ion, can also be used instead of calcium ion.

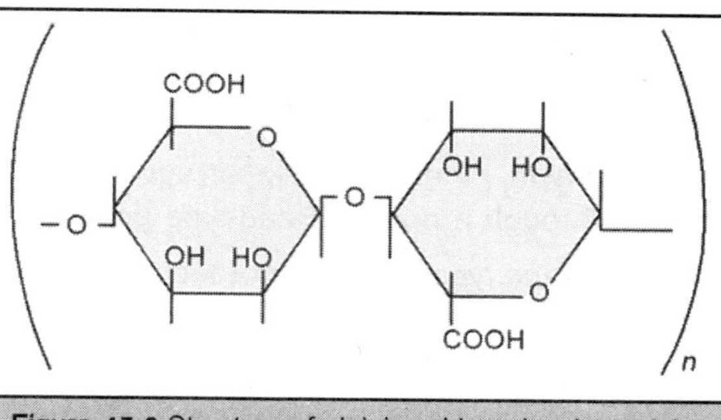

Figure 13.2 Structure of alginic acid (molecular weight, 50,000-200,000)

The treatment of formed calcium alginate gels with cationic polymers can improve gel resistivity against the solubilization by calcium-ion-trapping reagents. It is also reported that addition of polyacryl acid to sodium alginate in the entrapment by the procedure described can increase the mechanical strength of gels formed and prevent them from dissolving because of bridging polyacrylate chains with calcium ions. The method using alginate gel is widely applicable to immobilization of various useful biocatalysts, including very fragile ones such as plant cells and animal cells, because of the simplicity of the procedure, easy availability of sodium alginate, and mildness of immobilization conditions. It is worthy of note that the first report on immobilization of plant cells employed this method.

κ-Carrageenan Gel Method

κ-Carrageenan is a readily available nontoxic natural polysaccharide, isolated from seaweed. It is composed of a unit structure of β-D-galactose sulfate and 3,6-anhydro-α-D-galactose and is widely used in the food and cosmetic industries as a gelling, thickening, and stabilizing agent. κ-Carrageenan becomes a gel by cooling or in the presence of a gel-inducing reagent, such as ammonium ion, potassium ion, calcium ion, aluminum ion, magnesium ion, diamines, or water-miscible organic solvents. If the proper gel-inducing reagent is used to form a κ-carrageenan gel in the presence of cells, immobilized cells can be easily obtained. κ-Carrageenan gels can be easily solubilized in saline or in water, and a suspension of free cells can be obtained to investigate their characteristics. On the other hand, leakage of cells easily occurs by dissolution of gel when a gel-inducing reagent is not added to the reaction mixture.

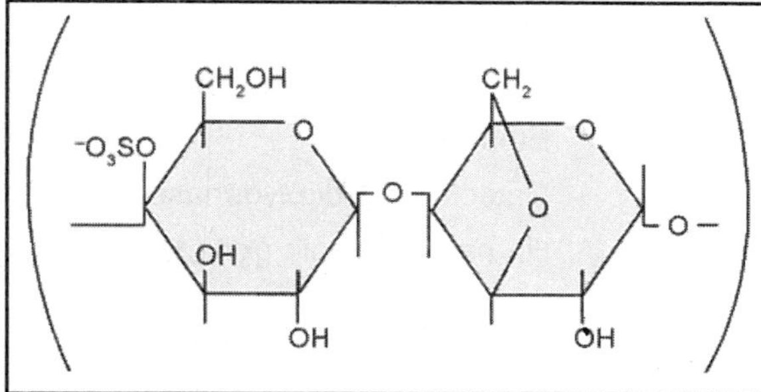

Figure 13.3 Structure of κ-carrageenan (molecular weight, 100,000-800,000).

In order to increase the stability of the κ-carrageenan-entrapped cells, treatment with glutaraldehyde and hexa-methylenediamine after entrapment is often effective. The conditions of immobilization by this method are also mild, and this method can be used for immobilization of various biocatalysts. This method has replaced the polyacrylamide gel method in industrial production of L-aspartate and L-malate by means of immobilized microbial cells. Another advantage of this method is that various shapes of immobilized cells can be prepared. The typical procedures for preparing various shapes of κ-carrageenan gel-entrapped cells are as follows:

Cubic type. A suspension of cells and κ-carrageenan dissolved in physiological saline are mixed and cooled to about 10°C to form a gel. After the gel is soaked in a cold potassium chloride solution in order to increase the gel strength, it is cut to give a cubic-type gel preparation.

Bead type. A mixture of κ-carrageenan and cells is dropped into a 0.3 M potassium chloride solution through a nozzle. Bead-type gels are obtained by this procedure.

Membrane type. A mixture of κ-carrageenan and cells is spread on a plate or sheet and soaked in a cold potassium chloride solution. Consequently, a membrane-type gel is obtained.

Table 13.1 Examples and applications of cells immobilized with κ-Carrageenan

Cells	*Products*
Bacillus subtilis	α-Amylase
Brevibacterium flavum	L-Malic acid
Corynebacterium dismutans	L-Alanine
Erwinia herbicola	L-DOPA
Escherichia coli	L-Aspartic acid
Penicillium urticae	Patulin
Propionibacterium sp.	Vitamin B_{12}
Pseudomonas dacunhae	L-Alanine
Saccharomyces bayanus	Ethanol
Serratia marcescens	L-Alginine, L-Isoleucine
Zymomonas mobilis	Ethanol
Recombinant *Escherichia coli*	Catechol-2,3-dioxygenase
Recombinant *Saccharomyces cerevisiae*	Human chorionic gonadotropin
Catharanthus roseus	Ajmalicine

Synthetic Resin Prepolymer Method

With the application of a variety of bioreactions, including synthesis, transformation, degradation, or analysis of various compounds having different chemical properties under various reaction conditions, gels having controlled physicochemical properties come to be required for entrapment of biocatalysts. Selection of suitable gel materials among natural polymers for each purpose is usually difficult. For this reason, entrapment methods using synthetic resin prepolymers, such as photo-cross-linkable resin prepolymers and urethane prepolymers, are promising. Synthetic resin pre-polymers having desired characteristics can be synthesized chemically in advance in the absence of biocatalysts. The methods are as follows:

Method adapted by Fukui and co-workers

Entrapment methods using photo-cross-linkable resin prepolymers with hydrophilic (ENT) or hydrophobic (ENTP) properties, cationic or anionic nature, and different chain lengths have been developed by Fukui and co-workers. ENT and ENTP can be prepared from hydroxyethylacrylate, isophorone diisocyanate, and poly(ethylene glycol) or poly(propylene glycol), respectively.

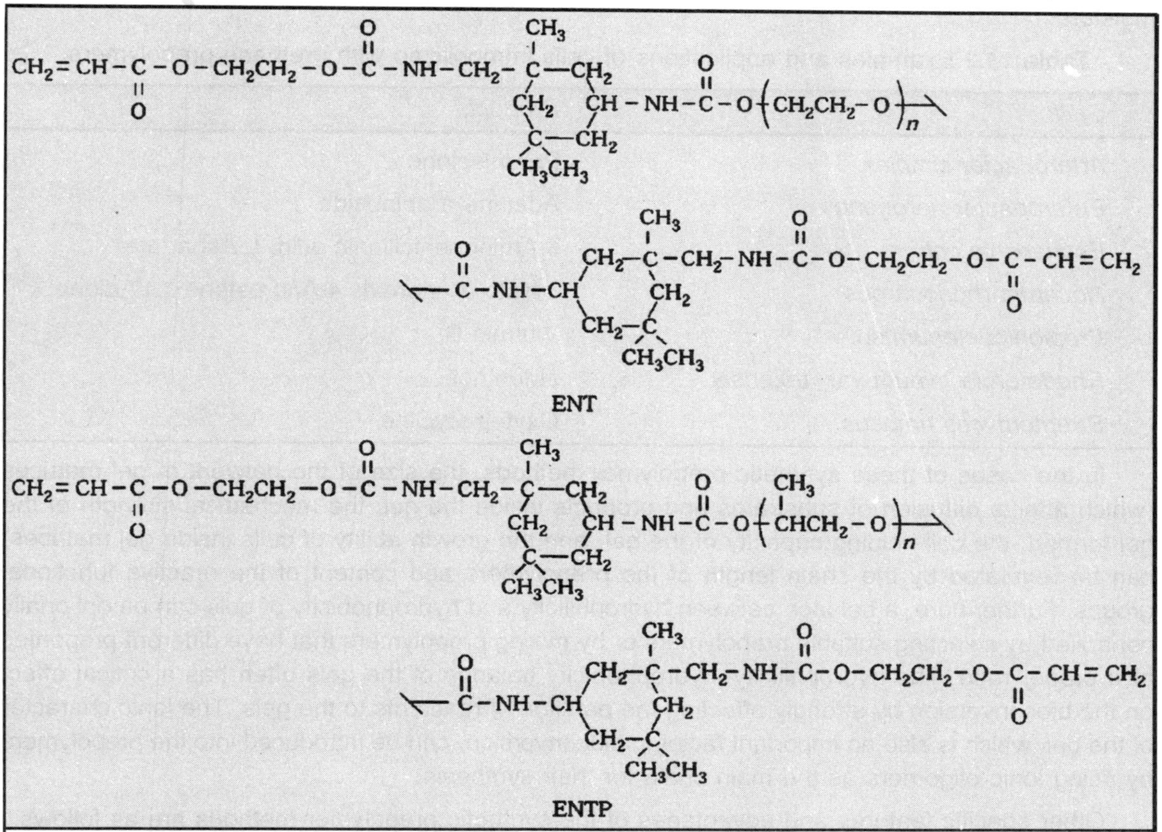

ENT

ENTP

Figure 13.4 Structure of typical photo-cross-inkable resin prepolymers.

A mixture of the prepolymers and cells, layered (about 0.5-mm thickness) on a transparent polyester sheet is gelled by irradiation with near-ultraviolet light (wavelength rage 300-400 nm) for several minutes in the presence of a proper sensitizer, such as benzoin ethyl ether. This method has been applied for the pilot-scale continuous ethanol fermentation by immobilized growing yeast cells. Photo-cross-linkable resin prepolymers can be stored in the dark and sterilized at 120°C.

Urethane prepolymer method

There are also various types of urethane prepolymers (PU), which have different hydrophilicities or hydrophobicities and chain length. Urethane prepolymers are synthesized from toluene diisocyanate and polyether diols composed of poly(ethylene glycol) and poly(propylene glycol) or poly(ethylene glycol) alone, although these are not commercially available. Isocyanate functional groups at the terminals of urethane prepolymers react with each other in the presence of water to form urea bonds and liberate carbon dioxide. Therefore, gel is easily formed only by mixing the prepolymer with an aqueous suspension of cells, and entrapment of cells can be easily carried out. Urethane prepolymers can also be stored in a desiccator and sterilized at 120°C, avoiding moisture.

Table 13.2 Examples and applications of cells immobilized with urethane prepolymers

Cells	*Products*
Arthrobacter simplex	Prednisolone
Enterobacter aerogenes	Adenine arabinoside
Escherichia coli	6-Aminopenicillanic acid, L-Aspartate
Nocardia rhodochrous	3-Keto-Δ^4-steroids 4-Androstene-3,17-dione
Propionibacterium sp.	Vitamin B_{12}
Rhodotorula minuta var. *texensis*	*l*-Menthol
Streptomyces rimosus	Oxytetracycline

In the cases of these synthetic prepolymer methods, the size of the network of gel matrices (which affects diffusion of substrates and products inside the gel, the mechanical strength of the gel formed, the cell-holding capacity of the gel, and the growth ability of cells inside gel matrices) can be regulated by the chain length of the prepolymers and content of the reactive functional groups. Furthermore, a balance between hydrophilicity and hydrophobicity of gels can be optionally controlled by selecting suitable prepolymers or by mixing prepolymers that have different properties at a proper ratio. The hydrophilicity–hydrophobicity balance of the gels often has a critical effect on the bioconversion by strongly affecting the partition of reactants to the gels. The ionic character of the gel, which is also an important factor in bioconversion, can be introduced into the prepolymers by using ionic oligomers as the main chain for their synthesis.

Other specific features and advantages of the synthetic prepolymer methods are as follows.

1. Procedures for entrapment are very simple and proceed under very mild conditions, and various types of biocatalysts can be immobilized by these methods.
2. Prepolymers do not contain monomers, which may have unfavorable effects on biocatalysts, such as cells. Therefore, the synthetic prepolymer method seems to be one of the most practically applicable immobilization methods for a wide variety of purposes.

14

EXPRESSION SYSTEMS

Since the advent of the recombinant DNA technology, a wide variety of expression systems for protein production have become available. These include systems based on bacteria, lower eukaryotes (yeast and fungi), invertebrates (cells or larvae), vertebrates (cells or whole animal), and plants (cells or whole plant). Of these, only the yeast and fungal expression systems have many of the attributes of both the bacterial and higher eucaryotic expression systems. Yeasts have the robust growth characteristics and ease of manipulation of a bacterial system and at the same time perform many post-translational modifications found in higher eukaryotes. The purpose of this article is to highlight the strategies one can apply to optimize protein expression in the *Pichia* yeast expression system.

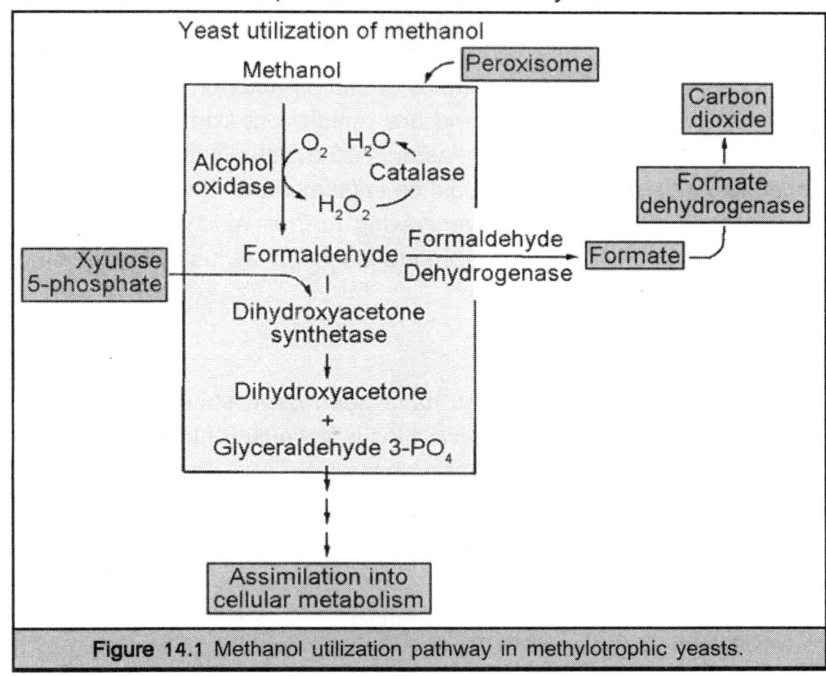

Figure 14.1 Methanol utilization pathway in methylotrophic yeasts.

WORK PROGRESS

Pichia pastoris is a methylotrophic yeast. It is able to use methanol for energy as well as for growth. It possesses a highly regulated pathway for the utilization of methanol. Synthesis of the enzymes of methanol metabolism increase rapidly once the cells are placed in the methanol medium. The most dramatic effect is seen for methanol oxidase, which accounts for 35% of the cellular protein in cells adapted to grow on methanol. An extensive proliferation of the peroxisomes, known to sequester methanol oxidase and dihydroxyacetone synthase, is also observed in methanol grown cells. *Pichia* was initially investigated at Phillips Petroleum Company as a potential source of single cell protein. Thus, a very efficient fermentation process (cell density > 130 g dry cell weight per L and biomass productivity > 10 g/L-h) was developed. Though impressive, this process could not compete with the economics of production of soy protein. After this setback, Phillips Petroleum Company directed its efforts in developing *Pichia* as an expression system for the production of recombinant proteins. This has turned out to be a worthwhile endeavor.

PICHIA EXPRESSION VECTORS

Typical *Pichia* expression vectors, for example, pHIL-A1, pHIL-D2, pHIL-D7 and pPIC9 are based on the strong methanol oxidase gene (*AOX1*) promoter. A wide variety of proteins have been produced in this system. The final yield of a protein expressed in *Pichia* is largely influenced by its inherent properties. Nevertheless, the production yield of a protein can be significantly enhanced by using a strategy that takes into account the multiple factors that influence protein expression.

Autonomous Replication

The expression cassette can be introduced into *Pichia* cells by way of chromosomal integration or autonomous replication. However, the integrative approach is preferable because it has the following advantages: (i) expression cassette stability; (ii) ability to generate clones that contain multiple copies of the expression cassette; (iii) control over the sites of integration (*HiS4* or *AOX1* loci); and (iv) ability to engineer different modes of integration (with or without the eviction of the *AOX1* coding sequences). Plasmids based on autonomous vectors such as pHIL-A1 are maintained in *Pichia* cells at low copies and are rapidly lost from the population of dividing cells. In a small number of cells, autonomous plasmids do eventually integrate. Thus, to arrive at a stable clone, a secondary selection screen should be implemented. This makes autonomous plasmids less attractive for protein expression. However, owing to their ability to transform *P. pastoris* spheroplasts at a very high frequency, the autonomous plasmids are useful for cloning genes in *Pichia* by functional complementation.

Integration

Both *AOX1* and *HIS4* sites have been used successfully for expression. Occasional loss of the expression cassette at the *HIS4* locus because of intra-chromosomal crossover between mutant *his4* and the good *HIS4* has been observed. Thus, the *AOX1* locus is the preferred site for stable expression.

Transformation

Transformation of a *P. pastoris his4* strain using linear DNA with the ends bearing homology to the 5′ and 3′ regions of the *AOX1* chromosomal locus (e.g., *Bgl*II digested pHIL-D7, or pPIC9 or *NotI*-digested pHIL-D2) results in the site-specific eviction of the *AOX1* structural gene. The eviction of the *AOX1* occurs at a frequency of 1 to 5 per 20 His^+ transformants. *AOX1*-deleted clones show a slower growth phenotype (Mut^-) on minimal methanol medium, as compared to *AOX1* intact clones, which have phenotypically normal growth characteristics (Mut^+) on minimal methanol medium.

The Mut^+ transformants arise presumably because of circularization of the linear DNA inside the yeast cell and subsequent integration into one or more of the *AOX1* or *HIS4* chromosomal loci. Thus, in a single transformation experiment, both Mut^+ and Mut^- transformants can be obtained. For intracellular expression, it is preferable to use Mut^- cells because they will have a lower level of alcohol oxidase protein and the expressed protein can be more readily purified. For secretion, either one of Mut^+ or Mut constructs can be used. There is no significant difference between these two types of host cells in the secretion of human serum albumin.

Clonal Variation

In many instances of *Pichia* expression of heterologous proteins, namely, expression of LACZ, hepatitis B surface antigen, invertase, and human serum albumin (HSA), a single copy of the expression cassette was sufficient for optimal production. Deliberately increasing the copy number had no significant effect on production in these instances. Yet, in other cases, multiple copies of the expression cassette (> 10) are a must for high-level expression. A dramatic effect of copy number on protein production in *P. pastoris* is seen in the production of human tumor necrosis factor (TNF), tetanus toxin fragment C, *Bordetella pertussus* pertactin P69, and mouse epidermal growth factor (EGF). In some rare instances, an increase in copy number has a negative effect on the production level. Because the effect of gene copy number on expression is unpredictable, it is necessary to examine the production level as a function of gene dosage.

The spheroplast method of transformation of *P. pastoris* results in transformants with a wide range of copy numbers. Analysis of as few as 100 individual clones for protein production is generally adequate to arrive at a good producer. If other methods of transformation (e.g., LiCl method or electroporation) that do not give rise to multicopy transformants at a high frequency are used, then more efficient screens can be used, such as colony hybridization (or dot blot analysis) with DNA probes or selection for multicopy integrants based on increased level of resistance to antibiotic G418 or Zeocin. Alternatively, multicopy constructs of the expression cassette can be obtained by transformation with DNA concatamers or by using multicopy construction vectors such as pAO815. This is not a method of choice because it permits production analysis over only a narrow range of gene dosage.

Noncoding Region

The nucleotide (nt) sequence and the length of the 5′ untranslated region (5′ UTR) can be detrimental to optimal protein translation. The leader length of the highly expressed *AOX1* mRNA is 114 nt long, and the sequence is A + U rich. For optimal synthesis of heterologous proteins, it is essential that the 5′ UTR of the expression cassette should closely resemble that of the *AOX1* mRNA. Ideally, it is preferable to make it identical to that of the *AOX1* mRNA. The expression level of HSA is increased more than 50-fold by adjusting the 5′ UTR to be closer to that of the *AOX1* mRNA. Expression plasmid such as pHIL-D7 can be used to readily make 5′ exact constructs.

Coding Region

The translation initiation codon AUG should be avoided in the 5′ UTR to ensure efficient translation of mRNA from the actual translation start site. The mRNA secondary structure around the initiator AUG may be adjusted, so that AUG is relatively free of secondary structure. This can be accomplished by redesigning the initial portion of the coding sequences with alternate codons.

Transcription

Genes with high A + T nucleotide clusters are inefficiently transcribed because of premature termination. One such sequence, ATTATTTTATAAA, present in HIV-gp120 has been identified to block transcription in *P. pastoris*, and when this stretch is altered to TTTCTTCTACAAG, the premature termination is abolished. Because there are many yet unidentified AT-rich stretches that act as transcriptional terminators, a general strategy would be to redesign the genes using *P. pastoris* preferred codons so as to have an A + T content in the range of 30 to 55%. By using this approach, it has been possible to construct strains for efficient production of tetanus toxin fragment C and *Bacillus sphaericus* mosquitocidal toxins (BSP1 and BSP2).

Secretory Proteins

If a protein can be secreted, it is the desired mode of protein production because of the ease of product recovery. Furthermore, certain normally secreted proteins, such as HSA and growth hormones, remain insoluble when expressed intracellularly. For a protein that is not normally secreted, it may be difficult to cause secretion. However, there are some encouraging results with respect to yeast secretion of cytochrome P-450 and rat liver elongation factor that suggest that at least in some instances it may be possible to secrete a normally intracellular protein. The choice of secretion signal is rather arbitrary. In several cases, (HSA, invertase, bovine lysozyme, etc.), the native signal sequence is adequate. We have found that the native secretion signals present in matrix metalloproteinases 1,2,3,9 (MMP-1,2,3,9) and tissue inhibitor of matrix metalloproteinases 1 (TIMP-1) are functional in the *P. pastoris* system.

If a native secretion signal sequence is not available, then a signal sequence based on the *S. cerevisiae* invertase (Suc2p), the *S. cerevisisae* α-mating factor (αMF), or the *P. pastoris* acid phosphatase (Pho1p) can be used. The pre-pro αMF signal works very efficiently, especially for the secretion of a large variety of proteins, including the smaller-sized products such as aprotinin, EGF, thrombomodulin fragment, blood factor XII, a fragment of amyloid β-protein, antibody single-chain Fv fragment, and ghilanten. In making protein fusions to the αMF signal, it is prudent to retain the Glu-Ala spacers adjacent to the Kex2-like protease cleavage site: (... V-S-S-L-E-Lys-Arg-^{vKex2p}Glu-Ala-^{vDapp}Glu-Ala-^{vDapp}fused protein).

The presence of the Glu-Ala spacers help to alleviate the steric interference by the fused protein, resulting in an efficient cleavage of the pro sequence by the *P. pastoris* Kex2-like protease. The Glu-Ala spacer sequence is subsequently cleared by a diamino peptidase (Dapp) to yield the protein of interest free of additional N-terminal amino acid residues. In spite of taking this precaution, we have noticed that the pro sequence is not cleaved when the *a*MF signal is fused with human brain-derived neurotrophic factor (BDNF) (mol. wt. ~14 kDa).

In this situation, BDNF is secreted into medium as a 26 to 30 kDa product, which is the expected result if the pro portion of the αMF signal is not cleaved and is heterogeneously glycosylated at the multiple glycosylation sites present in the αMF pro peptide. Acid phosphatase (Pho1p) secretion signal sequence, invertase secretion signal, and a hybrid sequence consisting of Pho1p secretion signal containing a Kex2 protease recognition sequence has also been used successfully for high level (>g/L) secretion of certain proteins.

Secretion Signals

In most of the examples, the relative efficacies of the different secretion signals for a given product have not been examined. Only in a limited number of cases has a rough comparison been made between native signal versus pre-pro αMF. For example, in the case of invertase secretion, both the extent of glycosylation and secretion kinetics are enhanced when the invertase signal sequence is substituted with the pre-pro αMF sequence (secretion $t_{1/2}$ of 10 min for pre-pro αMF signal, compared to 90 min for the native invertase signal). Thus, in this case, one would avoid using αMF signal if hyperglycosylation is not desired. However, if higher productivity (product yield per liter per hour) is desired, then it is preferable to use the pre-pro αMF signal. The secretion kinetics can also impact on the recovery of the intact protein, if the expressed protein is susceptible to proteolysis in the *Pichia* broth. For example, the yield of full length MMP-1 from the culture medium is significantly improved by using the αMF signal instead of the native signal.

Media Manipulations

In some instances, a secreted protein is rapidly degraded in the *Pichia* broth. The stability of a protein in the *Pichia* broth can be improved by manipulating the pH of the induction medium. The recommended pH range for experimentation is between 2.8 to 8.0. The pH control is best maintained in a fermenter, although it is possible to have a reasonable pH control even in shake flask cultures with the use of appropriate buffers (phosphate buffer for pH 5.7 to 8.0; alanine-HCl buffer for pH 2.5 to 3.6; citrate buffer for pH 3.0 to 6.0).

The product stability is further enhanced by addition of casamino acids (1 to 4%) or yeast extract (1%) plus peptone (2%) to the medium. For example, secretion of HSA was significantly improved by raising the pH of the medium from 5.2 to 6.0, and the yield was further enhanced by the addition of yeast extract and peptone. Production of mouse EGF was favored at pH 3.0 in the presence of casamino acids. Casamino acids are preferable to yeast extract plus peptone, because the peptide components of peptone (such as bovine collagen fragments) can interfere in product analysis and recovery. Addition of 5 mM EDTA to the medium also improves product stability. It should be noted that media manipulation can significantly alter the profile of protein components, so that previously unnoticed proteins can become evident. The susceptibility of a protein to proteolysis in the *Pichia* culture medium can be determined beforehand if sufficient quantity of protein is available and proper adjustments can be made to the culture medium.

Table 14.1 Partial list of heterologous proteins expressed in *P. pastoris*

Intracellular expression
E. coli β-galactosidase (LACZ)
Hepatitis B surface antigen (HBsAg)
Human tumor necrosis factor-α (TNF-α)
Human TNF-α analogues
Salmon growth hormone (SGH)
Streptokinase
Tetanus toxin fragment C
Bordetella pertussus pertactin P69
Human inteleukin 2
B. sphaericus mosquitocidal components, BSP1, BSP2, and BSP1 + BSP2
Superoxide dismutase (SOD)
Human serum albumin (HSA)
HIV gp120
Porcine leukocyte lipoxygenase
$p110^{RB}$ retinoblastoma tumor suppressor protein 919 Hamster prion protein
Spinach phosphoribulokinase
Spinach cytochrome C oxidase
Spinach glycolate oxidase
Dogfish shark cytochrome P450c17 (hydroxylase and lyase activities)
Algal phytochrome
Green fluorescent protein (GFP)
Arabidopsis chaperone protein ATJ2
Corn cytochrome *b* and cytochrome *c* reductases
Mouse thioether methyltransferase
Polyomavirus large T antigen
Rat liver mitochondrial carnitine palmitoyltransferase I and II
Mouse $5HT_{5A}$ serotonin receptor
Secretion using native secretion signal sequence
S. cerevisiae invertase (*SUC2*)
Human serum albumin (HSA)
Bovine lysozyme
Matrix metalloproteinases MMP-1,2,3, and 9
Tissue inhibitor of matrix metalloproteinases-1 (TIMP-1)
MMP-1,2,3, or 9 coexpressed with TIMP-1
Barley α-amylases 1 and 2
Cathepsin E

D-alanine carboxypeptidase (dacA) of *Bacillus stearothermophilus*

Pectate lyase pel B from *Fusarium solani f.* sp. *pisi*

IgE receptor (α subunit)

Secretion using pre–pro α matingfactor (αMF) secretion signal sequence

Aprotinin

S. cerevisiae invertase (*SUC2*)

Human epidermal growth factor (EGF)

Mouse epidermal growth factor (EGF)

Insulin-like growth factor-1 (IGF-1)

Kunitz protease inhibitor (nexin-2)

Human vascular endothelial factor (VEGEF)

κ-Bungaratoxin

Cathepsin L pro-peptide

Human fibroblast collagenase, matrix metalloproteinase-1 (MMP-1)

N-Lobe of human serum transferin

C. albican hyphal surface protein HWP1

Coffee bean α-galactosidase

Major histocompatibility (MHC) class II human leukocyte antigens DRα-Fos and DRβ-Jun

Human immunodeficiency virus envelope glycoprotein gp120

Ghilanten

Human factor XII

Cellulose binding domain (CBD)–factor X variant fusion protein

Cellulose binding domain (CBD)–steel factor fusion protein

Human gelatinase B (matrix metalloporiteinase-9, MMP-9) fragment

Measles viral glycoprotein

Rat liver elongation factor-2 (eEF-2)

Human growth hormone

Drosophila notch protein EGF-like domains

Type II fish antifreeze protein

Follicle stimulating hormone, FSH α and β chains

Secretion using acid phosphate secretion signal sequence

Human vascular endothelial factor (VEGEF)

Pectate lyase pelC from *Fusarium solani* f. sp. *pisi*

Catalytic domain of 92-kDa gelatinase (matrix metalloproteinase-9, MMP-9)

Angiotensin-converting enzyme (ACE) and its mutants

Rabbit plasma cholesteryl ester transfer protein (CETP)

β-Cryptogein elicitor protein secreted by *Phytophthora cryptogea*

Candida albicans aspartic proteinases

Angiostatin (an integral fragment of plasminogen)

Rabbit reticulocyte 15-lipoxygenase

A single chain antibody

Human monocyte chemotactic protein-3 (MCP-3)

Secretion using a hybrid of acid phosphatase signal sequence with a synthetic pro sequence and a KEX2 protease cleavage site

Tick anticoagulant peptide (TAP)

Secretion using invertase (SUC2) secretion signal sequence

A single-chain antibody

Porcine tumor necrosis factor-α (TNF-α)

Boophilus microplus Bm 86 antigen

Improvement

In addition to the media optimizations described in the previous section, the product yield can be further improved by using a protease-deficient *Pichia* strain. Two generally applicable techniques have been used to generate protease deficient strains. In one approach, a specific *Pichia* protease is knocked out by site-specific gene disruption, and in the other approach, a *Pichia* strain is engineered to express a specific protease inhibitor. Both types of protease-deficient strains have been found to be superior for expression of certain products. Even while using the protease deficient strains, it is essential to optimize the culture medium as discussed in the previous section. By using the gene disruption approach, three protease-deficient strains, SMD1168 (*his4, pep4*), SMD1165 (*his4, prb1*), and SMD1 163 (*his4, pep4, prb1*), that are defective in one or more of protease A (PEP4) or protease B (PRB1) have been constructed. Because a functional PEP4p is necessary for the maturation of PRB1p, the use of the single knockout strain SMD1168 alone should be adequate.

The production yield of insulin-like growth factor-1 and ghilanten are improved twofold to threefold by using SMD1168 as compared to the strain GS115, which has intact PEP4p. Co-expression of MMP-9 along with its inhibitor, TIMP-1, significantly improves the expression level of MMP-9. Furthermore, TIMP-1 expression results in overall stabilization of several *Pichia* secreted proteins. Thus, strains engineered to express protease inhibitors such as aprotinin and TIMP-1 should be useful for high-level expression of proteins that are susceptible to rapid degradation.

Coexpression

In many instances it is required or desirable to express two or more heterologous proteins at the same time. The fact that *Pichia* can stably integrate multiple copies of an expression cassette implies that it should be possible, at least in principle, to express numerous heterologous proteins at the same time in *Pichia*. The first example of simultaneous expression of two proteins is that of the construction of *Pichia* strains expressing high levels of BSP1 (43 kDa) and BSP2 (52 kDa) components of *Bacillus sphaericus* mosquitocidal toxin. Since then, there have been few other reports in literature for co-expression of two proteins in *Pichia*. Two generally applicable approaches have been used for coexpression. In one approach, *Pichia* is cotransformed with two expression cassettes, the resulting His+ transformants are screened for the presence of both the expression cassettes by PCR and subsequently examined for expression. In another approach, a *Pichia* expression plasmid is engineered with two expression cassettes so that the DNA used for transformation has both the expression cassettes present on the same fragment. In this situation, every His+ transformant is expected to contain both the expression cassettes. Plasmids such as pAO815 or pAO856 can also be used to readily construct expression plasmids for simultaneous expression of two or more products. One can also use sequential transformation to construct *Pichia* strains that coexpress two or more different proteins, because numerous transformation markers, namely His+, Arg+, Suc+, G418^R, and Zeocin^R markers are available for *Pichia*.

Protein Assembly

Several membrane receptors, such as major histocompatibility complex (MHC) class I molecules, MHC class II molecules, T-cell receptors, are multiprotein complexes. These receptors are comprised of α- and β-chains, each of which have a large NH_2-terminal extracellular portion, a trans-membrane region, and a short C-terminal cytoplasmic tail. Efforts to express soluble $\alpha\beta$-heterodimers of MHC class II molecules by coexpressing just the extracellular portions of α and β chains have proved impossible. The reason for this is that the transmembrane region of MHC class II α- and β-chains facilitate the proper assembly of $\alpha\beta$ heterodimer, and in their absence, assembly is hampered. Interestingly, this problem has been overcome by substituting the transmembrane regions of α- and β-chains with the leucine zipper dimerization motifs from the transcription factors Fos and Jun, which are known to assemble as stable, soluble heterodimers. The leucine zippers are characterized by five leucines that are spaced periodically at every seventh residue (heptad repeat); each heptad repeat contributes two turns of the α-helix (3.5 residues/turn). The leucine residues play a special role in dimerization, and they form the interface between the two α-helices in the coiled coil. The Fos-Jun dimer is soluble because of the charged amino acid residues on the outer surface of the coiled coil. The Fos/Jun leucine zipper dimerization domains may have general applicability in the expression of extracellular domains of multiprotein membrane receptors.

Repression

In some instances, the product being expressed may be toxic to *Pichia*. In such cases, a stepwise induction is suggested. First, the cell mass is built up using glucose as sole carbon source, taking care to see that glucose is not limiting. Under these conditions, very little product is made because of strong repression of the *AOX1* promoter by glucose. Subsequent to this, the product production phase is initiated with methanol feed. Whether a particular product is toxic to *P. pastoris* can be evaluated by comparing the growth of transformants in 1% sorbitol (or 100 mM alanine) media to that on sorbitol plus methanol or alanine plus methanol media. If the expressed protein is toxic, then the growth in the presence of methanol will be drastically impaired.

Induction of Protein Expression

Expression vectors for heterologous expression in *P. pastoris* are typically derived from *AOX1* regulatory sequences. One or more of the following induction strategies can be used for optimal protein expression.

Regular induction

The *AOX1* promoter can be maximally activated (derepression or induction) by growing cells on methanol as the sole source of carbon and energy in shake flasks, shake tubes, or plates at 30°C. The doubling time on methanol is ~6 h as compared to <2 h on glucose or glycerol medium. Even the cells in which the *AOX1* gene has been evicted (Mut^-) can grow on methanol. However, the growth rate is slow (doubling time, 18 to 24 h) because it is solely dependent on a secondary alcohol oxidase, *AOX2*, which is expressed only at low levels. In such cases the growth rate can be enhanced by supplementing the minimal methanol medium with 1% sorbitol or 100 mM alanine. The advantage of using sorbitol or alanine lies in the fact that these compounds do not interfere with the induction of *AOX1* promoter. Induction of expression on plates is carried out by streaking or plating cells on a methanol medium (MM) plate.

Gradual induction

For liquid cultures, cells are initially grown to an OD_{600nm} of 2 to 10 on a nonactivating carbon source such as glycerol or glucose (minimal glycerol yeast extract medium [MGY] or minimal dextrose medium [MD]) in shake flasks or tubes at 30°C. The cells are then harvested by centrifugation and shifted to MM and incubated with shaking at 30°C for 1 to 4 days. Induction of expression on plates is carried out by streaking or plating cells on a nitrocellulose or cellulose acetate membrane placed on an MD or MGY plate. Once colonies appear, the filter is lifted and placed on an MM plate.

Accumulation of proteins

The conditions described thus far work well for the intracellular accumulation of heterologous proteins, but are rather inadequate for secreted proteins. The following protocols work well for secreted proteins for both Mut^+ and Mut^- cells.

Shake tube cultures

First cells are grown to saturation in 10 mL buffered minimal glycerol yeast extract medium (BMGY) placed in a 50 mL tube (2 to 3 days). The OD_{600nm} of culture will be in the range of 10 to 20. The cells are harvested by centrifugation, and the cell pellet is resuspended with 2 mL of buffered minimal methanol–yeast extract medium (BMMY). The tube is covered with a sterile gauze (four layers of cheesecloth; USP type VII gauze, 20 ×12 mesh from Ultimed International Inc., Glendale Heights, Ill.) and incubated in the shaker for 4 days. Aliquots of supernatant can be analyzed for product secretion at desired time intervals. Fresh media should be added to account for loss of media caused by evaporation.

Shake flask cultures

Cells are grown in 1 L of BMGY in a 2-L flask. Cells are harvested and resuspended in 50 mL of BMMY in a 300- to 400-mL low baffle flask, covered with four layers of cheesecloth as described in the previous section, and incubated in a 30°C shaker for 2 to 4 days. The media supernatant is analyzed for product at different time intervals. Fresh BMMY media is added as necessary to prevent loss of liquid caused by evaporation.

Patched cultures

Cells are patched or plated on nitrocellulose filter (sterile), placed on a square or circular BMGY plate, and incubated at 30°C for 2 to 3 days. Once colonies have grown to 2 mm in size, the filter is removed and placed over a BMMY plate. The filter containing colonies is covered with a sterile nitrocellulose filter. If desired, a cellulose acetate filter can be interfaced between the two nitrocellulose filters and incubated at 30°C for 2 to 3 days. The secreted protein on the nitrocellulose filter can be analyzed by ELISA or another suitable procedure.

HETEROLOGOUS PROTEINS

The general methods for production of biomass as well as for the production of heterologous proteins using Mut^+ and Mut^- cells in both continuous and batch modes of fermentation are described here. The process can be scaled up (>1,000 L) or scaled down (0.2 L) as desired. Fermentation can be conducted over a wide pH range (3.0 to 6.5) at 30°C. For example, HSA secretion yield is improved more than threefold by using pH 5.85 compared to the generally used pH of 5.0. In the case of secretion of the V1 domain of CD4 (amino acid residues number 1 to number 106 of mature CD4), intact product was seen only at acidic pH (2.5 to 3.5). As high as a twofold to fourfold increase in secreted yields of human EGF and human insulin-like growth factor-1 are observed at pH 3.0.

Apparatus Required

A typical bench-top fermenter has a 2- to 20-L capacity, a 1-to 10-L operation volume, and monitors and controls for pH, dissolved oxygen (DO), agitator speed, temperature, air flow, pressure, foam, and weight. Sixfors, offered by Infors, UK Ltd., can bridge the gap between shake flasks and fermenters. They can be used for small-scale (typically 0.3 L) continuous, batch, or cascade processes where the growth from one vessel is passed into another.

Procedure

Bio-Rad's portable fermentation monitoring analyzer can be used on the spot to provide fast HPLC analysis of the concentration of methanol, glycerol, ethanol, glucose, acetic acid, lactic acid, fructose, maltotriose, and maltose. The equipment is compatible with automatic sampling and computerized data analysis. Analysis takes only 10 min. The conditions used are as follows:

1. *Instrument*: Bio-Rad's fermentation monitoring analyzer
2. *Column*: Fermentation monitoring column (150 x 7.8 mm)
3. *Sample*: Extracellular broth, prefiltered through a 0.22-μm syringe filter; sample volume: 10 to 20 μL
4. *Injector*: Rheodyne injector valve with a 10-μL injection loop
5. *Solvent*: 0.002 N H_2SO_4 in water at a flow rate of 0.8 mL/min (isocratic, so only one pump is required)
6. *Temperature*: 65°C
7. *Detector*: Refractive index detector

To determine the wet cell weight of the culture, 1 mL of the fermenter broth is centrifuged in a microfuge, and the cell pellet is weighed after carefully decanting the cell-free supernantant. To determine the washed dry cell weight of the culture, cells are harvested by centrifugation (3000 x *g* for 10 min at room temperature) from a known volume of the fermenter culture (typically 10 to 50 mL). The cell pellet is washed twice with water (10 to 50 mL), dried overnight at 100°C, and weighed. Alternatively, the dry cell weight can be readily determined in less than 10 min with the use of a moisture balance such as the Mettler LP16 Infrared Dryer. Mass transfer is determined by measuring the air flow to the fermenter and the composition of both the inlet and outlet air with a Perkin-Elmer gas analyzer. The O_2 transfer rate (OTR) is calculated as follows:

$$OTR = f(C_{in} - C_{out})/V$$

where f is flow rate L/h, C_{in} and C_{out} are the concentration in mmol of O_2 in the inlet and outlet gases, and V is the ungassed broth volume. OTR can also be established from yields using the equation

$$OTR = \mu X/Y \times O_2$$

where μ is the specific growth rate or dilution rate (h^{-1}), X is the cell density (g/L), and $Y \times O_2$ is the yield on O_2 (g cells/mmol O_2).

Heat transfer rate and heat load can be determined by measuring the temperature of the circulating water in and out of the fermenter heat exchanger. Productivity (P) for the continuous culture is determined using $P = DX$, where D is the dilution rate (h^{-1}), and X is the cell density. An increase in D and/or X will result in higher productivity. Because D also equals the specific growth rate μ, the dilution rate is limited by the microorganism's intrinsic characteristics. Once μ_{max} or D_{max} is reached, P can only be increased by increasing the cell density.

Table 14.2 Parameters determined for a high cell-density process in a 1,500-L fermenter

Substrate	*Methanol*
Substrate concentration (g/L)	263
Dilution rate (h^{-1})	0.11
pH	3.5
Temperature (°C)	30
Cell mass (g of cells/L)	105
Productivity (g/L h)	11.6
Yield (g of cells/g of methanol)	0.4
OTR (mmol O_2/Lh)	880
Oxygen consumption (g O_2/g cells)	2.42
Heat release	
kcal/Lh	109
kcal/mol O_2	123.9

Extracellular protein concentration (for secreted proteins) can be estimated by using a Lowry-type protein analysis on TCA-precipitated material and analyzed for specific product by SDS-PAGE and immunological methods. Aliquots of cells are lysed to prepare cell extracts for protein analysis.

Regular Culture

Fermentation is carried out in two steps, first in the batch mode on glycerol as the carbon source, followed by continuous mode on methanol-containing medium. Inoculum for the fermenter is prepared by growing cells in 1 L of MMD, YMPD, YMPGy, MGyB, MMGY, or YPD grown to an OD_{600nm} of 2 to 10 in a 2-L shake flask. This volume of inoculum is used for inoculating a 20-L fermenter containing 9 L of basal salt FM21 medium containing 5% glycerol (higher levels may be toxic to the cells) or 5 to 10% glucose and pH adjusted to 5.0 (by using 50% NH_4OH solution or NH_3 gas). Biotin stock (4 mL) and PTM1 trace mineral mix (11 mL) are also added. Fermentation is conducted until the glycerol or glucose (carbon source) is exhausted. During the run, dissolved O_2 is maintained at >20%, with the agitator speed set between 500 to 1,500 rpm and a vessel pressure of 2 to 3 psi.

Foaming is controlled through the addition of a 5% Struktol J673 or Mazu DF 37C. The cell yield expected for the batch phase on 5% glycerol under these conditions is 20 to 25 g of washed dry cell weight per liter. Continuous fermentation is established by feeding FM21-methanol (15% v/v) for Mut^+ cells or FM21-methanol (1% v/v) plus 15% glycerol, sorbitol, or alanine for Mut^- cells. The feed for continuous culture is also supplemented with PTM1 (1.1 mL of stock solution per liter) and biotin (0.4 mL of the stock solution per liter). Feed sterilization is carried out by filtration (Pall Ultipor disposable filter assembly DFA 4001 AR, 0.2 μm). Feed addition is achieved with a Milton-Roy positive-displacement metering pump. Continuous culture is performed as a chemostat under steady-state conditions where the dilution rate (*D*) is equal to the growth rate of the population. The growth rate of cells can be controlled by adjusting the flow rate of fresh medium into the fermenter. Typical *D* values range from 0.056 h to 0.11 h.

Dissolved oxygen is held in the 45 to 75% air saturation range by varying air flow, vessel pressure, or agitator speed. The maximum cell mass achievable under these conditions is around 80 g of washed dry cell weight per liter with a cell productivity of approximately 10 g/L h. If a higher cell mass is desired, the amount of carbon source in the feed can be raised (e.g., 25% methanol for Mut^+ and 1% methanol plus 25% glycerol, sorbitol, or alanine for Mut). This would also necessitate a proportional increase in the amount of minerals, trace elements, and biotin. As high as 110 g/L washed dry cell weight with 11.6 g/L h. productivity are achievable in this process. Highest induction of the *AOX1* promoter occurs using methanol (Mut^+ cells) or methanol plus sorbitol or alanine (Mut^-) as the carbon source in the feed. Intermediate levels of induction are seen with methanol plus glycerol feed.

Stepwise Culture

Fermentation is carried out in three steps: glycerol batch phase, glycerol fed-batch phase, and methanol-fed batch phase. One liter of inoculum is used for inoculating a 20-L fermenter containing 5 L of basal salt BSM medium with 5% v/v glycerol (higher levels may be toxic to the cells) or 5 to

10% glucose as carbon source; the pH of which is adjusted to 5.0 by using 50% NH_4OH solution or NH_3 gas. Biotin stock (40 mL) and PTM1 trace mineral mix (40 mL) are also added.

Fermentation is conducted as previously described until all the carbon source (glucose or glycerol) is completely consumed. This phase should take 18 to 24 h. After this, a fed-batch phase on glycerol is initiated with a 50% w/v glycerol feed (500 mL of 100% glycerol + 480 mL water + 10 mL each of PTM1 and biotin stock solution) at a feed rate of 18 mL/h/L initial fermentation volume with the aid of a peristaltic pump. Glycerol feeding is carried out for 4 h. The cell yield at this point will be in the range of 180 to 220 g/L of wet cells (equivalent to approximately 30 to 55 g washed dry cell weight per liter). This phase can be manipulated by varying the concentration of glycerol or the duration of fermentation to achieve optimal heterologous protein production in the methanol-fed batch phase.

Table 14.3 Cell yield ranges

Mut^- intracellular expression	50-100 g/L dry cell weight or 200-400 g/L wet cell weight
Mut^+ intracellular expression	30-80 g/L dry cell weight or 140-320 g/L wet cell weight
Mut^- secretion	20-40 g/L dry cell weigth or 80-160 g/L wet cell weight
Mut^+ secretion	12-80 g/L dry cell weight or 50-300 g/L wet cell weight

The fed-batch phase on methanol is initiated with a methanol feed (980 mL of 100% methanol + 10 mL of PTM1 + 10 mL of biotin stock solution) at a rate of 1 to 4 mL/L/h of the initial culture volume. The methanol-feed flow rate is adjusted such that with Mut^- strains, the methanol concentration in the fermenter is maintained in the 0.2 to 0.5% v/v level, whereas with the Mut^+ strains, methanol levels in the fermenter approach zero and the fermenter is run in a methanol-limited fashion.

Dissolved oxygen in both cases is maintained at the 20 to 70% range. If dissolved oxygen falls below 20%, the methanol feed is stopped, and nothing should be done to increase oxygen rates until an upward spike in the dissolved oxygen level is seen. At this point, adjustments (rpm, aeration, vessel pressure, oxygen feed) can be made. Maintaining the dissolved oxygen above 20% may be difficult depending on the OTR of the fermenter. With stainless steel vessels, the system can be pressurized up to 15 to 30 psi to increase OTR. Also, the oxygen feed (air plus oxygen mixture) at 0.1 to 0.3 vvm can be used to maintain adequate levels of dissolved oxygen. The methanol fed-batch phase generally lasts for 70 h.

However, it may prolong to more than 200 h if the feed rate is slowed down because of the fall in dissolved oxygen levels. Another factor that may also influence the fermentation time is the secretion rate of the heterologous protein. Longer times may be necessary to allow for accumulation of high levels of a slowly secreted protein in the broth. Depending on the product being produced, adjustments in the media (addition of casamino acids) and pH will have to be made to increase product yield. One technique that can be used to lower the pH is by setting the pH to the desired lower value (e.g., pH 3.0) and then letting the culture pH to decrease to the new set point of pH 3.0 as a result of cellular metabolism.

Multicyclic Production

Considerable time is expounded on setting up a fermentation run. It would make sense to reap as many cycles of product production as possible for each start-up. Continuous fermentation on methanol, which works well for biomass production, is largely not applicable for recombinant protein production, especially with Mut strains, because they grow poorly on methanol. Use of methanol plus glycerol mixed feed restores the cell productivity, with considerable level of protein expression; optimal level is not achievable because of partial repression of the *AOX1* promoter by glycerol. This problem can be overcome by using a sorbitol plus methanol mixed feed, because sorbitol is a nonrepressing carbon source. However, sorbitol is a poorer carbon source than glycerol and also needs supplementation with 1% each of malt extract, peptone, and yeast extract for improved growth. Nevertheless, this strategy has been used successfully to accomplish several cycles of MMP-2 production in a Biostat-B (Braun) bench-top fermenter.

In this approach, after recovery of 90% of the fermenter sample from the first fermentation cycle, the fermenter is rebooted with an appropriate volume of fermenter growth media containing sorbitol as carbon source and supplemented with 0.1% yeast extract and 0.2% peptone. After growth for 16 to 20 h, a fed-batch fermentation with sorbitol plus methanol nutrient feed is continued for another 72 h for MMP-2 production. The whole process of product recovery and rebooting the fermenter can be repeated as many times as one desires. By this approach, using a 4-L fermenter with one start-up, we were able to generate 25 L of MMP-2 broth in less than 4 weeks. This approach is also useful for Mut cells; it reduces overall methanol consumption because most of the growth is supported by sorbitol. Methanol can be added to the sorbitol feed at any desired point, to initiate induction of expression.

LARGE SCALE PRODUCTION

Problems generally encountered in protein expression may be overcome in most instances by taking due consideration of the factors that influence protein expression. Nearly every expression project has given new insight into the intricacies of the *Pichia* system. In spite of due consideration and careful planning, certain proteins are difficult to express in this system. Now that the expression system is readily available from Invitrogen, hundreds of investigators around the world are currently exploring the system, and their results will undoubtedly provide new insights as well as expand our knowledge base about the utility and limitations of the system. Within the near future, it is anticipated that *Pichia*-produced HSA will reach the market. It will make one think of what to do with the tens of thousands of tons of *Pichia* yeast that would become available as a by-product of HSA production plants.

It is tempting to speculate that the *Pichia* HSA strain would also be engineered to produce an intracellular insecticidal toxin, and thus the *Pichia* cells resulting from the HSA production can be used as a biopesticide. Several modifications have recently been made to the *P. pastoris* expression vectors by Invitrogen. These newer vectors (pPICZ A,B,C series, and pPICZα A,B,C series) are smaller in size and have Zeocin^R marker. Furthermore, they allow in-frame fusion to *myc* epitope and His_6 tags. It has been previously noticed that the addition of eight amino acid carboxy terminal tags to TNF rendered the tagged proteins completely insoluble under nondenaturing conditions, whereas the untagged TNF was completely soluble. Thus, a tag enthusiast should bear in mind that in some instances, a tag can bring about an undesirable change in the property of a protein.

Overexpression of ubiquitin (Ubi) seems to enhance the secretion of a human leukocyte protease inhibitor in *S. cerevisiae*. Ubi is recognized to be a normal secreted component of *P. pastoris*. Thus, it is presumable that the overexpression of Ubi in *P. pastoris* may also enhance heterologous protein secretion. A *P. pastoris* host strain that overexpresses Ubi can be engineered using one of the available Ubi cDNAs. Also, a Ubi expression plasmid can be introduced into any *P. pastoris* strain by using one of the dominant selection markers for transformation, such as invertase, *SUC2*, G418^R or Zeocin^R. Thus, the effect of Ubi overexpression on product secretion can be tested in any of the existing production strains.

A note of concern may be warranted while using zeocin^R for selecting *P. pastoris* transformants. Zeocin^R is a strong mutagen, and thus cells challenged with Zeocin*R* on a complex medium may acquire some undesirable change that may interfere with growth and thus reduce the overall productivity under the growth conditions in a fermenter. A major concern for large-scale use of *Pichia* in the production of biologicals has been the substrate methanol, which is toxic, inflammable, and volatile. With the existing technology in a typical production plant, one will have to use several tons of methanol for the production phase of the operation. It will certainly be an added advantage if the amount of methanol required can be reduced or even eliminated.

Two recent developments in this direction have been made. In one case, a mutant strain of *Pichia* that is able to turn on the *AOX1* promoter in the absence of methanol has been generated. This mutant still responds to repression by glucose and ethanol. In another instance, a *Pichia* strain in which both the alcohol oxidase structural genes *AOX1* and *AOX2* have been deleted has been constructed and shown to express heterologous protein LACZ in response to methanol induction. This novel alcohol oxidase null strain (*aox1, aox2*) uses considerably lower amounts of methanol for production of hetrologous protein. Another development that one can speculate is that *Pichia* will be engineered to express human post-translational modification enzymes (e.g., glycosyl transferases, amidating enzyme, vitamin K-γ carboxylase, etc.), so as to obtain *Pichia* strains that can be used to produce more authentic human proteins.

CULTURE MEDIA

Stock Solutions

Note: For filter sterilization of various solutions and liquids, filter wares (disposable or reusable types) equipped with cellulose acetate or cellulose nitrate membranes (pore size 0.2 to 0.22 μm) from one of the several manufacturers can be used. For filter sterilization of methanol and methanol-containing media, only cellulose acetate membranes (0.2 to 0.22 μm) are suitable, because methanol does not filter through cellulose nitrate membranes of pore size 0.2 μm.

- *10*x *YNB*. Dissolve 13.4 g of yeast nitrogen base without amino acids in 100 mL of water (heat if necessary) and filter sterilize. This solution can be stored for over a year at 4°C.
- *500*x *B*. Dissolve 20 mg of D-biotin in 100 mL of water and filter sterilize.

- *100x H.* Dissolve 400 mg L-histidine in 100 mL of water (heat if necessary) and filter sterilize.
- *10x D.* Dissolve 20 g of D-glucose in 100 mL water. Autoclave for 15 min or filter sterilize. Stores well for years at room temperature.
- *10x GY.* Mix 100 mL of glycerol with 90 mL of water. Filter sterilize. Stores well for years at room temperature.
- *10x M.* Mix 5 mL of methanol (100%) with 95 mL of water. Filter sterilize and store at 4°C.
- *100% methanol.* Filter sterilize pure methanol (100%). Store at room temperature in a fireproof cabinet.

Minimal Media Compositions

- *MD.* Mix 100 mL of 10x YNB, 2 mL of 500x B, and 100 mL of 10x D with 800 mL of autoclaved water (include 15 g bacto agar for plates).
- *MM.* Mix 100 mL of 10x YNB, 2 mL of 500x B, and 100 mL of 10x M with 800 mL of autoclaved water (include 15 g bacto agar for plates).
- *MGY.* Mix 100 mL of 10x YNB, 2 mL of 500x B, and 100 mL of 10x GY with 800 mL autoclaved water (include 15 g bacto agar for plates).

All these liquid media and plates store well for several weeks at 4°C. Minimal media with other carbon sources (such as D-sorbitol, D,L-alanine) are prepared by using the desired carbon source at 10 g/L in place of glucose in MD. Minimal media containing a mixture of carbon sources can also be prepared by combining two or more desired substrates in the growth medium.

Supplemental Minimal Media

Minimal media are supplemented with necessary supplemental nutrients such as amino acids depending on the specific requirement of a given strain. For example, *P. pastoris* strains GS115 and KM71, commonly used in molecular genetic manipulations, are auxotrophic for histidine. Such strains will grow in minimal media only in the presence of supplemental histidine. However, once transformed with *HIS4* (histidinol dehydrogenase gene) they readily grow in the absence of histidine. The composition of supplemental minimal histidine media (suitable for histidine auxotrophic strains such as GS115) is as follows. Other supplemental media can be prepared depending on the need of a particular strain in use.

- *MDH.* Mix 100 mL of 10x YNB, 2 mL of 500x B, 100 mL of 10x D, and 10 mL of 100x H with 790 mL of autoclaved water (include 15 g agar for plates).
- *MMH.* Mix 100 mL of 10x YNB, 2 mL of 500x B, 100 mL of 10x M, and 10 mL of 100x H with 790 mL of autoclaved water (include 15 g agar for plates).
- *MGyH.* Mix 100 mL of 10x YNB, 2 mL of 500x B, 100 mL of 10x GY, 10 mL of 100x H with 790 mL of autoclaved water (include 15 g agar for plates).
- *MGyB.* Dissolve 11.5 g KH_2PO_4, 2.66 g K_2HPO_4, 6.7 g YNB, pH 6.0, and 20 mL glycerol in 1 L water and autoclave.

All these liquid media and plates will store well for several weeks at 4°C. Supplemental minimal histidine media with other carbon sources is prepared by adding a similar amount of histidine as above to the minimal media with the desired carbon source.

Complex Media

- *YPD*. Dissolve 10 g of bacto yeast extract, 20 g of peptone, and 20 g of glucose in 1,000 mL of water (also include 15 g bacto agar for slants and plates) and autoclave for 20 min.
- *YMPD*. Dissolve 3 g of yeast extract, 3 g of malt extract, 5 g of peptone, and 10 g of glucose in 1 L of water and autoclave.
- *YMPGy*. Same as YMPD with the exception that 10 mL of 100% glycerol is used instead of 10 g of glucose.

Secretion Media

- *BMGY*. Mix 100 mL of 1 M potassium phosphate buffer, pH 6.0, 100 mL of 10x YNB, 2 mL of 500x biotin, and 10 mL of glycerol. Filter sterilize and add to an autoclaved solution of 10 g yeast extract and 20 g peptone in 788 mL water (15 g bacto agar is included for plates).
- *BMMY*. Same as BMGY, with the exception that 5 mL of methanol is added in the place of 10 mL glycerol.

Note: Yeast extract and peptone in the above media can be replaced by 1% casamino acids. The pH 6 suggested here may not be optimal for every secreted product. Experimentation with pH values in the range 2.5 to 8 (by using appropriate buffers) is suggested to determine the optimal pH for a particular product. Some suggested buffers are as follows:

1. Phosphate buffer for pH range 5.7 to 8.0
2. Alanine-HCl buffer for pH range 2.5 to 3.6
3. Aconitic acid-NaOH buffer for pH range 2.5 to 5.7
4. Citrate buffer for pH range 3.0 to 6.2

Avoid buffers such as succinate buffer, because succinate can serve as a carbon source and will repress activation of the *AOX1* promoter.

Fermentation Media

FM21 basal salt media

Composition is for 1 L final volume in water:

Phosphoric acid, H_3PO_4 (85%)	3.5 mL
Calcium sulfate, $CaSO_4 \cdot 2H_2O$	0.15 g
Potassium sulfate, K_2SO_4	2.4 g
Magnesium sulfate, $MgSO_4 \cdot 7H_2O$	1.95 g
Potassium hydroxide, KOH	0.65 g

Biotin stock solution

Biotin	0.2 g/L

PTM1 trace salts

Composition is for 1 L final volume in water:

Cupric sulfate ($CuSO_4 \cdot 5H_2O$)	6.0 g
Manganese sulfate ($MnSO_4 \cdot H_2O$)	3.0 g
Ferrous sulfate ($FeSO_4 \cdot 7H_2O$)	65.0 g
Zinc sulfate ($ZnSO_4 \cdot 7H_2O$)	20.0 g
Silfuric acid (H_2SO_4)	5.0 mL
Cobalt chloride ($CoCl_2 \cdot 6H_2O$)	0.5 g
Boric acid (H_3BO_3)	0.02 g
Sodium molybdate ($NaMoO_4 \cdot 2H_2O$)	0.2 g
Potassium iodide (KI)	0.1 g

BSM medium composition

Composition is for 1 L final volume in water:

Phosphoric acid, H_3PO_4 (85%)	26.0 mL
Calcium sulfate, $CaSO_4 \cdot 2H_2O$	0.9 g
Potassium sulfate, K_2SO_4	18.0 g
Magnesium sulfate, $MgSO_4 \cdot 7H_2O$	14.0 g
Potassium hydroxide, KOH	4.0 g

15

MECHANISM OF CELL CYCLE

The *cell cycle* is the name given to the process by which a cell "matures," synthesizes DNA, and divides to form daughter cells. Thus the cell cycle is a fundamental process with analogous mechanisms found in all cells, from the most primitive bacterium to higher animals and plants, from the unicellular to the most complex multicellular organism. Here we describe the mammalian cell cycle, high-light its importance in biotechnology, and describe some cell cycle applications in animal cell biotechnology. The cell cycle model (G_1-event model) currently accepted by authors of the major cell and molecular biology books describes the cell cycle as discrete phases controlled by cell cycle proteins interacting with each other and directing cellular events including DNA synthesis and cell division. However, the validity of this model and the interpretation of many experiments that underpin the current established view of the cell cycle and cell cycle regulation have been questioned.

The proposed alternative, the continuum model, suggests that the cell cycle does not exist as currently recognized but takes the form of a continuum from one cell division to the next, rather than discrete phases with discrete groups of regulatory proteins. A hypothetical initiator of DNA replication is synthesized by the cell during all phases of the cell cycle, and its concentration reaches a critical level that initiates DNA synthesis. Thus the cell can control its rate of division by regulating the amount of initiator. Nevertheless, until discussion of these competing models is concluded, with perhaps more work to determine the correctness and applicability of the continuum model, it is only reasonable to accept the weight of evidence for the established view of the cell cycle and its regulation. Thus the established model is summarized here and is the basis for the process biotechnology applications that are described. The pathways of cell cycle regulation are, unsurprisingly perhaps, shared with another important cell process: apoptosis, or "programmed cell death," which is, in effect, the opposite of the cell proliferation cycle. Because the cell cycle and apoptosis share some common pathways, regulatory imbalance can lead to cancer in multicellular organisms. Regulation of the cell cycle is also crucial in embryology and ontogeny, where features such as limb formation, tissue differentiation, and growth are a function of the balance between proliferation and apoptosis. Therefore, much of the cell cycle (and apoptosis) literature is from the field of biomedical science, particularly cancer and anticancer chemotherapy research. General texts, as well as much of what is written here, provide a consensus of information describing a hypothetical typical cell and thus, by implication at least, describe what is true of all eukaryotic cells.

CELL CYCLE

The cell cycle and its control mechanisms are highly conserved, so most of what is described in general terms for a "typical mammalian cell" is also true for other animal cells, such as insect and plant cells, although there is the obvious exception of the cell wall and its effect on cell division. The cell cycle and its control in yeast are essentially similar to but much simpler than the analogous mechanisms in higher eukaryotes; where one protein pair exerts a particular control function in a yeast, several may be required for the same function in an animal cell. As a result of this less complex system and the fortuitous discovery of cell cycle control mutants (cdc mutants) in which the cell cycle is inhibited at specific stages, many early discoveries about cell cycle control were made using the budding yeast *Saccharomyces cerevisiae* and the fission yeast *Schizosaccharomyces pombe*. Subsequent research has shown homology between cell cycle control proteins in yeast and those in mammalian cells. However, the bacterial cell cycle is markedly different from the eukaryotic cell cycle, and its control mechanisms are (to date) less well understood than the analogous processes in eukaryotes. Given a hypothetical typical bacterial cell, its structure and thus the function of its cell cycle are very different from those of a eukaryotic cell.

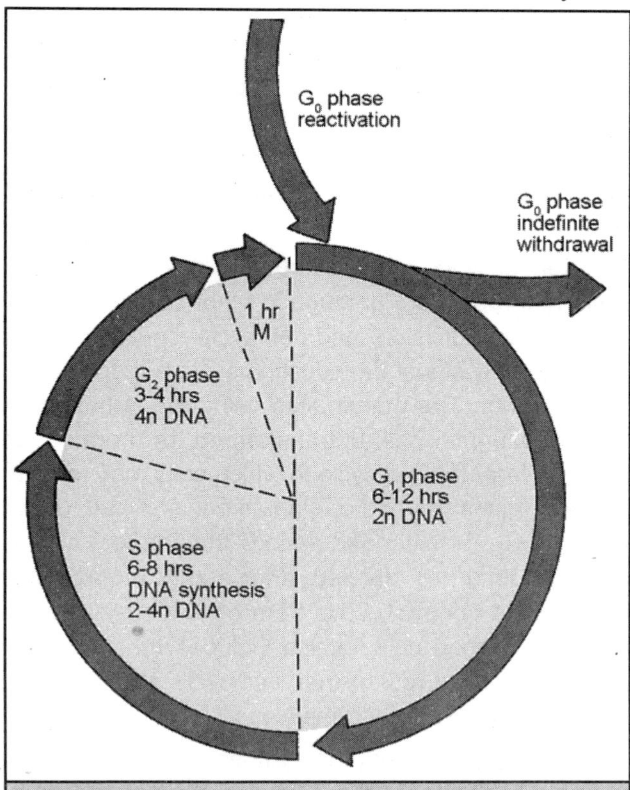

Figure 15.1 G_1 S and G_2 periods during an interphase of cell cycle.

Bacteria, like all cells, are enclosed in a cytoplasmic membrane, but in addition, they have a cell wall outside the cytoplasmic membrane. Within the cell there are no organelles, only free ribosomes and the nucleus, which is actually an area that contains DNA but is not segregated from the cytoplasm by a membrane. The entire genome is contained in a single, circular, double-stranded DNA chromosome; one strand is a closed circle and the other is open. There are no histone or similar DNA-associated proteins; instead there is a relatively high proportion of small, basic polyamines whose function may be to counteract the acidity of the DNA. The "joint" in the closed DNA strand, the origin of replication, is attached to a mesosome, an invagination of the cytoplasmic membrane. When DNA replication starts, the 5′ end of the open strand attaches to the cytoplasmic membrane and starts to form a second mesosome; meanwhile, the DNA unwinds in one direction from the origin of replication and splits into two single strands to form a replication fork. As the replication fork progresses along the chromosome, complementary strands of DNA are synthesized until two daughter chromosomes of double-stranded DNA are formed, each with an origin of replication attached to its mesosome. While the chromosome replicates, the cytoplasmic membrane and cell wall are synthesized in the region between the two mesosomes, enlarging the cell.

After chromosome replication, a septum forms across the middle of this region, finally separating the two daughter cells. However, it takes about 40 minutes to duplicate the single chromosome; hence, to allow doubling times as low as 20 minutes, each daughter chromosome must start to replicate in exactly the same way before the parent has finished replication. Thus instead of one there will be three origins of replication operating at once. In addition, when the doubling time of a culture is less than about 40 minutes, DNA synthesis is continuous, so it may be that DNA replication and cell division cycles are independent in bacteria. However, despite the essentially continuous DNA synthesis and cell division shown in the fast-growing cultures likely to be used in process biotechnology, it has recently been shown that slowly growing *Escherichia coli* cells (doubling time up to 5 hours) display cell cycle phases analogous to eukaryotic G_1, S, and G_2 phases, *vide infra,* followed by cell division. Furthermore it was shown that initiation of DNA replication was related to cell mass. Two genes, *seqA* and *dnaA,* were shown to be involved in the regulation of replication initiation; gene mutants exhibited decreased and increased mass, respectively, at initiation of replication.

Recent work with *Caulobacter* described a protein, CtrA, that has sequence homology with response regulatory proteins in other bacteria including *E. coli* and is involved in cell cycle control including cell division, and cell cycle–specific transcription. It was proposed that CtrA is part of a phosphorelay signal transduction system responsible for transcriptional-level control of bacterial cell cycle events. The mammalian cell cycle is best described by illustration. Starting from a recently divided daughter cell, untransformed, or "normal," cells may enter gap phase 0, or G_0. This is an offshoot from the cell cycle in which cells may remain viable and nondividing but retain the potential to divide given the appropriate stimulus. A cell, depending upon its type, can remain out of the cell cycle, in G_0, for extended periods but can return to gap phase 1, or G_1, due to cell maturation after a certain time has elapsed, after a critical mass/size is reached, or as a response to an external stimulus. In contrast, cells from continuous, sometimes called *established* or *immortal,* cell lines and transformed cells (which includes most animal cells used in biotechnology) do not enter G_0 after division but restart their cell cycle after mitosis by progressing directly to G_1. During G_1, cells prepare to duplicate themselves.

When a cell enters the synthesis, or S phase, it starts to precisely and accurately duplicate its genome by synthesizing DNA and the associated nuclear protein components, so that daughter cells carry a genome identical to each other and to the parent. During this time, ribosomes, mitochondria, endoplasmic reticulum, and all the other subcellular organelles are also replicated in preparation for cell division. Thus by the end of S phase, a normally diploid ($2n$) cell will be tetraploid ($4n$), containing enough DNA (and cellular components) for two cells. After S phase, the cycling cell enters the second gap, or G_2, phase. During G_2 the cell synthesizes the proteins required to direct chromosome segregation and cell division in the following mitosis, or M phase. After mitotic division the daughter cells return to G_1 at the beginning of a new cell cycle or exit the cycle to G_0. In a typical cell cycle of 24 hours' duration, G_1 would last 10 hours, S phase 9 hours, G_2 would last 4 hours, and mitosis 1 hour. However, the duration of the cell cycle varies both between cell lines and between individual cells within the same culture.

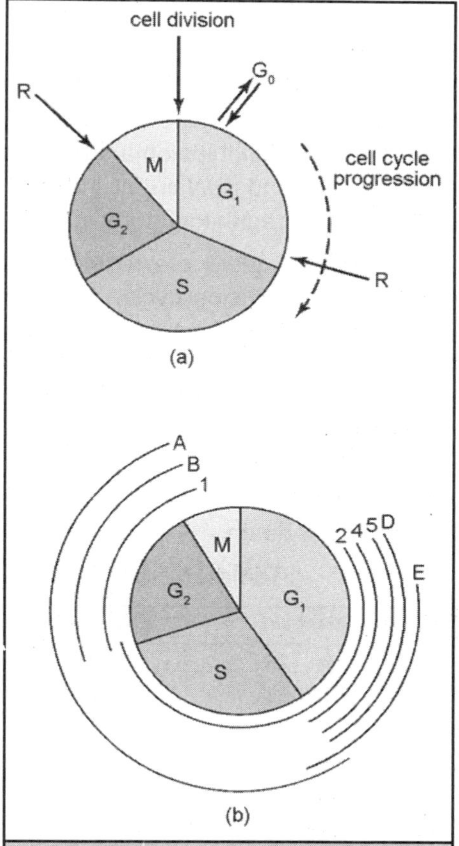

Figure 15.2 (a) A simple diagrammatic representation of the cell cycle. (b) A simple diagrammatic representation of cyclin and cyclin-dependent kinase activation during cell cycle.

Variation in cycle time is usually due either to cells exiting the cycle and arresting in G_0 or to a variation in the duration of G_1. It has been suggested that the intercellular variability of the post-mitotic early G_1 period is smaller than that seen with initiation of DNA replication (presynthetic late G_1). It is also postulated, by the same author, that cells that make the "yes or no" decision in the postmitotic early G_1 period about whether to proceed through the cell cycle also have the capacity to decide, in the presynthetic late G_1 period, when the cell will enter S phase. Heterogeneity in the duration of G_1 is the main reason why a synchronized culture loses synchrony with time to the extent that the culture is completely asynchronous within three or four cell cycles. This apparently simple cell cycle is subject to complex control mechanisms. Some of the molecular pathways have been understood only in recent years and there may be others that have not yet been described. Mammalian cells periodically transcribe cell cycle genes that encode proteins directly involved in cell cycle regulation. These include the cyclins and the cyclin-dependent kinase (Cdk) families.

The cyclin-dependent kinases are phosphorylating enzymes that are inactive when not associated with an appropriate cyclin; it is the associated cyclin that determines the substrate specificity of the complex. The activity of this heterodimeric complex is then regulated by phosphorylation/dephosphorylation of specific activation and/or inhibition sites, as well as by further binding with inhibitory proteins. Many of the cyclins, cyclin-dependent kinases, and related regulatory proteins are expressed or form active complexes only during particular cell cycle phases. Cdk4–cyclin D1 drives cells through mid G_1, Cdk2–cyclin E drives late G_1, Cdk2–cyclin A controls entry into S phase, and Cdk1–cyclin B drives the G_2–M transition; their phase-related activities are summarized. Others, including Cdk4 are expressed throughout the cell cycle. The noncyclic proteins exert their cell cycle–controlling effects by association with phase-specific cyclins or cyclin-dependent kinases, association with Cdk-cyclin inhibitors, and specific amino acid phosphorylation/dephosphorylation to form active complexes in specific phases. A family of genes encoding proteins related to the retinoblastoma susceptibility gene, *RB*, and genes for transcription factors including *BYMB, E2F1*, and *JUN* are all induced in late G_1 or early S phase and have been shown to interact with cell cycle regulatory complexes.

In addition, there are restriction points in the cell cycle that, once past, commit the cell to continuing the division cycle. There are also check points where, in effect, the cell ensures that the previous phase has been completed properly before proceeding or that there has been no DNA damage. To continue to mitosis before the genome is properly duplicated, for example, would be catastrophic for the cell; one if not both daughters would not be viable. It has been suggested that induction of the p53 tumor suppressor gene and its transcriptional activation of the Cdk inhibitor $p21^{WAF1/CIP1}$ with subsequent cell cycle arrest, following DNA damage, allows time for DNA repair before entering S phase. Activation of p53 in cells with unrepaired DNA damage may lead to apoptotic cell death.

Table 15.1 Cyclin-Cdk complexes, stability, and associated proteins

Cyclin	*Associated Cdk*	*Cell cycle activity*	*Associated proteins*	*Stability*	*Degradation*
A	Cdk2 and cdc2 (Cdk1)	S and $G_2 \rightarrow M$	p107, E2F, p21, PCNA	Unstable in mitosis	Ubiquitin
B1	cdc2 (Cdk1)	$G_2 \rightarrow M$	(p21), (PCNA)	Unstable in mitosis	Ubiquitin
B2		$G_2 \rightarrow M$	(p21), (PCNA)	Unstable in mitosis	Ubiquitin
C-type	CdkX	?	ND	ND	PEST
D1	Cdk4(2, 5, 6)	G_1	pRb, p21, p27, PCNA, p16, p15	Rapid turnover	PEST
D2	Cdk4(2, 5, 6)	G_1	pRb, p21, p27, PCNA, p16, p15	Rapid turnover	PEST
D3	Cdk4(2, 5, 6)	G_1	pRb, p21, p27, PCNA, p16, p15	Rapid turnover	PEST
E	Cdk2	G_1 and $G_1 \rightarrow S$	p107, E2F, p21, PCNA and p27 after TGFβ treatment	Rapid turnover	
F			ND	ND	
G			ND	ND	
H	Cdk7	All phases	ND	ND	ND

First Gap Phase or G_1

G_1 was originally thought to be a gap in the cell cycle—a resting phase while the cell recovered from division before starting a new cell cycle by synthesizing more DNA—although this is now known not to be so. Rather, complex control processes are preparing the cell to synthesize the cellular machinery required for DNA synthesis in S phase. There is also a "point of no return" near the end of G_1 called the restriction point (R), which is analogous to START in yeast. Once cells pass this point they are committed to the next stages of the cell cycle. Signals for passing this point include adequate cell size and a favorable environment, including the presence of growth factors. The transition at the R point from growth-factor dependence to growth- factor independence in normal cells, as opposed to cancer or transformed cells, is likely to reflect a stringent growth regulatory mechanism that is defective in the latter cell types. Coordination between cell cycle commitment, START (or R), and cell size, involving the G_1 cyclins, has been suggested as a regulatory mechanism for cell growth in yeast. During early G_1, immediately after G_0 cells are stimulated to reenter the cell cycle, a critical step occurs involving the induction of cell cycle–specific genes including *myc, cyclin D, and fos,* and Cdk genes. It is now apparent that Cdk 2, Cdk 4, and Cdk 5, despite constitutive expression of Cdk4 throughout the cell cycle, start to complex with cyclins D1, D2, and D3 and become active in early to mid-G_1 (possibly at a point when or soon after non-cycling cells return from G_0).

The level of activation of the G_1 Cdk-cyclin complexes peaks at about R; so it is possible that, by reaching a threshold activity, it is these complexes that allow cells to pass R. Cdk2 also forms active complexes with cyclin E starting from mid-G_1 and peaking later than the Cdk–cyclin D complexes, in other words, between R and the G_1–S phase transition. The primary substrate of the cyclin D complexes is the retinoblastoma tumor suppresser gene product pRB. Active, unphosphorylated pRB complexes with and inhibits the transcription factor E2F. When pRB is phosphorylated and inactivated, the complex dissociates, releasing active E2F; the cell cycle is not stopped at the G_1 checkpoint, and cells are allowed to proceed to S phase. E2F was originally characterized as an activator of the adenovirus early region 2 transcription unit. It is now recognized that E2F promotes the transcription of enzymes associated with DNA synthesis. It has been proposed that the cyclin E complex also phosphorylates pRB, with similar effects.

Synthesis Phase or S

The transition from G_1 to S phase is marked by the transcription and activation of the many synthetic enzymes required for cellular duplication, including DNA polymerase α, thymidine kinase, and dihydrofolate reductase (DHFR). This is followed by the exact duplication of the genetic material and replication of organelles and structural proteins. At the same time there is a marked change in regulatory proteins activated. At the onset of S phase, the activation of the G_1-phase Cdk-cyclin complexes decreases in the same temporal order as it increased, and the complexes dissociate. As Cdk2 disassociates from cyclins D and E, it complexes with cyclin A, which starts to be synthesized at about the G_1–S transition. The Cdk2–cyclin A complex retains its activity until the end of S phase.

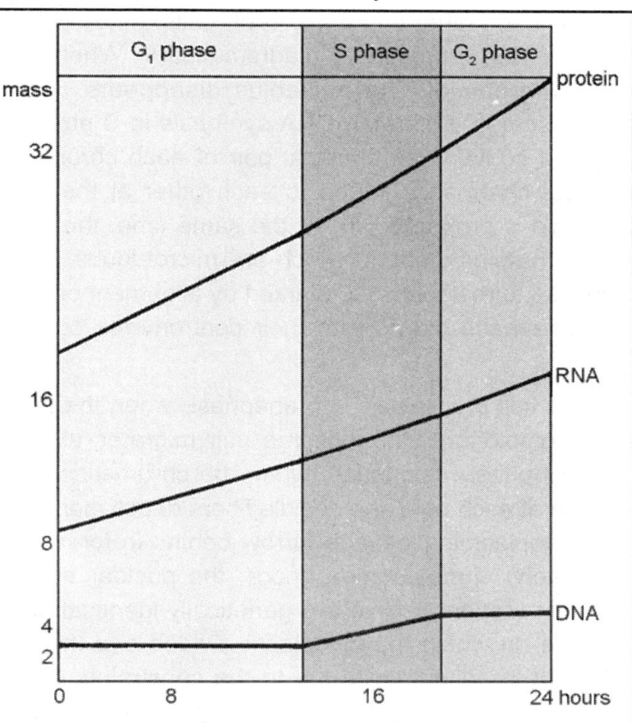

Figure 15.3 Synthesis of DNA, RNA and protein during cell cycle.

Disruption of cyclin A function inhibits DNA synthesis, suggesting that a threshold activity of the Cdk2–cyclin A complex is essential for the maintenance of S phase.

Second Gap Phase or G_2

As with G_1, G_2 was originally thought to be a gap in the cell cycle—a resting phase while the cell recovered from DNA synthesis before dividing. This has now been shown not to be so, although G_2 remains analogous to G_1 in that, once the appropriate control signals are received, genes encoding for the proteins responsible for cell division are transcribed and ready for cell division. At the end of S phase the activity of the Cdk2–cyclin A complex decreases as the components dissociate. Cyclin A and cyclin B then form complexes with Cdk1 (also often referred to as cdc2 because of its homology with the product of *S. pombe* cell division cycle gene 2). It is these complexes, increasing in activity throughout G_2 and continuing into mitosis, that mediate activation of the cell division machinery. At the transition point between G_2 and M is a second check point that, in effect, ensures that the cells are big enough, the environment is favorable, and DNA synthesis has been fully and accurately completed before the cell continues to mitosis and division.

Mitosis or M Phase

Mitosis is the division of one normally diploid, or $2n$, somatic cell into two genetically identical (both to each other and the parent) cells. Mitosis, like the cell cycle itself, displays distinct phases that are best represented diagramatically. When the cell leaves interphase at the end of G_2, it enters prophase. The nucleolus disappears (disperses), and nuclear DNA condenses into chromosomes. Following DNA synthesis in S phase, a normally diploid ($2n$) cell is now tetraploid ($4n$) and contains an identical pair of each chromosome, which in turn is organized as a pair of identical chromatids joined to each other at the centromere. During late prophase, the nuclear envelope is broken down. At the same time, the centrioles divide; one pair migrates to each cell pole with spindle fibers, which are microtubular structures, linking each pair. The transition from prophase to metaphase is marked by alignment of all the chromosome pairs on an equatorial plane, the metaphase plane, with their centromeres connected to each polar centriole pair by spindle fibers.

The cell progresses into anaphase when the chromosomes separate and one chromatid from each homologous chromosome pair migrates along the spindle fibers toward each centriole. In the next phase, telophase, when the chromatids (one from each homologous chromosome pair) arrange at each pole, the spindle fibers depolymerize, and cytokinesis (division of cytoplasm) begins as a cytoplasmic cleavage furrow begins to form. Further changes follow: the cytoplasm cleaves completely, chromosomes uncoil, the nuclear envelope reforms around each nucleus, and the nucleoli reappear to give two genetically identical daughter cells in interphase. The final telophase changes, in which the cytoplasm divides and the cells separate, are markedly different in plant cells and in animal cells due to the constraints of the rigid cellulose cell wall. As with other cell cycle phases, Cdk-cyclin complexes perform a major regulatory function in mitosis. The activity of Cdk1–cyclin A and Cdk1–cyclin B, which started in G_2, continues to increase until the middle of mitosis, at the metaphase–anaphase boundary. At this point the activity of both complexes declines abruptly to zero, allowing the cell to complete mitosis. It is also thought that the rapid decrease in G_2 cyclin-Cdk complexes is necessary for cells to start a new cell cycle at G_1.

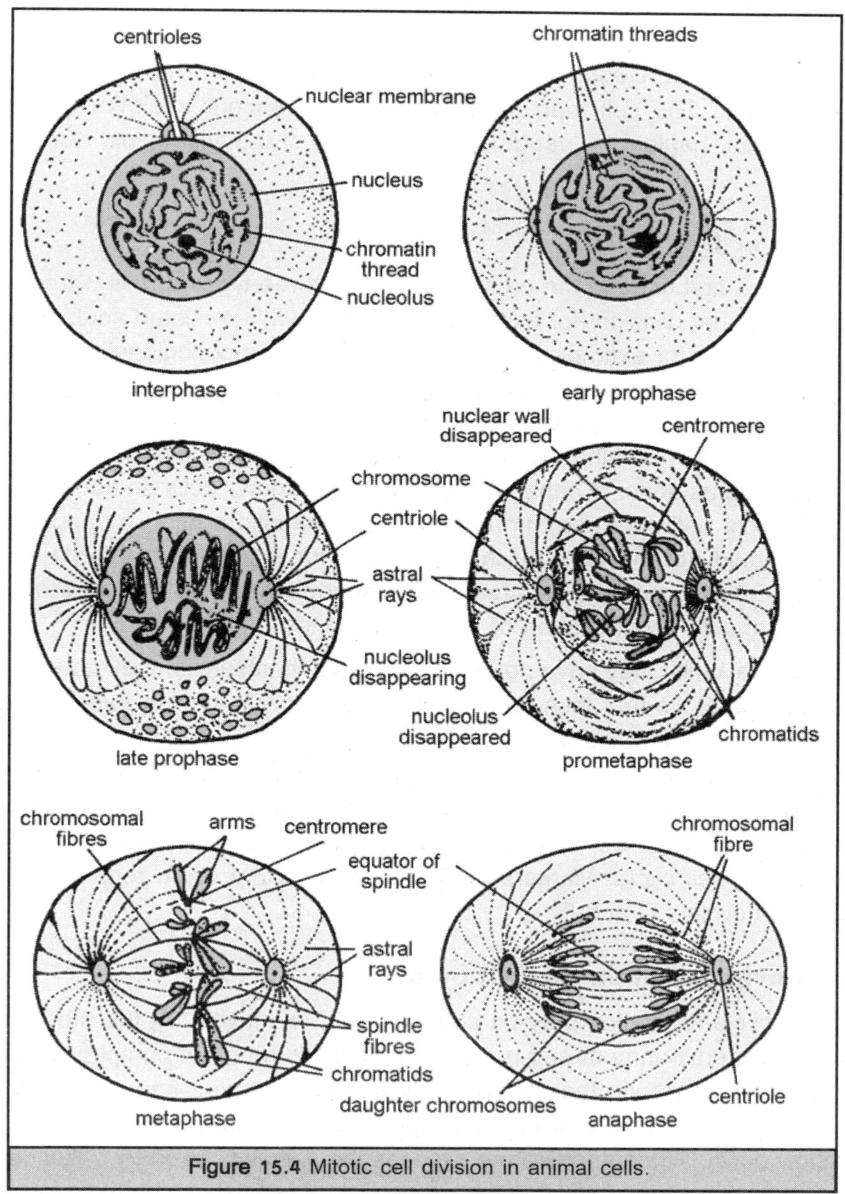

Figure 15.4 Mitotic cell division in animal cells.

Meiosis

Although unlikely to be encountered in process biotechnology, it is useful to point out that germ cells may undergo an alternative form of division to form zygotes, which may also be considered to be a form of terminal differentiation, called *meiosis*. Assuming a normally diploid ($2n$) cell, initial cell division occurs exactly as in mitosis until the end of anaphase, or in this case the first meiotic anaphase. At telophase the cell may divide or partially divide and enter a brief meiotic interphase comprising two identical diploid (2*n*) daughter nuclei (if not cells). Otherwise the cell progresses directly to the second meiotic prophase. Now there is only a single copy of each chromosome, still comprising two chromatids. During the second meiotic anaphase, the chromatids separate again to each of four poles. The cell then progresses to (the second meiotic) telophase and on to interphase to give four haploid (*n*) daughter cells instead of the two diploid (2*n*) cells produced by mitosis.

Zero Gap Phase or G_0

The term G_0 was first used in the early 1960s when three types of cell were recognized: static cells, incapable of further division (for example, a terminally differentiated adult neurone); renewing, constantly dividing cells (for example, bone marrow cells); and conditionally renewing cells, which do not normally divide but can respond to an appropriate extraordinary stimulus (for example liver cells starting to proliferate after partial hepatectomy). G_0 is used to describe this last category of cells, which have "opted out" of the cell cycle after a mitosis because of, for example, growth factor depletion. In this circumstance, in contrast, transformed or tumor cells do not enter G_0 but continue through the cell cycle until they die through nutrient exhaustion, often by apoptosis. It has been argued that G_0 should not be used to describe cells, for example, in the stationary phase of a hybridoma, or CHO batch cultures, which are in a pseudo-steady-state of growth and death due to nutrient depletion and/or accumulation of toxic metabolites.

Unlike the normal cells, they have not down-regulated themselves; they have been closed down by their environment. Cells that lose their ability to respond to growth factor depletion or do not exit the cell cycle to G_0 may have a defective G_1 program, which is different from the cyclin-Cdk4-pRB phosphorylation mechanism involved in R regulation in normal cells. However, G_0 is commonly used today, some claim inappropriately, to describe any cell (containing a G_1 amount of DNA) that is not expected to proceed to S phase for any reason, including the result of any means of artificial interference with the cell cycle. Markers of G_0 include reduced RNA, reduced ribosome content, and an absence of cyclins D and E.

Influence of Environment on the Cell Cycle

Extrinsic cell cycle regulation may be considered at two levels. At the more basic level is the effect of the physical environment and nutrient availability; the more complex level of extrinsic cell cycle control is through growth factors. The physical environment has a great effect on the cell cycle. When temperature is reduced from the optimum, the length of the cell cycle increases, and growth rates slow down. When temperature exceeds the optimum, even by a few degrees, profound changes in cell behavior occur, including the expression of heat shock proteins, and cells eventually die. Alternatively, if appropriate nutrients such as amino acids and glucose are not available in the medium, cell cycle duration and even progression will be influenced due to lack of energy and the precursors for protein and DNA synthesis. Serum starvation, in the presence of appropriate nutrients and physical conditions, may also affect the cell cycle, but this is more likely to be caused by the absence of specific growth factors that serum provides. Most, if not all, cell lines were originally grown in medium containing animal serum, usually newborn or fetal bovine serum. The main contribution to the cell cycle and, hence, growth of a culture from these undefined and heterogeneous supplements is the supply of growth factors that promote or enhance the cell cycle. Thus, because G_0 cells reenter the cell cycle as a result of serum addition, serum is regarded as a source of positive regulatory growth factors.

The presence of growth factors in the environment, their binding to specific transmembrane receptor proteins, and subsequent signal transduction effects changes in cellular processes, including the cell cycle. Expression of cyclin D, important in progression through G_1, is induced by growth factors. These effects are cell-type dependent because cells do not all express the same growth factor receptors. Furthermore, different cell types/lines may not only require different individual growth factors but may require a specific cocktail for optimal proliferation. Fibroblast growth, for example, requires growth factors including fibroblast growth factor (FGF), epidermal growth factor (EGF), and platelet-derived growth factor (PDGF).

Other growth factors that promote cell division include insulin-like growth factor (IGF) I and II, macrophage–granulocyte colony stimulating factor (G-CSF), and endothelial cell growth factor. Each growth factor stimulates a specific receptor on the plasma membrane, which in turn activates protein kinases, many of which phosphorylate tyrosine residues on their substrates, so they are generically called tyrosine kinases. Subsequent signal transduction by phosphorylation/ dephosphorylation cascades can lead to an increased rate of protein synthesis, influence the activity of DNA topoisomerase, or modify the glucose transport system—all with the similar effect of promoting cell division.

Transformed cells are known to have a reduced (or no) requirement for external growth factors compared with normal cells. However, through adaptation of cell lines and extensive development of medium formulation, many cells can now be grown in protein-free medium, indicating that it is possible to bring about cell proliferation and support long-term culture in media free of (extrinsic) growth factors, hormones, and other contaminating proteins. This absence of extraneous protein is particularly important in large-scale process biotechnology for facilitating downstream processing and product recovery. It is commonly accepted that the accumulation of toxic metabolites (ammonia and lactate, for example) inhibits culture proliferation. However, in addition to the positive growth-promoting (cell cycle–promoting) factors, there are also negative, growth-inhibiting factors, although these maybe the same growth factors displaying opposite effects under different conditions. Transforming growth factor $\beta 1$ (TGF-$\beta 1$), in addition to its strong positive stimulatory effect, has been shown to produce a density-independent negative response that prolongs G_1 in the NIH-3T3 mouse embryo cell cycle and has a similar effect on mortal human fibroblasts.

METHODOLOGY

Mitotic Index

There are three main approaches to describing the cell cycle. The first and most historically established approach is to determine the percentage of cells in mitotic cell division. This method, called the *mitotic index*, is simple and can be applied to attached cells grown on a microscope slide or any other appropriate substratum, as well as cells in suspension. There are many fixation and staining methods described in any textbook of cytology methods, from which a protocol appropriate to animal cell biotechnology can be selected. A generic protocol is as follows. Take cells grown on a glass microscope slide (or coverslip) or prepare a smear of suspended cells on a microscope slide, using an attachment agent such as polylysine if necessary. Fix the cells with a fixative appropriate to the proposed staining method. Stain the cells with either a standard cytological or histological stain that differentiates DNA, such as Feulgen staining, or alternatively, stain with a fluorescent DNA stain such as ethidium bromide (EB), Hoechst 33258, or DAPI. Examine stained slides by conventional or fluorescence microscopy, as appropriate; count the number of nuclei in mitosis and express the result as a percentage of the total.

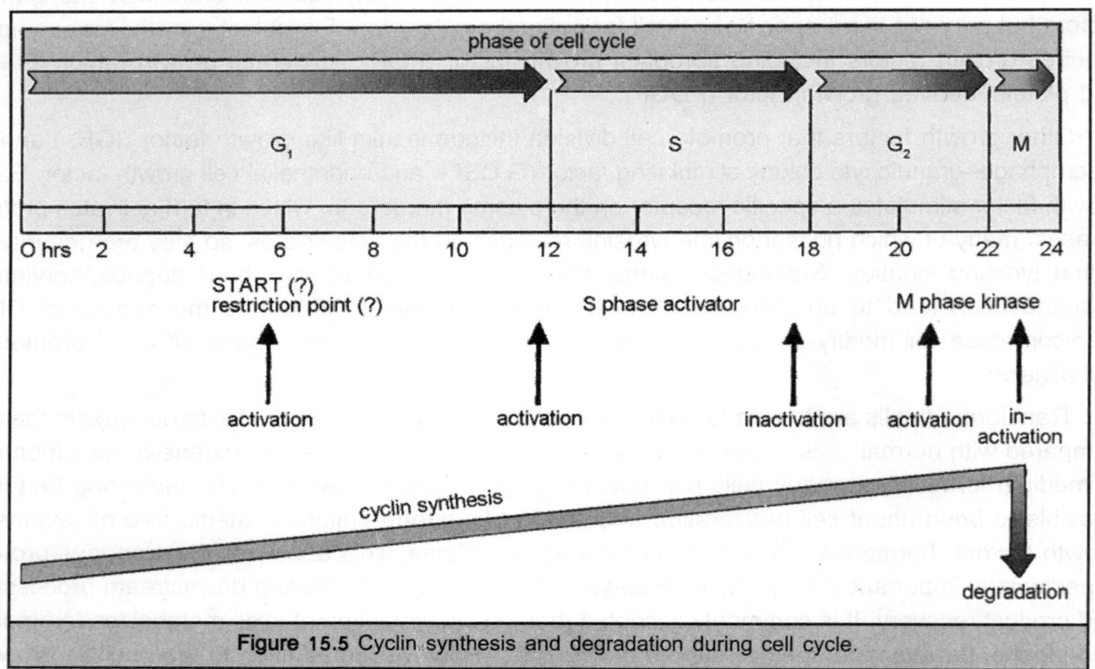

Figure 15.5 Cyclin synthesis and degradation during cell cycle.

A recent publication, investigating mitosis in a hydrodynamically stressed environment, described the use of the fluorescent DNA-intercalating stain propidium iodide (PI) with cells fixed with 3:1 ethanol:glacial acetic acid, followed by fluorescence microscopy. A major disadvantage to measuring the mitotic index is that, although it gives information about cell division in M phase, it can give no information about the other cell cycle phases or their associated processes.

Measuring the mitotic index also exhibits disadvantages common to all manual microscopic methods, including sampling error because a small proportion of the total population is examined, and interobserver variation resulting from subjective discrimination between mitotic and interphase cells. Furthermore this slow and labor-intensive method is not appropriate for online automation or use in integrated process control.

Nucleotide Uptake in S Phase

An alternative approach for describing the cell cycle is to document the incorporation of labeled nucleotide into nuclear DNA during S phase. This method relies on [^{3}H]thymidine incorporation followed by autoradiography and can be used in one of two ways: continuous labelling, in which the cells are exposed to the label for a fixed time, usually 24 or 48 h followed by analysis at that time; or pulse–chase labeling, in which the cells are exposed to the label for a short pulse (0.5-1 h), then analyzed at regular intervals for 24-48 h. Continuous labeling gives information about the growth fraction of a culture, which is the percentage of the cells that are in the cell cycle during the labeling period. However, this information is not likely to be of great use to the process biotechnologist because most cells under optimum conditions, when viability is close to 100%, will also have a growth fraction close to 100%.

A simple generic protocol for continuous labeling is as follows: [^{3}H]thymidine is added to the culture medium, and the culture continues for a specified period; cells are fixed and subjected to autoradiography; when the autoradiograph has developed, the number of radiolabeled nuclei is counted and expressed as a percentage of the total. In contrast, pulse–chase labeling is used to determine the average duration of each cell cycle phase. A generic protocol for [^{3}H]thymidine pulse–chase labeling is as follows: [^{3}H]thymidine is added to the culture; after a short period (1-2 h), cells are removed to fresh medium without [^{3}H]thymidine; the culture is sampled regularly at times greater than the length of one cell cycle (24-48 h); each sample is subjected to autoradiography; and the number of labeled mitoses is counted as a percentage of the total. The average duration of G_2, S, M, and the total cell cycle is measured from the plot of labeled mitoses against time, then G_1 is calculated. Details of using [^{3}H]thymidine incorporation for determining growth fraction and cell cycle phase duration are summarized elsewhere.

As with the mitotic index, any autoradiographic technique exhibits inherent disadvantages including sampling error because a small proportion of the total cell population is examined; the technique is slow, labor intensive, and not appropriate for on-line automation or use in integrated process control; and the use of radioisotopes requires additional containment and waste disposal facilities. In recent years the thymidine analogue bromodeoxyuridine (BrdU) has been used as a safer, nonisotopic alternative to [^{3}H]thymidine for DNA labeling. BrdU can be used in exactly the same way as [^{3}H]thymidine, as either a continuous or pulse–chase label. The only difference in the protocol is that BrdU incorporation is detected by immunostaining with fluorescent anti-BrdU antibodies followed by fluorescence microscopy instead of autoradiography, so the method still suffers the disadvantages of any manual microscopic method. Alternatively, BrdU incorporation can be measured by flow cytometry, *vide infra*.

Measuring DNA Content

The third and probably easiest approach to adopt for cell cycle analysis is to measure the DNA content of each cell in a population by flow cytometry and thus determine the proportion of cells in G_0/G_1, S, and G_2/M at the time of sampling. A major and obvious advantage to this approach is that heterogeneous nonsynchronized cultures can be analyzed easily and rapidly, with the useful bonus that sub-G_1 apoptotic populations can also be identified. Other advantages include minimizing sample error by analyzing a large number of cells, and minimal interobserver variation because of the objective discrimination between G_1, G_2, and S cells. Furthermore, the method is appropriate for near on-line automation and use in integrated process monitoring or control. Although this approach is not able to resolve G_0/G_1 or G_2/M because DNA content is the same, this is of little concern in most process biotechnology applications because the (transformed) cells commonly used do not enter G_0. Furthermore the G_2/M cell subpopulation can be resolved, if necessary, by measuring the mitotic index of the population to determine the proportion of cells in M phase.

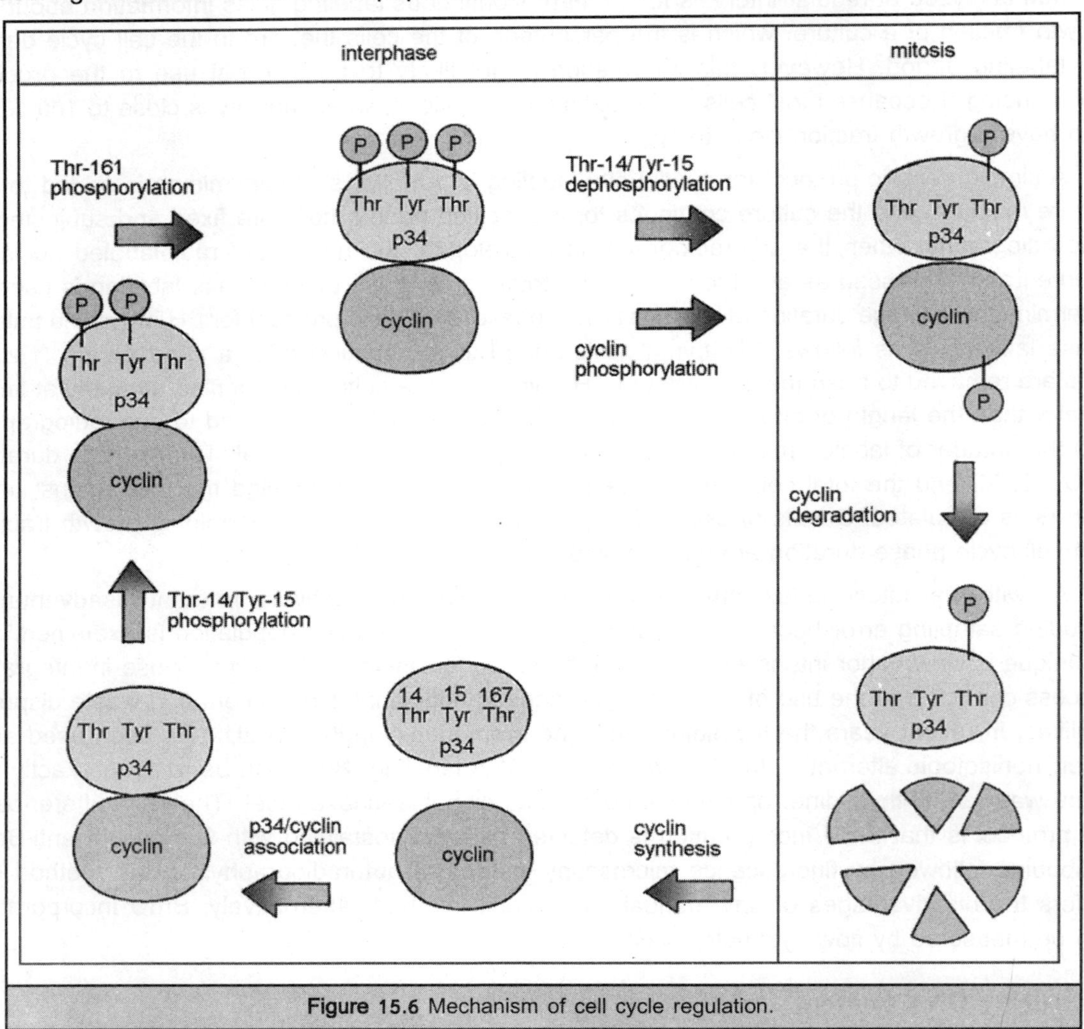

Figure 15.6 Mechanism of cell cycle regulation.

Flow Cytometry Cell Cycle Analysis

Flow cytometric cell cycle analysis by measuring DNA content may follow any one of many protocols. A generic protocol is as follows: fix/permeablize the cells; stain with a fluorescent DNA stain; measure stained DNA content by flow cytometry; and analyze data by simple gating or using dedicated cell cycle analysis software. Variations include pretreatment of cells; altered fixation/permablization protocols; post-fixation treatment of cells; different stains, although stains such as PI or EB, which intercalate into double-stranded nucleic acid, are commonly used; and methods for data analysis. A typical flow cytometric measurement of PI-stained DNA in ethanol-fixed recombinant CHO cells and the subsequent cell cycle analysis, using software based on a polynomial S-phase algorithm with an iterative nonlinear least squares fit, is shown. By fitting the G_1 and G_2 peaks as Gaussian distributions, and the S-phase distribution as a Gaussian broadened distribution, the cell cycle phase distribution can be deconvoluted. Flow cytometry can also be used with BrdU incorporation to measure the growth fraction or to determine average cell cycle phase times. For flow cytometric analysis, BrdU incorporation can be detected by either fluorescent immuno-staining or its quenching effect on Hoechst DNA stains. Thus BrdU incorporation can be analyzed simultaneously with DNA content.

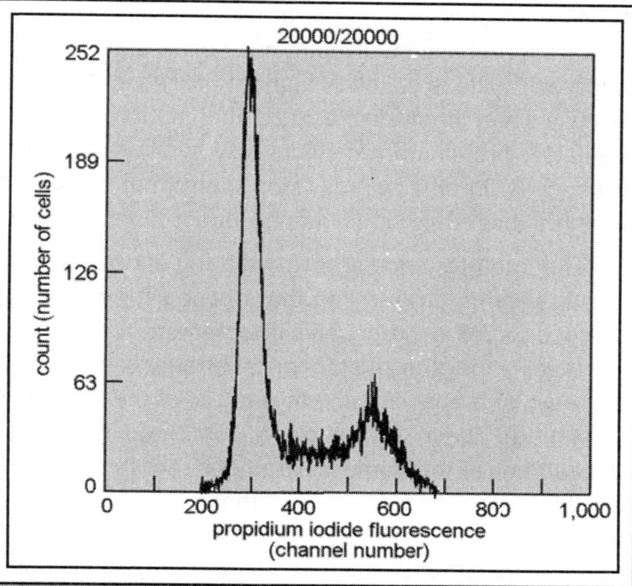

Figure 15.7 Typical flow cytometric measurement of fluorescent (propidium iodide-stained) DNA content in recombinant CHO cells.

The advantage of this method is that it can recognize arrested but non-DNA-synthesizing cells during S phase. Cell size and/or protein content can also be analyzed simultaneously to give further related information. Again, there are many variations on a theme, details of which and other flow cytometric methods for cell cycle analysis have been summarized elsewhere. Flow cytometry is also very useful for analyzing specific cell cycle–related proteins, following appropriate fluorescent immunostaining. For example, the absence of cyclin D or E in G_0 cells, shown by flow cytometry, has been used to distinguish G_0 from G_1 cells, and flow cytometric measurement of the Ki67 nuclear antigen has been shown to correlate with [^{3}H]thymidine and BrdU incorporation. Multiparameter flow cytometric analysis of G_1 and G_2 cyclin expression and DNA content has also been shown to be a useful tool in cell cycle analysis because G_1 cells can be differentiated from G_0 cells, and cells with unusual karyotypes (for example, G_1 tetraploid) can be distinguished from G_2 diploid cells. Flow cytometry can also be applied to investigating the relationship between the cell cycle and recombinant protein expression in economically important cell lines as a means of optimizing medium and environmental conditions for successful commercial operation.

SIGNIFICANCE OF THE CELL CYCLE

The cell cycle is the means by which a culture proliferates. Cell division and cell death, both active programmed cell death (apoptosis) and passive necrosis, are responsible for the change in the number of cells present. In addition to the cell cycle control genes, the transcription of genes coding for proteins involved in nucleotide and DNA synthesis is also cyclic. However, the progress of the cell cycle is costly in nutrient utilization and energy expenditure, particularly during S phase, which is when the nucleic acid content of the cell is duplicated, but also during other phases because regulatory proteins are continuously synthesized and degraded. In theory it should be possible to arrest cells, in other words, stop them from cycling, thus preventing the "wastage" of energy and nutrients that could otherwise be directed into product synthesis.

The maintenance energy model states that energy (as adenosine triphosphate [ATP]) consumption is divided into that needed for cell growth (cell cycle progression, as determined by increase in cell number) and that needed for cell maintenance. Based on this model it has been determined that the maintenance energy requirement of a hybridoma is 14.4×10^{-9} mmol ATP/(cell day) at a specific growth rate (μ) of 0.66 per day, corresponding to 62% of the energy (ATP) consumed. These values were confirmed by work that showed that the maintenance energy requirement of the same cells under slightly different nutrient and specific growth rate conditions was 11.4×10^{-9} and 12.0×10^{-9} mmol ATP/(cell day), again, corresponding to approximately 60% of the estimated ATP consumption rate. These observations agree with the value of 65%, which was determined as the proportion of energy consumption used for maintenance by mouse LS cells grown at $\mu = 0.60$ per day. Thus it is apparent that about two-thirds of the energy used by a culture is for its own maintenance, and this proportion is not likely to be readily altered.

However the remaining third, which is used for growth (and product elaboration), may be exploited. If the growth (cell cycle progression) requirement was removed, all of the remaining resources could be directed toward product elaboration, or alternatively, cultures would require up to one-third less energy and nutrient resources for the same product yield. Two main approaches have been used to prevent cells from cycling. The first is chemical blockade. For example, the addition of thymidine to culture medium causes cells to arrest at the G_1–S phase boundary, so cells accumulate in G_1, and specific productivity increases. However, this production advantage is short lived because the imbalanced growth state leads to culture death (usually by apoptosis) within a few days of the onset of the chemical blockade.

The second approach is not to arrest the cell cycle but to substantially reduce its progression by reducing the culture growth rate. Using a custom-built total cell recycle bioreactor system, where the growth rate of a mouse hybridoma was substantially reduced ($\mu < 0.05\ \mu_{max}$), Flickinger et al. showed that specific monoclonal antibody secretion remained constant as the specific growth rate decreased. This demonstrated that production could be uncoupled from growth (cell cycle progression), although there was no evidence that the energy saved by substantial (>95%) reduction of growth was diverted into production. However, using this method for production-scale processes may have the advantage of less medium (energy/nutrient) expenditure per unit product, although the total product yield per unit of production (from each bioreactor) would not be increased. Furthermore, questions of product fidelity, especially with respect to glycosylation, are raised when growth is limited by the stress of nutrient deprivation or limitation. Until recently, these were the only means available for controlling the cell cycle (i.e., preventing cycle progression). In 1997, Fussenegger et al. described a novel cytostatic process that enhances productivity. This was achieved by adopting a completely different approach compared with previous work. Metabolic engineering was used instead of the application of an external stress to the cells to prevent cell cycle progression. A series of transient CHO transfects was produced containing inducible dicistronic expression vectors, each of which contained a model product gene (human secretory alkaline phosphatase [hSEAP]) and a tumor-suppressor gene coding for an intracellular regulator of cell cycle progression (either p21, p27, or a mutant p53 [p53175P] that is said to have lost its proapoptotic properties while retaining its cell cycle regulation properties).

When the expression vector was induced, cells arrested and began to accumulate in G_1. The arrested cells showed a fourfold increase in hSEAP production compared with cells transfected with a control expression vector coding for the product only. Using stable dicistronic transfects, only p27 overexpression successfully induced cell cycle arrest (in G_1), but with it came a 10-fold increase in specific hSEAP productivity. Building on the success of these discistronic vectors, the tricistronic pBluescript-based pTRIDENT family of vectors was developed, allowing the possibility of a single stable transfection coding not only for a product but also for cell cycle control (cell cycle arrest) and enhanced culture survival using an anti-apoptosis oncogene such as *bcl-2* or *bcl-x_L*. Subsequent work with pTRIDENT vectors containing genes encoding for hSEAP with either p21 and the CCAAT/enhancer binding protein α (which induces and stabilizes p21) or p27 and Bcl-x_L showed even greater benefit, with specific productivities increased 15- and 30-fold, respectively, with respect to proliferation- competent controls.

Products and Cell Cycle

It is important not to confuse the relationship between cell cycle phase and productivity with the relationship between cell cycle arrest and productivity. The substantial increases in productivity from the metabolically engineered transfects just described are most likely to exist because the arrested cells do not need to devote resources to biomass production. This section addresses the situation in "simple" recombinant cells that may be grown in standard industrial processes such as batch, continuous, or draw-and-fill semicontinuous culture modes. Determining the relationship between cell cycle phase and productivity is important for developing automated on-line (or near online) process control strategies. These strategies will be based on models that, in addition to the basic parameters of pH, dO_2, cell density, and so forth, must incorporate information concerning cell state, including cell cycle phase and its effect on productivity. It has been shown that specific monoclonal antibody production by a synchronized mouse–mouse hybridoma culture is maximal during G_1; furthermore, the specific activity increases when the cell cycle is arrested and cells are maintained in late G_1. However, apparent specific productivity also increases when the culture declines (viability decreases); this is the result of substantial passive release of intracellular product into the culture medium from dead and dying cells.

Phase dependence was also found in recombinant CHO cells transfected with an expression vector containing *dhfr* and *lacZ* genes; the former was controlled by an SV40 promoter, and the latter was controlled by the constitutive hCMV promoter. In this study, the nonsecreted product, β-galactosidase, was produced almost exclusively during S phase. Furthermore, cell cycle arrest in S phase, by specific inhibition of DNA polymerase, did not inhibit transcription or translation and thus had no effect on β-galactosidase expression. However, the literature on the relationship between cell cycle and protein expression, summarized in Table 15.2, is inconsistent. Different researchers have shown that product expression (or the maximum rate of expression) is related to different cell cycle phases. In addition, others have shown that product expression is aphasic. These observations do not necessarily contradict each other and may be a true reflection of the systems used, because a variety of cells, products, and constructs were studied. Cell cycle–dependent product expression may not be universal and could vary with cell type or cell line, the construction of the expression vector, the nature of the recombinant gene expressed, or the promoter/enhancer used to drive product expression.

Table 15.2 Variation in phase expression of different products with cell line and promoter

Cell line	*Promoter*	*Product*	*Phase expression*
Mouse leukaemia	Endogenous	DHFR	Aphasic
CHO	Endogenous	DHFR	Aphasic
Human osteosarcoma	Endogenous	DHFR	Aphasic
Mouse embryo fibroblast	Endogenous	DHFR	Aphasic
Human prostate epithelium	Endogenous	PSAP	All except early G_1
Hybridoma	Endogenous	MAb	G_1/G_0 (mainly)
CHO	AMLP	DHFR	G_1 (mainly)
Hybridoma	Endogenous	MAb	G_1/S
Mouse plasmacytoma	Endogenous	Immunoglobulin	Late G_1/early S
Human lymphoid cell	Endogenous	Immunoglobulin	Late G_1/S
CHO	Endogenous	DHFR	G_1/S
Mouse leukemia	Endogenous	TMPS	S
CHO	CMV	β-Gal (*lacZ*)	S
CHO	SV40	tPA	S (mainly)
CHO	SV40	IFN-γ	S
Mouse L-cell	MMTV	β-Gal (*lacZ*)	S (predominantly)
CHO	Endogenous	DHFR	S
Rat adenocarcinoma	Endogenous	tPA	Shortly after S
Rat adenocarcinoma	Endogenous	uPA	Shortly after S
CHO	Endogenous	PA	G_2/M

Other variables may include posttranscriptional regulation (mRNA level), product gene copy number, and even the culture mode, (e.g., whether the cells are attached to a substrate or are free in suspension) may be relevant. However, there are two potential flaws in studying cell cycle-related productivity: synchronizing cells using chemical means, such as thymidine block or nutrient starvation, and comparing cells of different phases from the same culture that are separated in time. The two problems often go together, for example, following a culture for a day after release from thymidine block. The difficulty with the former is that the chemical stress used (what ever its nature) may, in addition to its desired effect of cell synchrony, directly cause perturbations in productivity that could wrongfully be ascribed to the cell cycle. The difficulty with the latter is simply that if two cell populations from a culture are separated in time, they are also separated in environmental conditions; recent work has shown that the effect of medium condition is dominant to any cell cycle effect on productivity. The effect of temporal separation is made more uncertain following stress synchronization; an early population may have only partially recovered from the stress, whereas a later population has fully recovered. Cells may be arrested in G_1 before the restriction point, but once past that point, they are committed to continue their cycle and divide unless prevented from doing so by catastrophe or artificial intervention. Therefore, one can presume that, once a cell enters S phase, under normal growth conditions, one single cell will in due course become two cells. This phenomenon can be used to predict culture growth. The growth-enhancing effect of peptone supplementation on hybridoma culture was predicted by cell cycle analysis 5 h after supplementation, although the effect on cell density was not apparent until 16 h later, 21 h after supplementation. A similar use of cell cycle analysis was made with CHO cells: the percentage of S-phase cells correlated with the specific growth rate 10.5 h later; the time difference was the average duration of S phase remaining plus the duration of G_2 and M for that population. Thus, the proportion of cells in S phase can be used to predict the subsequent proliferation rate of that culture.

Proliferation Predictor and Cell Cycle

The effect of hydrodynamic stress, as found in stirred and sparged bioreactors, is at least in part related to cell cycle phase, and there may be a role for hydrodynamic cell damage in the induction of apoptosis. It has been reported that DNA synthesis and the proportion of the cell population in S phase were increased under conditions of hydrodynamic stress (high sparging and agitation rates). However, the cell number remained low even though the S-phase population correlated with an increase in cell number. The observed reduction in growth rate was shown to be a consequence of a decline in the proportion of G_2- and M-phase cells, with a corresponding decrease in the cell size of the population, suggesting that the larger G_2 cells were particularly susceptible to hydrodynamic damage. Subsequently it was shown that mitotic cells, while also being susceptible to hydrodynamic damage due to their size, were even more sensitive to hydrodynamic damage than G_2 cells, perhaps due to cytoskeletal rearrangements, alteration of plasma membrane fluidity, or morphological changes during mitosis.

Cell Cycle and Apoptosis

Apoptosis (active or *programmed cell death*) is the means by which intrinsic or extrinsic stimuli activate a series of molecular cascades, triggering events that culminate in a particular morphologically, ultrastructurally, and biochemically defined process of cell death that is, in effect, the opposite of the cell cycle. Entry of a cell into apoptosis can be restricted to a particular cell cycle phase, although this cell cycle dependence does not appear to be universal. Cell cycle progression and apoptosis are not only opposite in effect, they may also be alternate outcomes from a single cell control mechanism; they are inextricably linked by common components in the signal transduction pathways for both processes. In addition, circumstantial evidence from the induction of apoptosis by DNA damage or cell cycle disruption and the existence of many oncogenes that both promote the cell cycle and induce apoptosis implies that there is an equilibrium between these two antagonistic processes in all eukaryotic cells.

16

EUKARYOTIC CELL CYCLE

The growth of eukaryotic cells is governed by a complex array of biochemical reactions that are collectively known as the cell cycle. The cell cycle controls are intimately coupled to the internal and external states of the cell and thus ultimately determine when a cell devotes the majority of its energy to growth, DNA replication, successful cell division, or cell maintenance. A recent explosion in the amount of research devoted to cell cycle controls has revealed that the activity of enzymes known as cyclin-dependent kinases (CDKs) regulates the order of cellular events that culminate in cell division. Furthermore, these CDKs are themselves subject to strict controls that govern their expression, activation, and degradation, giving a cell the freedom to respond to a variety of situations. An overview of the findings from the numerous biological studies that have been conducted on cell cycle control is presented here.

Additionally, because the cell cycle controls the most fundamental aspects of cell growth and division, it can be linked to the dynamics of many other cellular processes important to cell culture such as protein accumulation and secretion. Therefore, in order to achieve a more complete understanding of the cell cycle, a discussion of some of the investigations that have coupled the cell cycle to the dynamics of cellular process is presented. Indeed, the framework diagrammed here offers many opportunities for further research that can be used to engineer more efficient methods of cell culture.

BIOLOGY OF THE CELL CYCLE

The cell cycle comprises a group of complex interrelated events that exists to ensure that cells successfully divide to form two normal daughter cells. The eukaryotic cell cycle was originally divided into two stages, interphase and mitosis, as observed by light microscopy. Interphase was defined as the period between two successive rounds of cell division, whereas mitosis was identified as the period when the cell undergoes many structural changes, divides it chromosomes, and undergoes cell division. The advent of biochemical and genetic techniques later showed that interphase consists of many cellular processes necessary for the regulation of cell proliferation that can not be discerned under the light microscope.

The cell cycle was then partitioned into four separate phases: G_1 (or gap 1), DNA synthesis (S phase), G_2 (gap 2), and mitosis (M phase), where G_1 refers to the gap between mitosis and DNA synthesis, and G_2 to the gap occurring after DNA synthesis and prior to mitosis. Understanding the complex network of biochemical events that guide a cell through successful completion of cell division is of utmost importance in understanding how cells behave in response to different situations. There has been an explosion in the amount of cell cycle related research being done in the last 10 years. This is due, in part, to the fact that a wide variety cellular processes are thought to be somehow regulated by cell cycle events and proteins. Additionally, improper control of cell cycle processes in higher eukaryotes has been linked to the formation of cancerous tumors. Here we discuss the biochemical events that occur in the cell cycle and the control mechanism that maintains the temporal order of these events.

Zero and First Mitotic Gap Phase

Upon initial inspection, it would seem that the G_1 phase of the cell cycle is a particularly uneventful time period (or gap) between the more spectacular events that occur during mitosis and DNA synthesis. Closer analysis of this interval, however, reveals that the cell is closely monitoring its environmental and internal state to determine whether the cell has sufficient resources and an intact genome to safely complete the rest of the cell cycle. In contrast to the other phases of the cell cycle, G_1 is highly variable between different cell lines and growth conditions. For example, if certain factors are not available, some cells do not progress toward S phase, but instead may enter a resting state called the G_0 phase, or quiescence.

During this time, the cell undergoes a significant reduction in RNA and protein synthesis and will remain in this arrested state, with unduplicated DNA content, until the cells receive the proper stimuli. The sequence of events that takes place during the G_1 phase has been deduced by observing how cells respond when cultured in different environments. For instance, when cultured cells are deprived of serum containing growth factors, the synthesis of new proteins is drastically reduced and a portion of newborn cells cease to advance through the cell cycle. Through observations made using time-lapse cinematography on a culture of Swiss 3T3 cells, it was determined that only those cells that exited from mitosis less than 4 h prior to growth factor deprivation were arrested in G_0, while all other cells continued to mitosis as normal. This observation led to the division of G_1 into two sub-phases.

The first subphase consists of cells that are still dependent on growth factors for cell cycle progression and are referred to as G_1 postmitotic (G_{1pm}) cells, whereas cells that can proceed through the rest of the cycle are called G_1 presynthesis phase (G_{1ps}) cells. The transition point where cells pass from G_{1pm} to G_{1ps} is commonly called the restriction point in mammalian cells and is analogous to "Start" in yeast. During G_{1pm}, progression through the cycle is highly dependent on growth factors and amino acids that are necessary for the production of certain regulatory proteins (called cyclins) that are discussed later. It is thought that the purpose of the G_{1ps} period is to allow cells to grow, perhaps so that later in the cycle, the cell can focus its energy on other cellular processes such as

DNA replication in S phase and reorganization of the cellular infrastructure during mitosis. This also explains the fact that the G_{1ps} portion of G_1 accounts for nearly all the temporal variability in the cell cycle both within a population and between cell lines. For example, after having passed the restriction point, larger cells may proceed almost immediately to the S phase, while smaller cells may linger in G_{1ps} for up to 10 h before beginning the transition to S phase.

Additionally, although growth factors are not required to promote S phase entry once past the restriction point, the length of G_{1ps} is significantly increased when factors such as insulin and insulin-like growth factor (IGF) are lacking in the environment, presumably due to a decrease in the net protein synthesis. Under these conditions, cells undergo mitosis with a significantly reduced protein content, suggesting that different growth factors play an important role in growth and progression through the cell cycle. A culture of 3T3 cells arrested in G_0 by serum deprivation can reenter the cell cycle and proceed through the restriction point only after stimulation by the addition of platelet-derived growth factor (PDGF); this process is referred to as competence. Factors that initiate competence, however, vary between cell lines, and several other growth factor/cell-line combinations have been described. The biochemical effects of competency factors such as PDGF are discussed later. Time-lapse cinematography also revealed that cell cultures that lack serum do not contain cells arrested all at one point, but at numerous times throughout G_{1pm}. Additionally, when stimulated to proliferate, arrested cells require an extra 8 h in G_1 before proceeding to S phase, presumably so the cells can recover from G_0.

Synthesis Phase

The S phase is the period when the cell replicates its entire genome so that each daughter cell will receive a complete, intact set of chromosomes. Due to the immense size of mammalian genomes, multiple bidirectional replication sites are necessary for the synthesis of new DNA to be accomplished in such a short time (approximately 8 h in most cell lines). In addition to chromosome duplication, the cell must also synthesize proteins such as histones for the formation of nucleosomes and the chromosomal scaffolding. The replication of DNA has been shown to be a highly ordered process. This was observed by culturing cells with a thymidine analog called bromodeoxyuridine (BrdU). Cells in the process of synthesizing DNA take up BrdU and integrate it into newly formed DNA. Cells can then be separated by their position in S phase, based on the amount of BrdU in the DNA, to determine the exact time of replication of certain genes. These studies revealed several aspects of the replication of certain genes.

First, individual genes within a cell line are replicated at defined periods in the S phase, with many of the more active genes replicated earlier than others. Furthermore, newly replicated DNA is packed into nucleosomes a short distance behind the replication fork. It is also very important that the cells not replicate the same chromosome more than once; the mechanism that ensures that this does not happen is known as the re-replication block. Early cell fusion experiments on human HeLa cells showed that, prior to S phase, nuclei could be induced to undergo DNA synthesis early, whereas nuclei that have already undergone a round of DNA synthesis do not undergo another round of DNA synthesis. Later cell fusion experiments showed that nuclei with replicated DNA can undergo another round of DNA synthesis before mitosis after the nuclear membrane has been permeablized and then repaired. A model involving a replication-licensing factor (RLF) was then proposed to describe these phenomena. DNA would be replicated only if RLF was bound to the chromatin to be broken down by the replication machinery during the next S phase.

During mitosis, the nuclear envelope breaks down, allowing new RLF to access the chromosomes and bind to them, thus licensing the chromosomes to undergo a single round of DNA replication. Support for this model was gained when two protein complexes, called RLF-M and RLF-B, were shown to be required to modify chromatin before replication is allowed to proceed. These protein complexes were shown to bind to chromatin in late mitosis and be gradually removed during S phase. The re-replication block model described earlier has recently been modified to accommodate a more detailed understanding of the proteins and mechanisms involved. Chromosomes in G_1 must first be "primed" by the establishment of a pre-replication complex (pre-RC) containing an origin-recognition complex (ORC), an RLF complex, and other proteins. After initiation of DNA synthesis, the pre-RC is disrupted to become the post-RC, which still contains the ORC. The ORC may be some sort of "docking" factor that allows other components of the pre-RC to assemble on the DNA.

Second Mitotic Gap Phase

Cells enter the G_2 phase upon completion of DNA synthesis, and this period lasts between 2 and 4 h, culminating with the onset of chromosome condensation. During this time, duplication of the centriole is completed, and cells produce many mRNAs and proteins that will be used during mitosis. Cells may also use this period of time before mitosis to ensure that the cell is sufficiently large and is ready to undergo cell division. Mitosis is generally the shortest phase of the cell cycle, lasting only 15 min to 1 h, depending on the cell line. However, visual inspection of a cell culture reveals that this is a very physically active period for the cell in that it goes through many structural changes in preparation for cell division. Mitosis is traditionally divided into six stages that describe the mechanical events that occur during this period: prophase, prometaphase, metaphase, anaphase, telophase, and cytokinesis.

The beginning of mitosis, known as prophase, is marked by condensation of the chromosomes into compact structures that allow sister chromatids to be more easily separated. The cytoskeletal microtubules begin to disassemble, and the mitotic spindle, a weblike array of microtubules that mediates the events in mitosis, begins to form. Also during this phase, the Golgi apparatus breaks up into tiny vesicles, and pinocytosis and RNA synthesis stop. Prometaphase begins with the disruption of the nuclear envelope, followed by the attachment of the chromosomes to the microtubules that make up the mitotic spindle. The microtubules attach to protein structures on the DNA called kinetochores, which as we shall see are very important in regulating progression through mitosis.

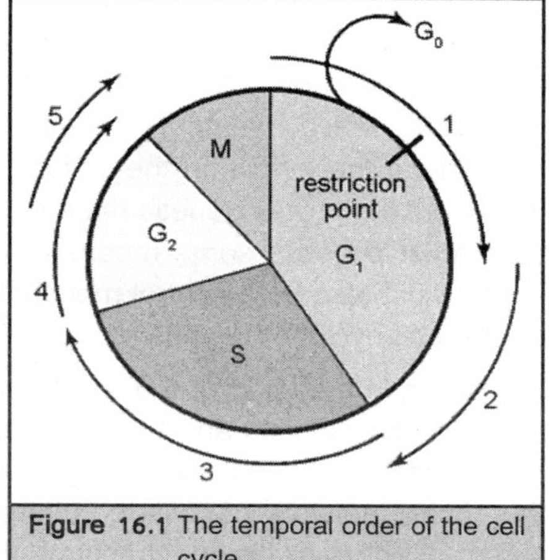

Figure 16.1 The temporal order of the cell cycle.

During metaphase, the mitotic spindle arranges all the chromosomes in a single plane across the middle of the cell, called the metaphase plate. After alignment of the chromosomes, the cell enters anaphase, which is when sister chromatids are pulled to opposite sides of the cell by the microtubules so that there is a complete genome on either side of the cell. Microtubules not attached to kinetochores then elongate, pushing the spindle poles further apart. Telophase then begins, resulting in the formation of two nuclei, one around each set of chromosomes on opposite sides of the cell. The cell then begins the process of cell division, called cytokinesis, when a ring of actin and myosin that has formed around the center of the cell begins to contract. Upon completion of this process, two new daughter cells are formed, each containing a complete, intact genome and cytoplasmic organelles sufficient to support further growth and proliferation.

Table 16.1 Activity of the major cyclin–CDK complexes

Complex	*Cyclin–CDK complex*	*Cell cycle phase*	*Function*
1	Cyclin D–CDK4	G_0, early to mid-G_1	Stimulates exit from G_0 and progression through the restriction point; phosphorylates pRb
2	Cyclin E–CDK2	G_1–S transition	Triggers S phase entry; activates cdc25A
3	Cyclin A–CDK2	S	Phosphorylates DNA synthesis machinery; inhibits E2F/DP1 transcription factor
4	Cyclin A–CDK1	G_2, M	Triggers formation of mitotic spindle
5	Cyclin B–CDK1	G_2–M transition, M	Triggers entry into M; responsible for many structural changes in mitosis
6	Cyclin H–CDK7	All cell cycle phases?	Cyclin activating kinase (CAK); phosphorylates other cyclin–CDK complexes

REGULATION

Successful completion of cell division requires that the cell reside in an environment capable of supporting growth, and that the cell cycle proceed as normal. As discussed previously, cells require the presence of growth factors and essential amino acids to support the synthesis of new proteins. In addition, the immediate environment must not be overcrowded; daughter cells must have room to grow and subsequently divide. If any of these conditions is not met, cells enter the G_0 phase, which eventually results in cell death if environmental conditions are not improved. Similarly, if the events in the cell cycle do not occur in the proper sequence, daughter cells may not inherit the proper genetic material or organelles to properly function, and they will die. Therefore, cells must possess some means of sensing whether the environment is able to support growth. There must also be a control mechanism in place that regulates the sequence of events that make up the cell cycle.

An understanding of the mechanisms that control cell growth is of interest to many scientists including oncologists and biochemical engineers. Oncologists may be able to use the knowledge of the events regulating the cell cycle to develop therapies against certain types of cancers, whereas the control network of cells in culture can be engineered to reduce the culture's dependence on serum. Central to the mechanisms that control the events of the cell cycle are the checkpoints.

Checkpoints are points in the cell cycle where certain criteria must be met before the cell cycle can proceed any further. For example, the restriction point in G_1 is a checkpoint because cells not cultured in the presence of growth factors are arrested in G_0 and not allowed to proceed through the cell cycle until growth factors are supplied.

Indeed, several other checkpoints that control the order of the cell cycle phases and the events within each phase exist throughout the cell cycle. For instance, cells with unduplicated DNA do not enter mitosis, and chromosomes are replicated only once in each cycle. Additionally, cells with damaged DNA do not undergo mitosis, rather they become arrested in G_2 to allow the DNA repair machinery to correct any defects in the genetic material. Controls that mediate the order of events in mitosis also exist. For example, entry to metaphase is prevented until the chromosomes are lined up at the metaphase plate. As we will see, complexes made up of two classes of proteins are involved in regulating all these events throughout the cell cycle. These two classes of proteins are cyclins and cyclin-dependent kinases (CDKs).

Molecular Regulation

Cyclins were originally defined as molecules that accumulated throughout the cell cycle in marine invertebrate eggs and were then degraded during each mitosis. Several years later, the amino acid sequences of various cyclins were compared, and a consensus sequence of approximately 100 amino acids was found and dubbed the "cyclin box". Cyclins are now defined on the basis of the presence of the cyclin box in the native sequence. Currently, 12 cyclins have been identified in mammalian cells (cyclins A–H, with two A-type, three B-type, and three D-type cyclins). The cyclin box domain facilitates binding to CDKs, whose kinase activity is dependent on being bound to a cyclin molecule and are thus defined as proteins that require binding to cyclins in order to attain activity. There are seven known CDKs in animal cells (named CDK1–CDK7). Different cyclins may interact with different CDKs at specific points in the cell cycle, and each of these complexes is specific for substrates with differing functions. Because several of the complexes have similar substrate specificity in vitro, the change in specificity is thought to be due not only to changes in the affinity toward different substrates, but also to localization of complexes in distinct parts of the cell by various cyclins. We first briefly discuss the different cyclin–CDK complexes that form throughout the cell cycle and then review the regulation of these complexes. The major studies that resulted in the discovery of G_1 cyclins were performed in budding yeast. It was shown that three cyclin molecules, Cln1, Cln2, and Cln3, have important functions in advancing cells through the restriction point (called Start in yeast).

In addition, inactivation of all three molecules is required to cause G_1 arrest of the cell cycle. By searching for factors that could complement the function of Clns in yeast, several mammalian cyclins were isolated: cyclin C, cyclin D, and cyclin E. Other studies revealed several other G_1 cyclins, cyclins G and F, however, neither the roles that these cyclins play nor their CDK partners are known. It has been shown that cyclin C mRNA synthesis reaches its peak in mid-G_1, and cyclin G mRNA synthesis starts soon after growth-factor stimulation of quiescent cells. Conversely, cyclins D and E are fairly well characterized and have been the focus of much attention because defects in control of these proteins have been implicated as a cause of cancer. Although D-type cyclins have been shown to associate with CDK2, CDK5, and CDK6, they seem to interact primarily with CDK4. Transcription of various combinations of D-type cyclins (cyclins D1, D2, and D3) is stimulated by different mitogens (growth factors), depending on the cell type, and their synthesis throughout the rest of the cell cycle is maintained by the presence of growth factors.

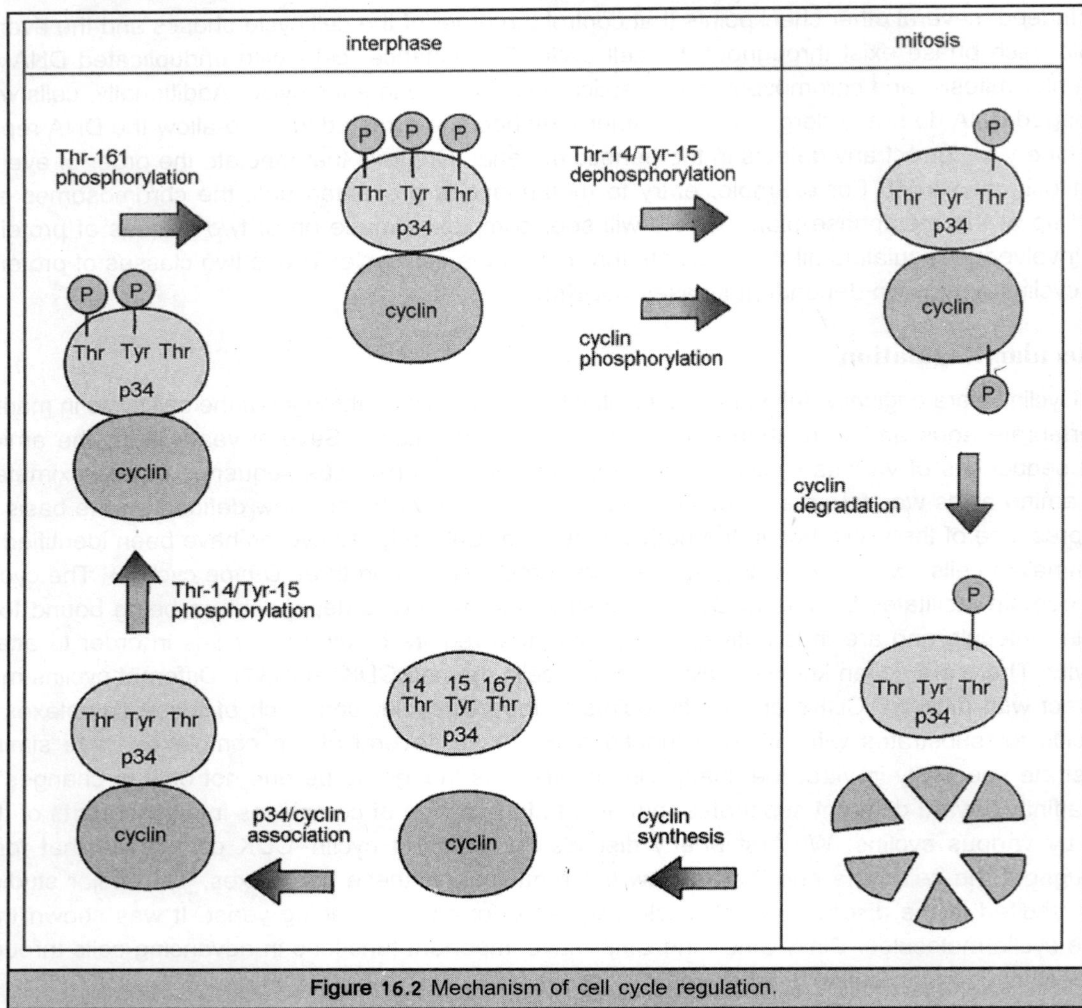

Figure 16.2 Mechanism of cell cycle regulation.

Removal of growth factors halts the synthesis of D cyclins, resulting in a rapid decrease in their levels because they have very short half-lives. Furthermore, in cycling cells, the level of cyclin D hardly varies, and only a small peak showing its accumulation is observed. As noted previously, growth-factor deprivation has no effect on cells that have passed through the G_1 restriction point, and these cells continue to cycle until their progeny become arrested in early G_1. Additionally, microinjecting cells with cyclin D1–specific antibodies results in their cell cycle arrest only during early to mid-G_1, whereas overexpression of cyclin D1 results in a reduction in the length of G_1, the size of the cell, and growth-factor dependency. These findings indicate that the primary function of

D-type cyclins is to link the cell cycle machinery with the extra-cellular environment, which permits the cell to become committed to DNA synthesis when in a favorable environment or become arrested in G_0 under unfavorable conditions. Cyclin E is expressed soon after cyclin D expression in G_0 cells stimulated with growth factors, and its level peaks late in G_1. It has been shown to interact with a single catalytic subunit, CDK2.

In contrast to D-type cyclins, cyclin E is synthesized periodically and is rapidly degraded soon after entry into S phase, thus adhering to the original definition of a cyclin. Similar to cyclin D, cyclin E function has been shown to be essential for progression through G_1, because microinjection of anti–cyclin E antibodies prevents cells from entering S phase, whereas its overexpression results in acceleration toward S phase. This acceleration is, however, balanced by an increase in the lengths of the S and G_2 phases so that the length of the entire cycle remains unchanged. Because an increased amount of cyclin E advanced progression to the S phase early, it is possible that the cyclin E-CDK2 complex is responsible for initiating the synthesis of proteins that are used during DNA synthesis. This hypothesis accounts for the increase in the length of the S and G_2 phases following increased expression of cyclin E. These findings, when taken together, suggest that the cyclin E–CDK2 complex is involved in regulating the passage from G_1 into S. Upon entry into S phase, cyclin E is degraded and cyclin A takes its place as the partner of CDK2.

Transcription of cyclin A is begun immediately after entry into S phase, and its inhibition by microinjection of anti-cyclin A antibodies allows, at most, only 10% of the DNA to be replicated and may also disrupt the temporal relationship between DNA synthesis and mitosis. As a cell progresses into the G_2 phase, cyclin A is no longer associated with CDK2, however, it forms a new complex with CDK1 (also known as cdc2 due to its homology with the budding yeast kinase of the same name) to regulate progression through G_2 and into mitosis. It has been shown that CDK1 forms complexes with both cyclin A and cyclin B near the G_2–M transition. These complexes are located in different parts of the cell because cyclin A accumulates in the nucleus, and cyclin B resides in the cytoplasm until the breakdown of the nuclear envelope during mitosis. Once they have completed their respective functions, both cyclins are then rapidly degraded later in mitosis. Experiments done in fission yeast suggest that the cell monitors the presence of cyclin B, because cells undergo another round of DNA replication when cyclin B–CDK1 complexes are not present. This implies that the level of cyclin B–CDK1 complexes, both active and inactive, are constantly monitored, and the presence of these complexes maintains cells in G_2. Although this theory has not been investigated in other eukaryotes, it is not entirely unlikely given the high degree of conservation of cell cycle controls among eukaryotic cells.

Cyclin–CDK complex regulation

Three different mechanisms allow tight control over the myriad of cyclin–CDK complexes: (a) cyclin transcription and degradation, (b) CDK phosphorylation and dephosphorylation, and (c) binding of the cyclin–CDK complexes to various inhibitory proteins. These three forms of cyclin–CDK regulation interact to form an intricate regulatory network that allows cells to temporally control events throughout the cell cycle and to ensure that these processes are successfully completed. Intracellular concentrations of different cyclins, and thus the activity of cyclin–CDK complexes, are regulated by synthesis and degradation. There are two different pathways by which cyclins can be degraded, depending on the type of cyclin. Cyclins D and E, which function in early G_1 and the G_1–S transition, respectively, are inherently unstable and thus have short half-lives of approximately 30 min. The instability of these cyclins has been attributed to the presence of a PEST sequence near the C-terminus of the cyclin box. In contrast, cyclins A and B, which function in the S, G_2, and M phases, are stable throughout interphase, even when they are not bound with CDKs. However, these cyclins are degraded during mitosis in the proteosomes through a ubiquitin-mediated pathway. These proteins contain a sequence of amino acids near the amino terminus, called the destruction box, that marks them for destruction.

Degradation of these proteins is also regulated so that cyclin–CDK complexes that are necessary for cell cycle progression are not inactivated before their specific tasks are successfully completed. For example, it has been shown that proteolysis of cyclin A and B does not occur in *Xenopus* egg extracts until purified cyclin B–CDK1 complexes are added. This suggests that the proteolysis pathway is initiated by active cyclin B–CDK1 complexes. Conversely, the addition of purified cyclinA–CDK1 complexes to the extracts had the opposite effect of delaying cyclin degradation triggered by cyclin B–CDK1 kinases. The fact that only cyclin B–CDK1 complexes can activate proteolysis may explain why cyclin degradation does not occur until correct assembly of the mitotic spindle is achieved in metaphase. The exact mechanism of regulation of cyclin degradation is unknown, but it has been hypothesized that cyclin B–CDK1 activates a cyclin-specific enzyme that catalyzes the conjugation of ubiquitin to cyclins. Cyclin–CDK complexes are also regulated by phosphorylation and dephosphorylation. Specifically, phosphorylation on T160/161 (depending on the CDK) is necessary for activation of the complex, whereas phosphorylation of T14 and Y15 residues inhibits cyclin-CDK activity. The T160 residue is located on a T loop, which is conserved in all CDKs. When not phosphorylated, this loop prevents kinase activity by covering the substrate binding domain. Phosphorylation at this site serves two purposes: it increases the affinity between the cyclin and the CDK and therefore stabilizes the complex, and it alters the conformation of the T loop, allowing access to a substrate.

The kinase that is responsible for phosphorylation of this residue on several CDKs is also a cyclin–CDK complex. This complex, commonly referred to as CDK-activating kinase (CAK), is made up of cyclin H, CDK 7, and a third 32-kDa protein. Although the level of CAK activity has not been shown to vary in a cell cycle–dependent manner, it is possible that cyclin H is regulated by its synthesis and degradation. The kinase responsible for activating CAK has not been identified, but it has been shown not to be an autophosphorylation process. Further studies done with this complex have shown that it may play a role in activating the RNA polymerase II C-terminus domain kinase of transcription factor IIH. This suggests that cyclin H and CDK7 may have a more diverse role than merely activating other CDKs. Cyclin–CDK complexes are also subject to negative regulation through the phosphorylation of the T14 and Y15 residues.

These residues are found within the ATP binding site of the enzyme. Phosphorylation of the Y15 residue inhibits kinase activity by interfering with the transfer of a phosphate group to a bound substrate, whereas phosphorylation of the T14 residue acts by interfering with binding of ATP. The regulation of phosphorylation on these residues is not yet well defined in animal cells. In addition, only the process of tyrosine phosphorylation is known in yeast because phosphorylation of the threonine residue is not necessary in this organism. In yeast, the tyrosine residue is phosphorylated by the Wee1 and Mik1 kinases. The Wee1 kinase is in turn regulated by the Nim1 kinase, which inhibits Wee1 activity. The kinase responsible for phosphorylation of the Y15 residue in animal cells has been isolated and is closely related to Mik1.

The dephosphorylation of these residues, which is necessary for cyclin–CDK activation, is performed by the phosphatases Cdc25A, B, and C. Cdc25A has been shown to be activated by the cyclin E–CDK2 complex in early S phase, which then activates cyclin A–CDK2 complexes. The Cdc25C phosphatase activates cyclin B–CDK1 complexes at the transition from G_2 to mitosis; however, the function of cdc25B is as yet unknown. The activity of cyclin–CDK complexes is also controlled by certain proteins known as cyclin-dependent kinase inhibitors (CKIs or CDIs) that bind to these complexes and repress their activation. Several different inhibitors have been isolated from mammalian cells that interact with either different cyclin–CDK complexes or CDKs themselves. These proteins are commonly named according to their molecular weights, such as the 21-kDa inhibitor known as p21. Other cyclin–CDK inhibitors are p27, p15, and p16. These inhibitors are intricately involved in regulating progression through the G_1 phase, and have thus been the subject of recent reviews.

Immunoprecipitation studies have revealed a very interesting fact: cyclin D1 is commonly found among a quaternary complex of proteins that includes either CDK4 or CDK6, p21, and proliferating-cell nuclear antigen (PCNA). The p21 protein acts as an inhibitor by binding directly to cyclin B–CDK1, cyclin E–CDK2, cyclin A–CDK2, and cyclin D–CDK4 complexes. These cyclin–CDK complexes containing p21 are, oddly enough, still enzymatically active in vitro, which directly conflicts with studies showing that p21 is an inhibitor of CDK activity and is able to induce cell cycle arrest. It was later demonstrated that active cyclin–CDK–p21 complexes isolated from cycling cells could be inactivated by the addition of more p21, suggesting that the stoichiometry of p21 binding to these complexes is very important. For example, in cycling cells, there is only enough p21 present so that each cyclin–CDK complex has a single p21 molecule and thus remains active. However, when the amount of p21 is increased or the cyclin–CDK pool depleted, the ratio of p21 to cyclin–CDK complexes increases so that more than one p21 molecule binds to each complex, resulting in inhibition of the complex's kinase activity. We explore the circumstances under which this sort of inhibition becomes important in a later section. A second inhibitor that has been identified is called p27. p27 and p21 have been shown to have highly similar amino acid sequences near their amino terminus domains. In addition, they also share the same range of specificity for cyclin–CDK complexes.

The cell cycle arrest of certain cell lines in response to some environmental signals such as contact inhibition and transforming growth factor β (TGF-β) has been shown to be due, in part, to inactivation of cyclin–CDK complexes by this inhibitor, and the exact mechanism of this type of arrest will be described shortly. A recent discovery has given some insight into one of the mechanisms by which this protein inhibits cyclin–CDK activity. Cyclin–CDK complexes that have not yet been activated by CAK phosphorylation can be bound by p27, which then blocks the CAK's access to the threonine residue found on the CDK subunit. Presumably, another mechanism exists for inhibition of active complexes. The main function of the p27 inhibitor seems to be to prevent the G–S transition from occurring when the need arises by inhibiting the activity of cyclin E–CDK2 complexes.

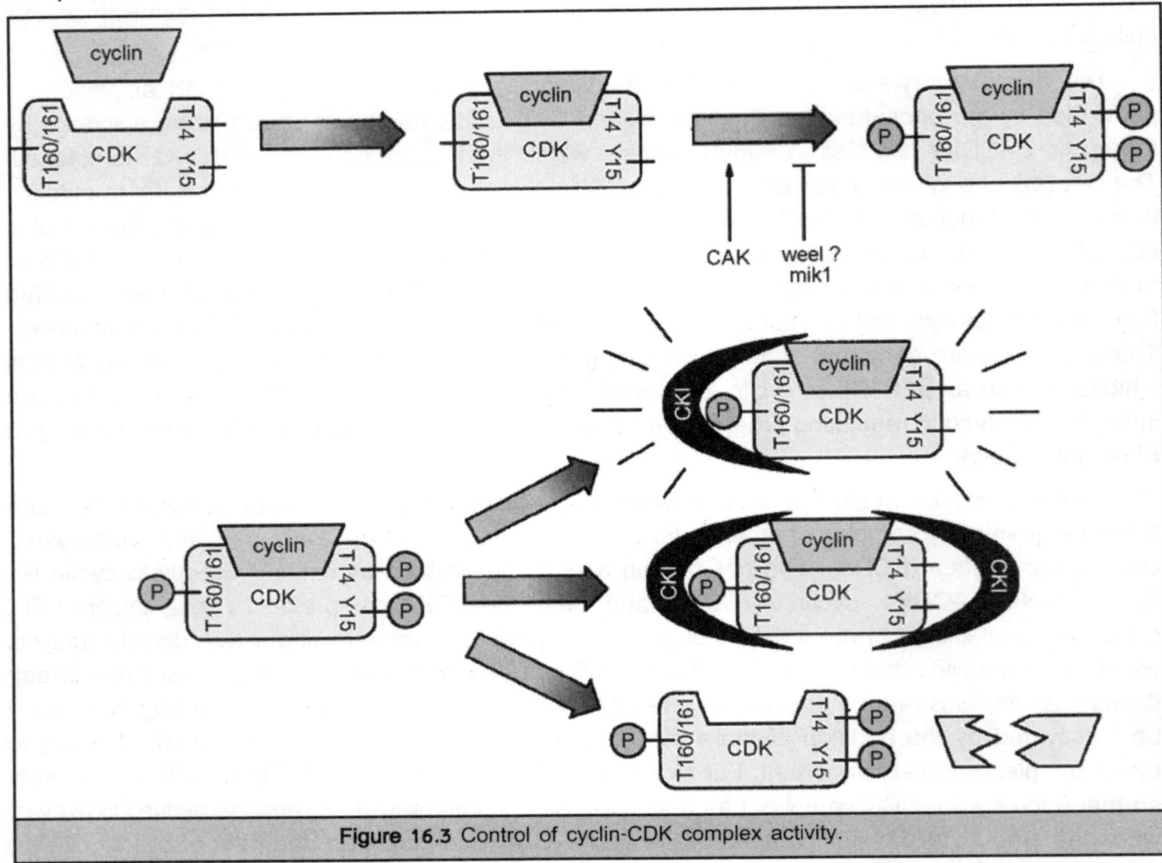

Figure 16.3 Control of cyclin-CDK complex activity.

In normal cycling cells, the p27 inhibitor seems to be dispersed among the different cyclin–CDK complexes that are active during G_1, therefore it does not have a significant impact on any essential events. However, in response to stimulation by TGF-β, it has been suggested that the expression of CDK4 is down-regulated, resulting in an increased amount of p27 bound to cyclin E–CDK2 complexes. The p16 inhibitor was originally found associated with CDK4 in a transformed cell line. Unlike the p21 and p27 inhibitors, p16 was found to associate specifically with the cyclin D CDKs, specifically, CDK4 and CDK6. It competes with cyclin D for the CDK subunits and may even act to disrupt cyclin D–CDK complexes. This inhibitor is thought to play an important role in regulating growth because the p16 gene has been shown to be deleted in many tumor-cell lines. Examination of TGF-β-arrested cells led to the discovery of another inhibitor, p15, because it was shown to be bound to CDK4. Similar to p16, p15 binds to CDK4 and CDK6, competes with cyclin D, and may be involved in arresting cells treated with TGF-β in G_1. TGF-β stimulates the synthesis of p15 so that its level is increased approximately 30 times. This allows p15 to sequester the CDK4 complexes so that a much larger amount of the p27 inhibitor binds to and inactivates cyclin E–CDK2 complexes, resulting in G_1 arrest.

Action of Major Regulatory Proteins

Now that we have discussed the events that occur in the cell cycle and defined the major regulatory proteins, we are in a position to examine the mechanisms that control cell cycle progression. Indeed, control of the activity of the cyclin–CDK complexes is at the heart of cell cycle regulation because these complexes either directly or indirectly act on cellular constituents that are capable of bringing about changes in intracellular events. This implies that the proper regulation of these cyclin–CDK complexes is essential for maintaining the temporal integrity of cell cycle events. This is accomplished by repressing the activity of the appropriate kinase through different combinations of cyclin degradation, phosphorylation/dephosphorylation of specific CDK residues, and action of CDK inhibitors. These controls not only maintain temporal stability, but also ensure that DNA synthesis is not carried out when DNA is damaged or that mitosis does not occur in the presence of damaged or misaligned chromosomes.

Influence of extracellular environment

An important property in normal cells is their ability to sense the state of the extracellular environment, such as the presence of various growth or inhibitory factors or contact inhibition. This communication between the cell and its surroundings takes place in the G_0 and G_1 phases, and the cells respond by becoming quiescent, undergoing programmed cell death (known as apoptosis), or continuing with the cell cycle. The interaction with the environment ceases when the cell passes the restriction point, after which the cell is said to be committed to DNA synthesis. We now consider the machinery responsible for directing the path that the cell will traverse in response to signals from the environment. Several lines of evidence indicate that the product of the retinoblastoma gene, pRb, is responsible for linking signals from the extracellular domain. Specifically, mutation or deletion of this gene is found in many cancerous cell lines that have lost their ability to regulate proliferation. Additionally, pRb accumulation was found to coincide with suppression of growth in TGF-β-induced arrest, whereas inhibition of pRb results in a resumption of cell cycle progression.

Mammalian DNA viruses, whose main purpose is to activate quiescent cells so that they can use the cell's replication machinery to duplicate their own DNA, have been found to seek out the hypophosphorylated form of pRb. This led to the belief that pRb is in its active state when it is under phosphorylated, and phosphorylation of this protein results in its inactivation. This is supported by the fact that hypophosphorylated pRb is present in G_0 cells and is phosphorylated later during the G_1 phase. In its active form, pRb binds to several proteins including a heterodimer of proteins from the E2F and DP family of proteins. This heterodimer, called DRTF1/E2F, is a transcription factor that stimulates the transcription of numerous proteins necessary for cell cycle progression, such as CDK1, cyclin A, and a variety of proteins that are indispensable during DNA synthesis. In order to further characterize G_1 events, it is necessary to find out what is responsible for inactivating pRb. A likely candidate seems to be the cyclin D–CDK4 complex. The presence of this active kinase is highly dependent on the presence of growth factors that stimulate the synthesis of cyclin

D, and thus, the activity of that complex appears during the same period when pRb is inactivated. Moreover, cells that lack functional pRb do not require the presence of cyclin D–CDK4 to proceed through the cell cycle, implying that the primary function of this complex is to inactivate pRb. Indeed the belief that pRb is a substrate of this complex was confirmed because cyclin D–CDK4 is capable of binding to and phosphorylating pRb in vitro.

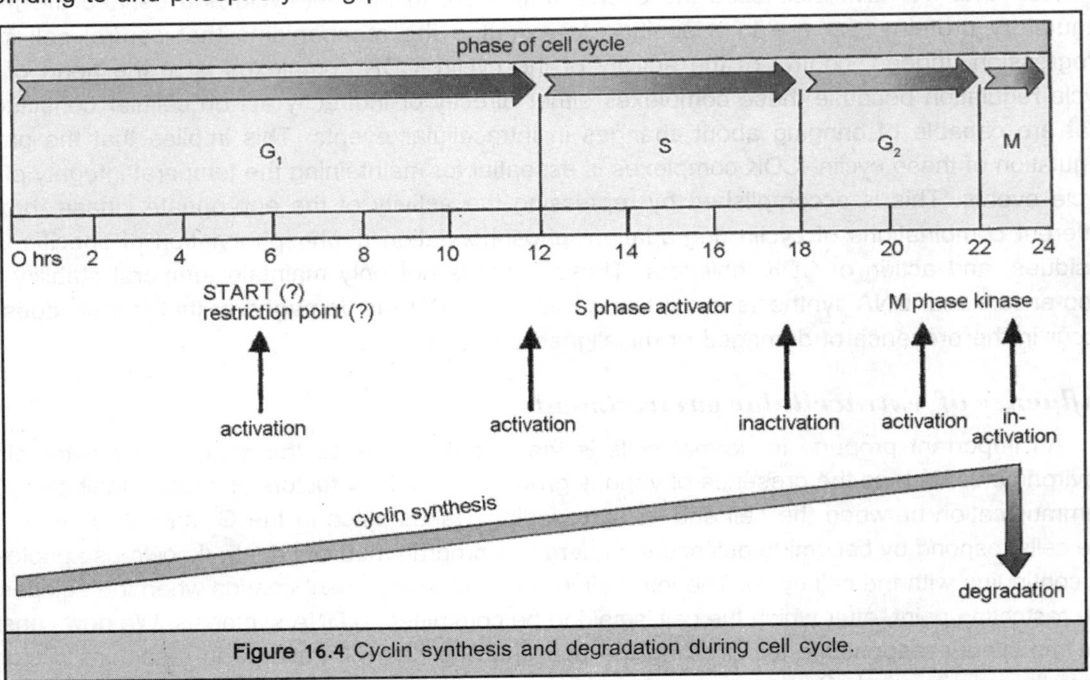

Figure 16.4 Cyclin synthesis and degradation during cell cycle.

The following feedback loop, involving the levels of cyclin D and the functional state of pRb, has been proposed. Cells that lack functional pRb have drastically reduced levels of cyclin D; however, addition of pRb to these cells results in the synthesis of cyclin D. Active kinase complexes are then allowed to form, which act by phosphorylating pRb. The hyperphosphorylated form of pRb down-regulates the levels of cyclin D during late G_1, causing the inactivation of CDK4, which in turn allows pRb to be activated once again by the action of phosphatases, and the cycle begins anew upon completion of mitosis. Because the CDK inhibitors p21 and p27 are constitutively expressed in cycling cells, these molecules are present throughout G_1. This indicates that in order for cells to advance past the restriction point, the level of cyclin D–CDK4 complexes must increase until it exceeds the number of inhibitor molecules present so that there is at most one CKI per cyclin–CDK complex to allow kinase activation.

This model not only satisfies the timing of cyclin D–CDK4 activation, but suggests a dual role for this complex. Its primary function is to phosphorylate pRb to promote the transcription of important genes, while its secondary role is to sequester the CKIs so that the cyclin E–CDK2 complex, which is the next kinase to be active in the cell cycle, cannot be inhibited by these molecules, thus allowing the transition to S phase to occur. As mentioned previously, the activity of cyclin E–CDK2 complexes during the cell cycle has been narrowed down to the transition from G_1 to S. Some of the possible substrates for this complex include the so-called pocket proteins, which include pRb, p107, and p130. The term *pocket protein* refers to those proteins involved in binding to certain proteins, such as transcription factors, and releasing them after activation of the pocket protein by the action of a CDK kinase. The cyclin E–CDK4 complex is localized in the nucleus where it is bound to E2F, p107, and p130 (although it has not been shown to phosphorylate these proteins), suggesting that it too plays a role in regulation of transcription. The E2F subunit that interacts with cyclin E–CDK4 has been shown to differ from the E2F found associated with pRb, which hints that they have different properties. One confirmed substrate of cyclin E–CDK2 is the Cdc25A phosphatase that is involved in activating CDKs by cleaving the inhibitory phosphates on both a tyrosine and a threonine residue.

The cdc25A phosphatase has been proposed to activate both cyclin E–CDK2 and cyclin A–CDK2 complexes, creating a positive feedback loop between cyclin E–CDK2 and Cdc25A. If this feedback loop does indeed exist, it has been hypothesized that the activation of Cdc25A occurs in response to the completion of some G_1 event such as the duplication of the centrosome or the accumulation of certain DNA synthesis proteins. Experiments with yeast suggest that proteolysis may be involved in activating the cyclin E–CDK2 homologue that plays an important role in the G_1–S transition. It has been shown that a protein called Cdc34 is necessary for the G_1–S transition to take place, possibly because it is responsible for initiating the degradation of certain CKIs. Functional homologues of the Cdc34 protein have been isolated in vertebrate cells and are thought to stimulate the degradation of the p27 inhibitor. This, in addition to isolation of CKIs by the cyclin D–CDK4 complex, could possibly aid in beginning the onset of the synthesis phase. In summary, a cell becomes arrested in G_0 in response to some extracellular signal that triggers a decrease in

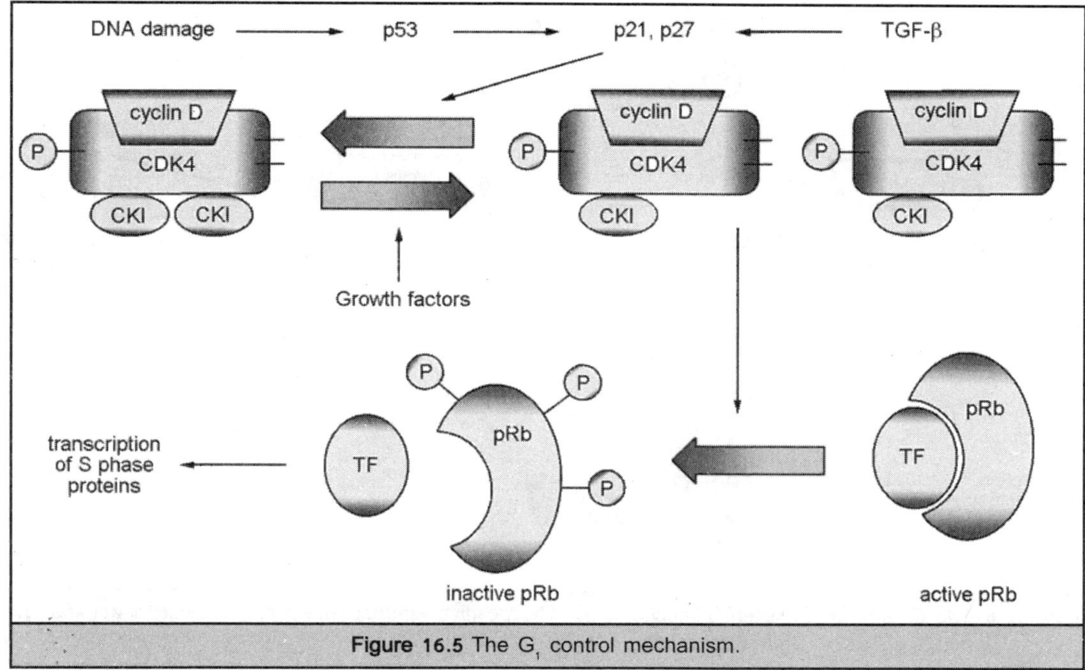

Figure 16.5 The G_1 control mechanism.

the level of cyclin D below a threshold level of CKIs. This allows p21 and p27 to bind to cyclin D–CDK4 complexes in a 2:1 ratio, resulting in an inhibition of kinase activity.

Following stimulation of cyclin D synthesis due to the presence of growth factors, there is an increase in cyclin D–CDK4 kinase activity because there is no longer enough inhibitor present to inactivate all the complexes. This active CDK kinase then serves two purposes: (a) to inactivate pRb, allowing the release of transcription factors that induce the synthesis of other cyclins and DNA synthesis proteins, and (b) to isolate the CKIs to allow activation of the cyclin E–CDK2 complex. This complex then stimulates the release of additional transcription factors and initiates a positive feedback loop that triggers the onset of the DNA synthesis phase. Once cells move beyond the restriction point, they are committed to entering the DNA synthesis phase. Cells do, however, have the power to delay the onset of later events to ensure the proper coordination of important processes. For example, failure to repair damaged DNA greatly increases the occurrence of genetic mutations, which can have disastrous consequences for the cell. Thus, an additional checkpoint exists in late

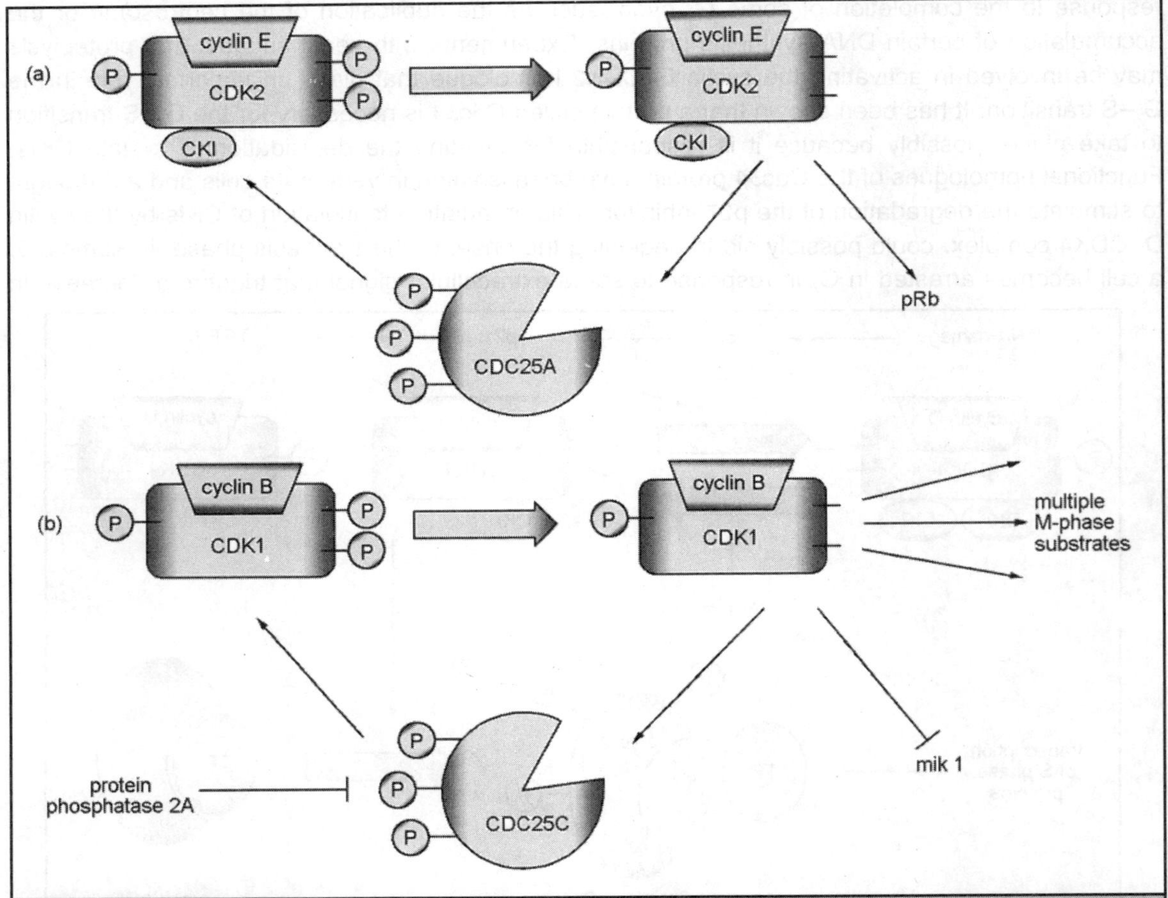

Figure 16.6 The S- and M-phase entry mechanisms. The positive feedback mechanisms that trigger the entry into S phase (a) and mitosis (b).

G_1 phase, which ensures that the DNA is free from damage and ready to be replicated. We will now see that cells possess the means to sense when there is damage to the genetic material and then halt cell cycle progression to allow the DNA machinery time to repair any defects in the genetic material. It has been determined that the there is a single protein, p53, that is held accountable for inducing G_1 arrest in response to DNA damage.

By inducing DNA damage in normal cells, it was determined that cells responded by increasing the level of p53 and the CKI p21. The increase in p21 then leads to the inhibition of both cyclin D–CDK4 and cyclin E–CDK2 complexes to prevent entry into the S phase. Additionally, p21 also functions to oversee the proper repair of DNA by interaction with PCNA. PCNA is required for both DNA synthesis and repair. When bound with p21, PCNA is no longer able to interact with polymerase-δ, thus preventing DNA synthesis, however, the DNA repair function remains unaffected by the inhibitor. Therefore, p21 must inhibit both CDK activity and interaction between PCNA and polymerase-*d* in order for DNA repair to be carried out before advancing to the S phase.

Progression of synthesis S phase and G_2 phase

Shortly after the cyclin E–CDK2 complex triggers entry into the S phase, cyclin E is degraded and CDK2 interacts with cyclin A. There is evidence that cyclin E may be degraded in a ubiquitin-mediated pathway, as in budding yeast, but this has not been confirmed. The cyclin A–CDK2 complex is localized in the nucleus where it associates with p107, p130, and E2F. Interacting with E2F allows the kinase to phosphorylate the DP1 subunit of the transcription factor complex, which inhibits its DNA binding capability. This suggests that when cyclin A synthesis begins in early S phase, it aids in halting the transcription of genes triggered by the E2F complex in the late G_1 phase. No other substrates of this cyclin–CDK complex have been identified except for helicase RF-A, which is one component of the DNA synthesis machinery. Upon completion of DNA synthesis, CDK2 is degraded and cyclin A forms a new complex with CDK1. It is not known if this complex has any function in the G_2 phase; however, its activity has been implicated in triggering initial events that occur in mitosis.

During G_2, cyclin B begins associating with CDK1 in the cytoplasm and is transported to the nucleus just prior to mitosis. Mitosis begins when the cyclin B–CDK1 complex is activated, and it has been suggested that a checkpoint mechanism exists that prevents activation of this complex until duplication of the centriole has been completed. One model that describes this transition involves the regulation of Cdc25C activity by down-regulating protein phosphatase 2A (PP2A), which controls Cdc25C activity. Signals could be initiated from the DNA replication machinery or from the proper organization of chromatin, which would inhibit the action of PP2A. This allows Cdc25C to become activated and dephosphorylate the inhibitory residues on CDK1, causing its activation. A positive feedback loop is then initiated, whereby active cyclin B–CDK1 complexes activate more Cdc25C and, in budding yeast, inactivate the Wee1 kinase by phosphorylizing a residue that inhibits its action. This mechanism brings about a sharp transition into mitosis almost immediately after a small amount of CDK1 is activated. Another mechanism exists that allows cells with damaged DNA to become arrested in G2. In response to DNA damage, transcription of the CHK1 kinase is initiated. This kinase targets Cdc25C and phosphorylates the serine-216 residue, which allows formation of a binding site for the 14-3-3 protein that inhibits Cdc25C activity. This then allows the DNA repair machinery to work because the cyclin B–CDK1 complex can not be activated by cdc25C.

Stepwise events in mitosis

The point when the cyclin A–CDK1 complex is most active occurs just after the cell enters into mitosis, which coincides with nucleation of the centrosomes and the formation of the long microtubular extensions of the mitotic spindle. Indeed it has been shown that cyclin A–CDK1 activity has a profound affect on the nucleating ability of the centrosomes in *Xenopus* extracts. Once spindle formation is complete, cyclin A begins to be degraded and the activity of cyclin B–CDK1 complexes takes over. This complex is either indirectly or directly involved in regulating almost all of the structural reorganization that takes place during mitosis. Chromosome condensation begins in prophase when cyclin B–CDK1 phosphorylates histone H1, transcription factors, and several other proteins. Phosphorylation of these proteins is thought to weaken the interaction between these proteins and the DNA, allowing the chromosomes to pack more tightly. In addition, after formation of the mitotic spindle is complete, the cyclin B–CDK1 complex reduces the nucleating ability of the centrosomes, which causes the microtubules to shorten to a length suitable for the alignment and separation of the chromosomes.

The breakdown of the nuclear envelope, which marks the beginning of prometaphase, and the reorganization of the cytoskeleton are also brought about by phosphorylation. The nuclear lamina are present in a hypophosphorylated form in an intact nucleus, and when phosphorylated by the cyclin B–CDK1 complex, the lamina network breaks down, contributing to nuclear disassembly. Lamina phosphorylation itself is not sufficient to achieve disruption of the nucleus, however, and these mechanisms remain unclear. Caldesmon has also been shown to be a substrate of the mitotic CDK complex, and microtubular reorganization ensues after its phosphorylation. Caldesmon binds to actin and calmodulin, and its phosphorylation weakens its affinity for actin, causing the microfilaments to dissociate. This process causes the cells to take on a round shape that is characteristic of anchorage- dependent cells undergoing mitosis. The activity of the cyclin B–CDK1 complex is also responsible for repressing the formation of the contractile bundle until the cell enters anaphase. The CDK1 complex phosphorylates myosin, which prevents its association with actin filaments so the contractile bundle is unable to form. This inhibitory function of the CDK complex stays in effect until cyclin B is destroyed in the ubiquitin degradation pathway.

The activity of the cyclin B–CDK1 complex is, ironically, responsible for the initiation of its own destruction, which stays in effect until this pathway is shut down in late G_1, possibly by the action of the cyclin E–CDK2 complex. The final regulatory mechanism that exists in the cell cycle is a checkpoint that prevents the onset of anaphase until several other structural events have taken place in the cell. Some of these events include the assembly of a bipolar spindle, attachment of the chromosomes to the spindles through the kinetochore, and the migration of the chromosomes to the metaphase plate. This suggests that proper completion of these processes is somehow monitored to ensure that anaphase does not begin prematurely. Several experiments suggest that one of two events signals that the cell is ready to undergo the chromosome separation of cell division. These events are either the attachment of the chromosomes to the mitotic spindle or the presence of tension on the kinetochores due to their attachment to a bipolar spindle. Because all the structural reorganization processes that occur in mitosis depend on the successful completion of the other events, failure to complete any one of these processes could result in the failure of the signaling event to occur. This would then stimulate a cascade of phosphorylation events, culminating in cell cycle arrest in mitosis.

Mathematical Modeling

Understanding the complex array of biochemical events we have outlined can be difficult at times. Moreover, our knowledge of the controls responsible for cell cycle regulation is rapidly becoming more and more complicated. This complexity makes it extremely difficult to understand the mechanisms involved through logical thinking alone. Mathematical modeling has been used for the purpose of explaining the various physiological events that are commonly observed in different situations, and the events observed in *Xenopus* oocytes and yeast cells. The cell cycle controls discussed in this review have been described for animal cells. However, the majority of initial experiments on the cell cycle were carried out in yeast cells and in *Xenopus* oocyte extracts. The components that make up the control systems in these organisms are largely identical to those of animal cells, and many findings resulting from studies on these organisms have been extrapolated to apply to animal cells. An initial review of the literature would suggest that the control systems in these organisms are different; however, this is not the case—only the terminology used to represent the components differs. The cell cycle nomenclature used in these organisms has been reviewed elsewhere. Modeling of the molecular controls that are thought to control the cell cycle in *Xenopus* oocytes and yeast cells has been done in order to determine the effect of changing the activity of various components of the regulatory network. We now believe that progression through the cell cycle is governed by the activity of cyclin–CDK complexes, in other words, a sharp rise in the complex's activity results in a transition to the next phase.

The aim of the molecular modeling done by Novak and Tyson was to determine whether the control mechanisms presented here could be used to describe the observed behavior of cells in a variety of situations, or more simply stated, to compare theory with experiment. In order to do this, a set of equations (one for each component of the system) was generated, based on the mechanisms described, to describe the rate of change of each component as a function of time by translating the proposed control mechanisms into simple reaction rate equations. The resulting set of differential equations was then solved using numerical techniques; the activity of the cyclin–CDK complexes could be followed with time, and the sharp oscillations in complex activity could be varied. Because sets of equations

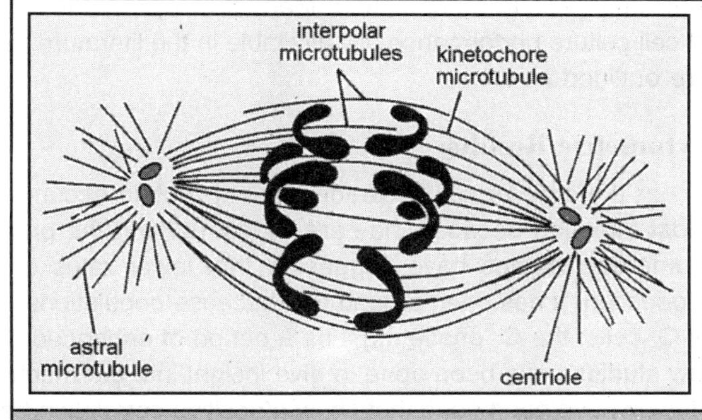

Figure 16.7 Spindle apparatus showing the three types of microtribules.

commonly contain more than one solution (depending on rate constants and initial conditions used), phase-plane plots were constructed. Observation of limit-cycle-type behavior in these plots indicates an oscillatory solution that is characteristic of proposed cell cycle control. This type of model is easily solvable by today's personal computers and has been successfully implemented to describe the observed behavior of a variety of yeast cell cycle mutants and embryo and oocyte maturation.

LARGE SCALE CULTURES OF ANIMAL CELLS

The use of large-scale cultures of animal cells is becoming increasingly important as the demand for their biological products grows. Animal cell cultures are used for the production of many different products of diagnostic and therapeutic value, including vaccines, monoclonal antibodies, and in some cases, the cells themselves (for example, to replace the function of damaged tissue). Determining the culture conditions that optimize the production and secretion of proteins is therefore of utmost importance in meeting these demands. Consequently, the key to designing such systems is not only determining the kinetics of growth, secretion, and proliferation, but also understanding how the cell cycle regulates certain cellular processes. The heterogeneous nature of cell populations has made it extremely difficult to obtain these data. Various mathematical models have been developed to describe cell behavior under different conditions; however, they are, for the most part, unable to account for the heterogeneous nature of cell populations and usually do not provide insight into variations introduced by the cell cycle.

One model used to describe cell growth within a heterogeneous population of cells, known as population balances, has been developed. However, the implementation of this approach has been hampered by the difficulty in obtaining the single-cell parameters necessary for its use. These problems are being tackled through the development and improvement of tools such as flow cytometry and fluorescent markers, which allow the examination of single-cell parameters within a heterogeneous population. Flow cytometry is a very sensitive tool that allows one to simultaneously obtain several single-cell parameters related to growth physiology from a large number of cells in the population. Collecting these data allows one to determine the size, DNA content (and thus cell cycle position), and protein content (when the protein is conjugated to a fluorescent marker) of individual cells in a population. Many flow cytometric studies relating the cell cycle to optimization of cell culture performance are available in the literature. Several of these analyses and their findings are outlined next.

Cytometric Results

In the area of cell cycle–dependent protein accumulation, hybridoma cells have received the most attention because they are widely used in the production of monoclonal antibodies (MAbs). Numerous studies have suggested that lower rates of growth result in higher levels of protein production. It has been argued that because populations growing at slow rates have a larger fraction of G_1 cells, the G_1 phase must be a period of enhanced protein production and secretion. However, few studies have been done to give insight into the mechanism for this behavior. Additionally, other studies have been done that show a positive correlation between growth rate and protein production. These variations in response observed by different researchers may be explained by the fact that the overall growth rate of a culture depends on both cell growth and cell death. This means that the same overall growth can be observed in two cultures with different rates of cell loss and cell growth. The conclusion that can be made from the examination of these different investigations is that it is important to be aware of the different methods used to quantify growth and protein production in cell cultures because they may result in conflicting results. Al-Rubeai and colleagues used flow cytometric results obtained from synchronous cultures and concluded that the rates of both immunoglobulin synthesis and secretion are at a maximum at the G_1–S transition. The synchronization of the culture was achieved by the addition of thymidine, which arrests cells prior to entry into the S phase. This type of analysis presents another problem, however, because factors used for cell cycle arrest have been shown to alter cell metabolism; therefore these results should be viewed with care.

A more complete series of studies on rates of MAb production during the cell cycle was performed by Kromenaker and Srienc. The first study involved using flow cytometric data obtained in asynchronous cultures and applying these data to population balances to find the single-cell rates of protein accumulation in various cell cycle phases for both antibody-producing and non-antibody-producing cell lines. It was found that during the early G_1, producer cells with low protein content have a specific rate of accumulation of protein approximately 3–5 times greater than the specific rate observed for the remainder of G_1 cells. This indicates that the rate of protein synthesis is much greater than the rates of protein degradation and secretion in G_1. The S phase of the cell cycle was shown to a have a low specific rate of protein accumulation that remained relatively steady throughout this period. Extending this analysis to the G_2 + M phase yielded a negative specific rate of protein accumulation, indicating that protein is lost due to secretion and degradation during this period of the cell cycle. From these findings, it was concluded that the disruption of the secretion system that occurs during the division of the cell accounts for the extremely high rate of accumulation observed in early G_1 because the transport system probably has not had time to recover from its disruption. A later report by Kromenaker and Srienc described the relation between growth rate and secretion rates during the cell cycle as determined using single-cell data and population balances. Lactic acid was used as a growth suppressor, and the concentrations were changed to vary the specific growth rate. This study brought several previously unconsidered factors to light.

By analysis of the DNA distribution of the populations at different growth rates, it was found that the lengths of all the cell cycle phases were increased by the same factor that the specific growth rate was increased. This is in direct contrast to other studies that suggest that the increase in productivity is due to the increase in the length of the most productive phase of the cell cycle. Secondly, the application of population balances in this study revealed that the net specific rate of antibody synthesis is independent of growth rate. Additionally, the specific secretion rate of G_2 + M cells is greater than the rate in the other two phases and remains constant with decreasing growth rates. In contrast, the specific secretion rates for G_1 and S increase with decreasing growth rate. These findings are consistent with the hypothesis that the disruption of the secretion pathway during cell division results in lower secretion throughout G_1, until the Golgi apparatus has had sufficient time to reassemble. During this time, antibodies accumulate in the cell until they can be secreted during G_2 + M, before the Golgi apparatus is disrupted once again. Thus, at lower growth rates, more time is spent in G_1, so an increased secretion rate is observed because the cells have had time to recover from cell division. A subsequent study tracked the amount of cell surface antibodies as a function of DNA content. Because prior experiments determined that the amount of cell surface antibodies observed on the membrane is a good measure of the rate of secretion, these data were used to determine the pattern of secretion during the cell cycle.

The results obtained from this method also agreed with the hypothesized regulation of secretion already described. Understanding the regulation of foreign genes in animal cells is also of great interest because recombinant DNA is commonly used to express proteins of interest in immortal cell lines such as Chinese hamster ovary (CHO) cells. This provided the motivation for studies in which the expression of foreign genes in animal cells was tracked throughout the cell cycle. One study tracked the secretion of tissue-like plasminogen activator (tPA) in recombinant CHO cultures synchronized by fluorescence-activated cell sorting (FACS) based upon DNA content. This study shows that secretion of tPA reaches its maximum near late S phase, while the increase in this protein is maximum in G_1.

These results are fairly similar to those obtained by Kromenaker and Srienc using different methods. Gu and colleagues also measured the expression of foreign β-galactosidase during the cell cycles in recombinant CHO cells and in recombinant mouse cells and hypothesize that control of foreign gene expression is promoter dependent. These particular studies, however, showed that the use of either the cytomegalovirus (CMV) promoter or the mouse mammary tumor virus (MMTV) promoter causes the production of β-galactosidase mainly in S phase.

Behavioral Studies

There is a steady increase in the number of reports in the literature that look to the cell cycle to explain various phenomena such as differing growth rates between cultures and cell death due to shear stress in bioreactors. For example, the effect of different inoculation sources was examined by Al-Rubeai et al. Two cultures that were inoculated with cells from either the exponential or stationary growth phase were examined using flow cytometry. The culture inoculated with cells from the exponential phase grew at a much greater rate than the other culture and therefore reached the death phase more quickly due to nutrient exhaustion and buildup of toxins. Analysis of the cell cycle, cell size, and mitochondrial activity of the cultures revealed differences among the cultures. For instance, the slower-growing culture was found to have a larger proportion of its cells in G_1, while the culture with the greater growth rate had a larger fraction of its cells in S. Because G_1 cells are generally smaller than the S-phase cells, the culture with a larger fraction of cells synthesizing new DNA uses more nutrients than the one with smaller cells.

Additionally, cultures with elevated mitochondrial activity were also shown to have a greater capacity for proliferation. These findings lead the researchers to conclude that examination of these parameters can provide a means of predicting how the culture will behave and a means of explaining variations in the growth patterns of cultures. It is also of great interest to determine how cells behave in different culture systems. For cells in continuous culture, it has been found that the fraction of cells in the G_1 phase decreases with increasing growth rate, while the fraction of cells in the S phase increases.

As expected, the G_1 phase accounts for most of the resulting variability in the lengths of the cell cycle phases; however, the S phase is affected slightly more than the G_2 and M phases. A more detailed analysis of the cell cycle was performed in cultures containing microcarriers, and the fraction of contact-inhibited cells was determined and compared to a model. Microcarriers are commonly employed in cultures of anchorage-dependent cells, such as CHO cells, because these types of cells must be attached to a surface in order to proliferate. When the culture grows to the point where there is no longer any surface area available for newborn cells, contact inhibition occurs, and a portion of the cells cease to proliferate even when provided with an ample supply of nutrients. Analysis of this system yielded the distribution of cells in the cell cycle during the lag phase, exponential growth, and confluency.

During the lag phase, the culture is predominantly in the G_1 and/or G_0 phases and rapidly enters the other cell cycle phases upon reaching exponential growth. As the culture reaches confluency, the culture gradually shifts back to predominantly the G_1 and/or G_0 phases; however, a portion of the cells continue cycling. These findings are similar to those of suspension cultures. Furthermore, a method was developed to determine the fraction of cells affected by contact inhibition, and the results showed good agreement with a model that had been developed. Furthermore, when culturing animal cells in large-scale bioreactors, the cells can be injured due to the shear stress that results from agitation and sparging.

There have been several studies that examined the mechanism of cell damage and sought to minimize injury to the cells through the use of various shear-protecting agents such as Pluronic F-68 and serum. However, there have been relatively few studies that examine the effect that these forces may have on cell metabolism and growth. It is evident that shear forces have an effect on the overall growth rate of the culture due to an increase in cell death.

Lakhotia and coworkers analyzed the rate of DNA synthesis and the cell cycle phase distributions of cultures grown in small-scale bioreactors with gas sparging and varying agitation intensities to determine the effect that these forces may have on growth kinetics of the surviving cells. Measurement of the kinetics was accomplished by measuring the uptake of BrdU and the DNA content distributions of the surviving cells with the aid of flow cytometry. They determined that, in cultures exposed to high agitation, the fraction of cells in the S phase increased by up to 45%, while the fraction of cells in G_1 decreased by up to 50%, compared to control cultures. The simultaneous analysis of the DNA synthesis rate suggests that the change in the cell cycle phase distribution is due to an increase in the DNA synthesis rate.

Additionally, when cultured in the absence of shear stress, the apparent specific growth rate of the cells previously exposed to the shear forces was initially higher than and later the same as that of control cells. It was also shown that an increase in the agitation intensity resulted in a further increase of the DNA synthesis rate. By subjecting cells to extreme, short-term stresses induced by turbulent capillary flow, it was determined that larger cells are more susceptible to damage than smaller cells. This conclusion was made based on the preferential disappearance of S and G_2 cells when passed through the capillary. Another possible explanation for these results is that an apoptotic response triggered by the shear forces caused a decrease in the size of the cells, but verification of this hypothesis is difficult.

Bioprocessing Technology

Once a more complete picture of how the cell cycle regulates cell physiology and metabolism has been painted, it can possibly be manipulated in culture in such a way that cells will perform desired processes at an optimal level. For example, controlling the regulatory mechanisms of cell growth and proliferation, or controlling the rate at which cells advance through the cell cycle can lead to the enhancement of protein production, improved economic feasibility, and ease of downstream processing. Indeed there have been several studies published to date that address these issues and will undoubtedly lead to further examination of directing cellular operations through the engineering of the cell cycle in cell cultures.

Cells in culture are subjected to many different forms of stress that may ultimately lead to cell death. Cell death can be caused by external factors such as hydrodynamic forces that may cause the cell to rupture. Additionally, the response to certain stimuli such as limitation of glutamine, glucose, or serum or the buildup of toxins may result in programmed cell death, called apoptosis, which is regulated by some of the same pathways as is the cell cycle. When in culture, cells reach their maximum cell density in a few days, after which the process of programmed death is induced in the majority of the cells.

The discovery that the product of the protooncogene *bcl-2* is able to suppress apoptosis lead to its use in cell culture to prolong the productive period of the culture. Several recent studies show that cells expressing this gene increased MAb production by 40% and prolonged the viable culture time by up to 4 days by delaying the onset of apoptosis for 2 days. Additionally, it was determined that the increase in MAb production was equally due to both the prolonged survival of cells and the increase in protein production per cell. The expression of the *bcl-2* gene also allows cells to be cultured with a significant decrease in serum supplementation. Specifically, *bcl-2* overexpressing cells could maintain high viable cell concentrations for up to 4 days, while the control culture lasted only 1 day when supplemented with 0.5% serum. Moreover, cells cultures supplemented with 9% serum with additional 2% serum fed after 4 days sustained productivity for 10 days, while the control culture underwent apoptosis after 4 days. These experiments show that the cell cycle controls can be bypassed when expression of the appropriate gene is induced in culture.

A similar approach has been used in CHO cells to bypass the cell cycle controls that require the presence of growth factors for efficient proliferation. The observation that cells stimulated by growth factors possess elevated cyclin E levels led researchers to transfect the CHO cells with a cyclin E expression vector. The expression of this gene in cells cultured in protein-free media resulted in cell proliferation and levels of cyclin E, cell morphologies, and cell cycle phase distributions comparable to those obtained in growth-factor-stimulated cells. An approach related to the one just described employed expression of the transcription factor E2F-1 in protein-free media to bypass the growth-factor requirements for proliferation.

Although the desired effect of eliminating growth-factor dependency of the cells was achieved in both cases, the two approaches had different effects on several cellular properties. For instance, the cells expressing cyclin E rapidly round up and move into suspension, whereas the E2F-1 cells adhered more tightly to the surface. Additionally, the same doubling time was observed for both cell types; however, the cyclin E cells resided in G_1 for approximately 20 min longer, while the S phase was reduced by 2 h as compared with the E2F-1 cells. Finally, the expression of certain proteins varied between the two cell types. The expression of cyclin A was shown to be 2 times higher in the E2F-expressing cells, and these same cells expressed a greater number of proteins at higher levels than wild- type cells than did the cyclin E–expressing cells. These results present two different ways of circumventing growth-factor-dependent controls and show different effects on cellular processes. Additionally, these findings may give some insight into the roles that these factors play in the regulation of the cell cycle.

Another recent study involved increasing protein production in cultured cells by arresting cells in G_1. This was done by transfecting CHO cells with a gene coding for either p21, p27, or p53175P controlled by a tetracycline- regulated promoter. The p53 protein used in this study is a mutant that has lost its apoptotic function, resulting in a culture that undergoes G_1 arrest rather than programmed cell death. The culture was allowed to proliferate until it reached the desired cell density, after which expression of the cell cycle arrest molecules was stimulated, thus blocking proliferation. This sort of cytostatic process has several advantages over other processes involving uncontrolled growth of cultures, including slower rates of nutrient exhaustion, higher levels of protein production, and low genetic drift.

Treatment of Genetic Disorders

A firm grasp of the events that control the cell cycle will also yield significant advances toward the treatment of various illnesses including cancer and genetic disorders. The connection between the cell cycle and some cancers has been reviewed elsewhere. Many genetic disorders that alter the way cells function result from either the absence of an enzyme or the presence of a flawed enzyme. This situation may be remedied by the introduction of the appropriate genes into a target cell, a procedure known as gene therapy. Although cells that have incorporated the desired gene into their genome have displayed the desired effects, this technique is limited by low rates of gene transfer. Numerous studies have attempted to uncover the factors that determine transfer efficiency, resulting in the hypothesis that gene transfer is cell cycle dependent. More specifically, the most widely used method for gene transfer is recombinant retroviruses, and genes introduced into cells in this manner are not integrated into the host's genome until mitosis.

It has been postulated that nuclear envelope breakdown is necessary for the viral integration complexes to act. Based on these studies, Andreatis and Palsson suggested that transfer efficiency depends both on cell cycle and retroviral life cycle events and developed a mathematical model describing the interplay between the two. This model aided in the design of experiments that allow the direct measure of intracellular retroviral decay. Furthermore, they developed a method to monitor the cell cycle dependence of gene transfer by using the transduction of β-galactosidase as a marker. Thus it was determined that almost all transduced cells show up in the early G_1 phase, supporting the hypothesis that only mitotic cells are susceptible to gene transfer. These studies reveal the importance of the kinetics of both retroviral decay and cell cycle progression of the target cells in the efficiency of gene therapy.

SUMMARY

An overview of the mechanisms by which cells respond to different environmental stimuli has been presented. It is evident that even after the commitment to synthesize new DNA is made, the cell possesses the ability to monitor and direct internal events. For instance, the transcription and activation of key enzymes are regulated by cell cycle controls to ensure that necessary cellular processes are carried out successfully. The existence of this type of control suggests that almost every cellular process is somehow linked to the cell cycle machinery to guarantee the proper coordination of events throughout the cell cycle. A failure to maintain the order of these events may result in cell death or a disruption of normal cell behavior. Thus, a more complete understanding of factors that have an effect on cell cycle regulation would allow one to gain insight into better directing the cellular processes bound to cell cycle control. This fact is characterized by the recent increase in the number of studies that have focused on better understanding the components involved in directing cell division and their influence on fundamental cellular processes. It seems that obtaining a better understanding of the cell cycle and understanding how the cell cycle directs fundamental components of cell growth, such as protein production and secretion, may provide the missing link that will allow researchers to not only improve cell culture, but develop treatments for many types of ailments and dys-functions such as cancer, which are indubitably cell cycle related.

BIOPROCESSING PHENYLALANINE

Phenylalanine which is an amino acid is currently manufactured worldwide at an annual scale exceeding 12,000 metric tons. The vast majority of this material is L-phenylalanine, synthesized specifically for incorporation into the dipeptide sweetener aspartame. Production of L-phenylalanine grew rapidly during the 1980s in parallel with the demand for aspartame and has grown steadily in recent years. L-Phenylalanine is also used in additional food and medical applications and is in increasing demand in pharmaceutical development as a chiral intermediate or as a precursor of chiral auxiliaries such as benzyloxazolidinone. In contrast, only a few metric tons of D-phenylalanine are used annually, also in pharmaceutical drug development, although this application is increasing. During the past two decades, a number of manufacturers have developed chemoenzymatic and purely biological routes for L-phenylalanine manufacture.

Table 17.1 Large scale production of L-phenylalanine

Manufacturer	*L-Phenylalanine processes developed*
Ajinomoto	Microbial fermentation
Biotechnica	Microbial fermentation
Degussa/Rexim	Enzymatic resolution (aminoacylase)
DSM	Enzymatic resolution (amino acid amidase)
Genex	Biotransformation (phenylalanine ammonia lyase)
Kyowa Hakko	Microbial fermentation
Miwon	Microbial fermentation
Nutrasweet Co.	Microbial fermentation, biotransformation (aminotransferase)
PEI	Biotransformation (aminotransferase)
Tanabe Seiyaku	Enzymatic resolution, biotransformation (aminoacylase, aminotransferase)

Chemoenzymatic syntheses have relied on resolution technology, applied to chemically prepared racemic phenylalanine derivatives or DL-5-monosubstituted hydantoins and biotransformation routes from chemically prepared a chiral precursors of L-phenylalanine. Purely biological routes have concentrated on direct synthesis of L-phenylalanine through microbial fermentative means. Today, the largest and most commercially successful synthetic routes to L-phenylalanine production use large-scale fermentation of bacterial strains that overproduce L-phenylalanine. Such microorganisms have been isolated through a combination of classical mutagenesis selection procedures and molecular genetic manipulation, the latter deriving from a detailed understanding of the molecular biology of the pathways involved in bacterial L-phenylalanine biosynthesis. Conversely, the present commercial production of D-phenylalanine relies solely on chemoenzymatic resolution.

MICROBIAL PRODUCTION OF L-PHENYLALANINE

Efforts to develop L-phenylalanine-overproducing organisms have been vigorously pursued by Nutrasweet Company, Ajinomoto, Kyowa Hakko Kogyo, and others. The focus has centered on bacterial strains that have previously demonstrated the ability to overproduce other amino acids. Such organisms include principally the coryneform bacteria, *Brevibacterium flavum* and *Corynebacterium glutamicum* used in L-glutamic acid production. In addition, *Escherichia coli* has been extensively studied in L-phenylalanine manufacture because of the detailed characterization of the molecular genetics and biochemistry of its aromatic amino acid pathways and its amenability to recombinant DNA methodology. The biochemical pathway that results in the synthesis of L-phenylalanine from chorismate is identical in each of these organisms. In each case, the L-phenylalanine biosynthetic pathway comprises three enzymatic steps from chorismic acid, the product of the common aromatic pathway.

The precursors of the common aromatic pathway, phosphoenolpyruvate (PEP) and erythrose-4-phosphate, derive from the glycolytic and pentose phosphate pathways of sugar metabolism, respectively. Although the intermediate compounds in both pathways are identical in each of the organisms, there are differences in the organization and regulation of the genes and enzymes involved. Nevertheless, the principal points of pathway regulation are very similar in each of the three bacteria. Conversely, tyrosine biosynthesis, which is also carried out in three biosynthetic steps from chorismate, proceeds through different intermediates in *E. coli* than the coryneform organisms. Regulation of phenylalanine biosynthesis occurs both in the common aromatic pathway and in the terminal phenylalanine pathway, and the vast majority of efforts to deregulate phenylalanine biosynthesis have focused on two specific rate-limiting enzymatic steps.

These are the steps of the aromatic and the phenylalanine pathways carried out, respectively, by the enzymes 3-deoxy-D-arabino-heptulosonate-7-phosphate (DAHP) synthase and prephenate dehydratase. Classical mutagenesis approaches using toxic amino acid analogues and the molecular cloning of the genes encoding these enzymes have led to very significant increases in the capability of host strains to overproduce L-phenylalanine. This has resulted from the elimination of the regulatory mechanisms controlling enzyme synthesis and specific activity. The cellular mechanisms that govern the activity of these particular enzymes are complex and illustrate many of the sophisticated means by which bacteria control gene expression and enzyme activity. Chorismate mutase, the first step in the phenylalanine-specific pathway, and the shikimate kinase activity of the common aromatic pathway are subject to a lesser degree of regulation and have also been characterized in detail.

Production through Mutagenesis

Classical methods of strain improvement have been widely applied in the development of phenylalanine over-producing organisms. Tyrosine auxotrophs have frequently been used in efforts to increase phenylalanine production through mutagenesis. These strains often already overproduce phenylalanine because of the overlapping nature of tyrosine and phenylalanine biosynthetic regulation. Limiting tyrosine availability leads to partial genetic and allosteric deregulation of common biosynthetic steps. Such strains have been subjected to a variety of mutagenesis procedures to further increase the overall titer and the efficiency of phenylalanine production. In general, this has involved the identification of mutants that display resistance to toxic analogues of phenylalanine or tyrosine, such as β-2-thienyl-DL-alanine and *p*-fluoro-DL-phenylalanine. Such mutants can be readily identified on selective plates in which the analogue is present in the growth medium. The mutations responsible for the phenylalanine overproduction have frequently been located in the genes encoding the enzymatic activities of DAHP synthase, chorismate mutase, and prephenate dehydratase. In turn, this has prompted molecular genetic approaches to further increase phenylalanine production through the isolation and in vitro manipulation of these genes, as described next.

Regulatory Activities

The activity of DAHP synthase commits carbon from intermediary metabolism to the common aromatic pathway converting equimolar amounts of PEP and E4P to DAHP. In *E. coli* there are three isoenzymes of DAHP synthase of comparable catalytic activity encoded by the genes *aroF, aroG,* and *aroH*. Enzyme activity is regulated by the aromatic amino acids tyrosine, phenylalanine, and tryptophan, respectively. In each case, regulation is mediated both by repression of gene transcription and by allosteric feedback inhibition of the enzyme, though to different degrees. The *aroF* gene lies in an operon with *tyrA,* which encodes the bifunctional protein chorismate mutase–prephenate dehydrogenase (CMPO). Both genes are regulated by the TyrR repressor protein complexed with tyrosine. The *aroF* gene product accounts for 80% of the total DAHP synthase activity in wild-type *E. coli* cells. The *aroG* gene is repressed by the TyrR repressor protein complexed with phenylalanine and tryptophan. Repression of *aroH* is mediated by tryptophan and the TrpR repressor protein.

The gene products of *aroF* and *aroG* are almost completely feedback inhibited respectively by low concentrations of tyrosine or phenylalanine, whereas the *aroH* gene product is subject to maximally 40% feedback inhibition by tryptophan. In *B. flavum,* DAHP synthase forms a bifunctional enzyme complex with chorismate mutase and is feedback inhibited by tyrosine and phenylalanine synergistically but not by tryptophan. Similarly, in *C. glutamicum,* DAHP synthase is inhibited most significantly by phenylalanine and tyrosine acting in concert, but unlike *B. flavum* reportedly does not show tyrosine-mediated repression of transcription. Many examples of analogue-resistant mutants of these organisms display reduced sensitivity of DAHP synthase to feedback inhibition. In *E. coli,* the genes encoding the DAHP synthase isoenzymes have been characterized and sequenced. The mechanism of feedback inhibition of the *aroF, aroG,* and *aroH* isoenzymes has been studied in considerable detail, and variants of the *aroF* gene on plasmid vectors have been used to increase phenylalanine overproduction. Simple replacement of transcriptional control sequences with powerful constitutive or inducible promoter regions and the use of high copy number plasmids has readily enabled overproduction of the enzyme, and reduction of tyrosine-mediated feedback inhibition has been described using resistance to the aromatic amino acid analogues β-2-thienyl-DL-alanine and *p*-fluoro-DL-phenylalanine.

Control of CMPD

The three enzymatic steps by which chorismate is converted to phenylalanine appear to be identical among *C. glutamicum, B. flavum,* and *E. coli,* although only in *E. coli* have detailed reports appeared upon the characterization of the genes involved. In each case, the principal regulatory step is that catalyzed by prephenate dehydratase. In *E. coli,* chorismate is converted first to prephenate and then to phenyl pyruvate by the action of a bifunctional enzyme, CMPD, encoded by the pheA gene. The final step, in which phenyl pyruvate is converted to L-phenylalanine, is carried out predominantly by the aromatic aminotransferase encoded by *tyrB*. However, both the aspartate aminotransferae, encoded by *aspC* and the branched-chain aminotransferase encoded by *ilvE,* can efficiently catalyse this reaction. Phenylalanine biosynthesis is regulated by control of CMPD through phenylalanine-mediated attenuation of *pheA* transcription and by feedback inhibition of the prephenate dehydratase and chorismate mutase activities of the enzyme. Inhibition is most pronounced upon the prephenate dehydratase activity, with almost total inhibition observed at micromolar phenylalanine concentrations. Chorismate mutase activity in contrast is maximally inhibited by only 40%.

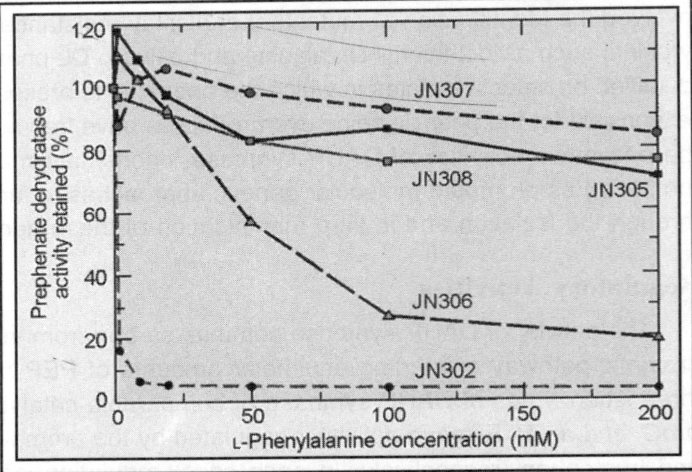

Figure 17.1 L-Phenylalanine-mediated feedback inhibition of wildtype *E. coli* K12 prephenate dehydratase and four inhibition resistant enzyme variants.

In *B. flavum,* prephenate dehydratase and chorismate mutase are encoded by distinct genes. Prephenate dehydratase is again the principal point of regulation, with the enzyme subject to feedback inhibition, but not transcriptional repression, by phenylalanine. Chorismate mutase, which in this organism forms a bifunctional complex with DAHP synthase, is maximally inhibited by phenylalanine and tyrosine to 65%, but this is significantly diminished by the presence of very low levels of tryptophan. Expression of chorismate mutase is repressed by tyrosine. The final step is carried out by at least one transaminase. Similarly, in *C. glutamicum*, the activities are encoded in distinct genes, with prephenate dehydratase again being more strongly feedback-inhibited by phenylalanine. The only transcriptional repression reported is that of chorismate mutase by phenylalanine. The *C. glutamicum* genes encoding prephenate dehydratase and chorismate mutase have been isolated and cloned from analogue-resistant mutants of *C. glutamicum* and used along with the cloned DAHP synthase gene to augment L-phenylalanine biosynthesis in overproducing strains of *C. glutamicum*. Many publications and patents have described successful efforts to reduce and eliminate regulation of prephenate dehydratase activity in phenylalanine overproducing organisms. As with DAHP synthase, the majority of the reported efforts have focused on the *E. coli* enzyme encoded by the *pheA* gene that is transcribed convergently with the tyrosine operon on the *E. coli* chromosome.

The detailed characterization of the *pheA* regulatory region has facilitated expression of the gene in a variety of transcriptional configurations, leading to elevated expression of CMPD, and a number of mutations in *pheA* that affect phenylalanine-mediated feedback inhibition have been described. Increased expression of the gene is readily achieved by cloning *pheA* onto multicopy plasmid vectors and deletion of the nucleotide sequences comprising the transcription attenuator. Most mutations that affect the allosteric regulation of the enzyme by phenylalanine have been identified through resistance to phenylalanine analogues such as β-2-thienylalanine, but there are examples of feedback-resistant mutations arising through insertional mutagenesis and gene truncation. Two regions of the enzyme in particular have been shown to reduce feedback inhibition to different degrees. Mutations at position Trp338 in the peptide sequence desensitize the enzyme to levels of phenylalanine in the 2 to 5 mM range but are insufficient to confer resistance to higher concentrations of L-phenylalanine. Mutations in the region of residues 304 to 310 confer almost total resistance to feedback inhibition at L-phenylalanine concentrations of at least 200 mM. It is not clear if the mechanism of resistance is similar in either case, but the difference is significant to commercial application because overproducing organisms readily achieve extracellular concentrations of L-phenylalanine over 200 mM.

Aromatic Biosynthetic Pathways

Rate-limiting steps in the common aromatic and phenylalanine biosynthetic pathways are obvious targets in the development of phenylalanine overproducing organisms. However, the detailed biochemical and genetic characterization of *E. coli* has enabled additional areas of its metabolic function to be specifically manipulated to determine their effect on aromatic pathway throughput. Besides efforts to eliminate additional lesser points of aromatic pathway regulation, attempts have been made to enhance phenylalanine production by increasing the supply of aromatic pathway precursors and by facilitating exodus of L-phenylalanine from the cell. The precursors of the common aromatic pathway D-erythrose 4-phosphate and PEP are the respective products of the pentose phosphate and glycolytic pathways. Precursor supply in aromatic amino acid biosynthesis has been reviewed very recently.

Theoretical analyses of the pathway and the cellular roles of these metabolites suggest that the production of aromatic compounds is likely to be limited by PEP availability because PEP is involved in a number of cellular processes, including the generation of metabolic energy through the citric acid cycle and the transport of glucose into the cell by the phosphotransferase system. Strategies to reduce the drain of PEP by these processes have included mutation of sugar transport systems to reduce PEP-dependent glucose transport and modulation of the activities of pyruvate kinase, PEP synthase, and PEP carboxylase, which regulate PEP flux to pyruvate, oxaloacetate, and the citric acid cycle. Similarly, the availability of E4P has been increased by altering the levels of transketolase, the enzyme responsible for E4P biosynthesis. In general, these efforts have been successful in directing additional flux of PEP of E4P into the aromatic pathway, although their effect has not proved to be as predictable as the deregulation of rate-limiting pathway steps, and their overall impact on L-phenylalanine overproduction has not been well characterized.

Commercial Production of Phenylalanine

Exodus of phenylalanine is of manifest importance in the large-scale production of phenylalanine because the amino acid is typically recovered only from the extracellular medium and washed cells. Lysis of cells to recover additional phenylalanine is not generally practical, and so L-phenylalanine remaining within the cells after washing is usually lost. Because cell biomass is extremely high in large-scale fermentations, this can represent up to 5% of total phenylalanine produced. Additionally, because L-phenylalanine is known to regulate its own biosynthesis at multiple points through feedback inhibition, attenuation, and TyrR-mediated repression, methods to reduce intracellular concentrations of phenylalanine through reduced uptake or increased exodus throughout the fermentation have been sought. Studies have addressed the means by which L-phenylalanine is both taken up and excreted by bacterial cells and whether this can be altered to increase efflux. In overproducing strains, it is likely that most phenylalanine leaves the cell by passive diffusion but specific uptake and exodus pathways also exist. In *E. coli,* L-phenylalanine is taken up by at least two permeases, one specific for phenylalanine encoded by *pheP* and the other encoded by *aroP* a general aromatic amino acid permease responsible for the transport of tyrosine and tryptophan in addition to phenylalanine.

The genes encoding the permeases have been cloned and sequenced, and *E. coli* mutants deficient in either system have been characterized. The effect of an *aroP* mutation has been evaluated in L-phenylalanine overproduction. In addition, specific export systems have been identified that although not completely characterized have been successfully used to increase exodus of L-phenylalanine from strains of *E. coli.* In one such system, the C in invertase of bacteriophage P1 has been shown to induce a metastable phenylalanine hypersecreting phenotype upon a wild-type strain of *E. coli.* The phenotype can be stabilized by transient introduction of the invertase on a temperature- sensitive plasmid replicon and is sufficient to establish L-phenylalanine overproduction in the absence of any alterations to the normal biosynthetic regulation. The incremental gains made from the various levels at which phenylalanine biosynthesis has been addressed has led to the high-efficiency production strains currently in use today.

Almost all large-scale L-phenylalanine manufacturing processes in present operation are fermentations using bacterial strains such as those described in this article in which classical strain development or molecular genetics have been extensively applied to bring about phenylalanine overproduction. The low substrate costs and the economics of scale associated with this approach have resulted in significant economic advantages. However, a number of alternate biotransformation and chemoenzymatic resolution strategies have been extensively developed that have also demonstrated commercially viable levels of L-phenylalanine synthesis.

ROUTES OF L-PHENYLALANINE

Role of L-Phenylalanine Ammonia Lyase

The use of L-phenylalanine ammonia lyase (PAL) in various overproducing strains of the yeast *Rhodoturula* was developed by Tanabe Seiyaku Co. In this process, fermented cells of yeast strains *R. glutinis*, *R. rubra* or *R. graminis* were recovered and washed before bioreaction under batch or immobilized conditions with *trans*-cinnamic acid and ammonia. The elevated pH and high concentration of ammonia used under the process conditions enabled the normally catabolic PAL reaction to favor formation of L-phenylalanine. Significant development work was carried out to determine the appropriate conditions for optimal growth and maximal induction of the PAL enzyme and the appropriate process conditions for retention of PAL activity.

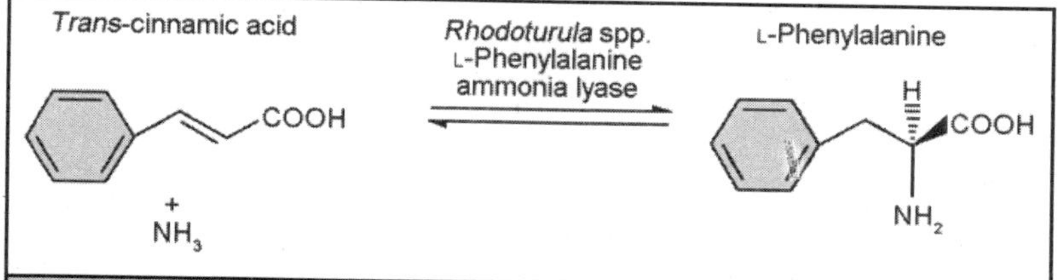

Figure 17.2 L-Phenylalanine-ammonia-lyase-catalyzed synthesis of L-phenylalanine from trans-cinnamic acid.

The principal drawbacks of the PAL process lay in substrate inhibition of the enzyme, low enzyme-specific activity, and enzyme instability. Screening regimes carried out to address these limitations resulted in the identification of an *R. graminis* strain that displayed greater enzyme specific activity and stability than the *R. rubra* and *R. glutinis* strains used previously. Subsequent mutagenesis and selection procedures carried out on this strain using the cinnamic acid analogue phenylpropiolic acid led to a further increase in PAL activity through a significant elevation in PAL gene expression. Such strain and process development strategies enabled the Genex process to achieve yields of L-phenylalanine in excess of 50 g/L, with close to 90% conversion of substrate to product. Nevertheless, the biotransformation could not favorably compete with the economics of fermentative production, and it is not presently used in L-phenylalanine production.

Role of Aromatic Aminotransferase

A number of reports from PEI, Nutrasweet, and Allelix, among others, have described the use of aminotransferases (transaminases) to produce L-phenylalanine. The reaction proceeds by a pingpong mechanism whereby the amino group of a donor amino acid, typically aspartic acid, is exchanged with the keto group of phenylpyruvic acid, yielding L-phenylalanine and the by-product oxaloacetate. Amino-transferases are ubiquitous in nature, and microorganisms typically possess multiple aminotransferases that are involved in the biosynthesis of a number of amino acids, often catalyzing the terminal step from the keto acid precursor. The enzymes use pyridoxal 5′-phosphate as cofactor and typically possess high catalytic rates and stereoselectivity with broad substrate specificities. This is exemplified by the multiple transaminases in *E. coli* that can each catalyze the synthesis of L-tyrosine, L-aspartate, and L-valine and L-leucine in addition to L-phenylalanine.

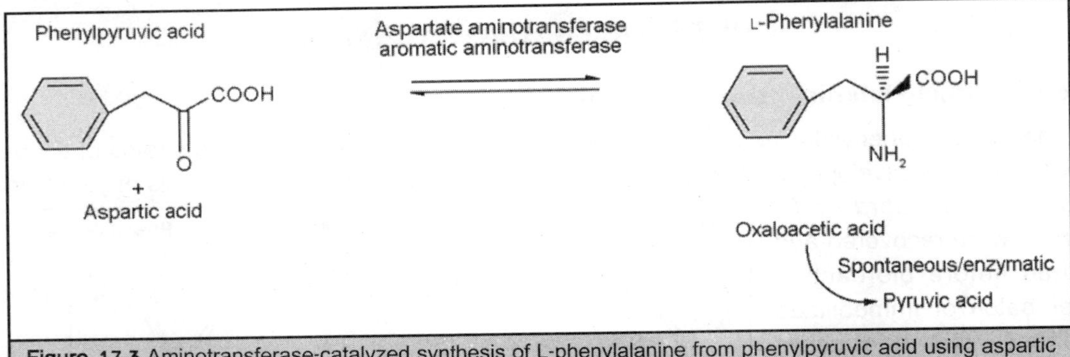

Figure 17.3 Aminotransferase-catalyzed synthesis of L-phenylalanine from phenylpyruvic acid using aspartic acid as the amino donor. Decarboxylation of oxaloacetic acid derives the reaction toward completion.

Many aminotransferase encoding genes have been cloned and used to overproduce the enzymes. Significant sequence homology often exists among the aminotransferases of individual strains as well as between the corresponding enzymes of microbial and mammalian origin. In addition, the tertiary structures of a number of aminotransferases have been determined, furthering the mechanistic understanding of the enzyme and prompting mutagenesis approaches to altering enzyme properties. Aminotransferases from many microorganisms have been applied in L-phenylalanine biosynthesis. In the case of *E. coli,* the highly homologous aspartate and aromatic aminotransferases encoded respectively by the *aspC* and *tyrB* genes have both been developed and applied in this bioconversion. The aspartate aminotransferase has been generally preferred, having shown greater throughput and increased stability over the *tyrB* gene product. A number of L-phenylalanine biosynthetic processes involving aminotransferases have been described using either isolated enzyme or whole cells in batched or immobilized systems.

In the mid 1980s, one such immobilized process was commercialized by PEI using the *E. coli* aspartate aminotransferase. In this case, the PPA substrate was economically derived from chemically synthesized 5-benzilidene hydantoin. The highest yields in the transaminase synthesis are generally obtained from batch processes at the expense of the greater catalyst stability of the continuous systems. The cost of catalyst regeneration in batch systems is a drawback of the aminotransferase approach, as is the need for two primary metabolites as substrates. The principal drawback, however, lies in the reaction equilibrium, which being close to 1 limits the yield of the reaction to 50% unless the keto acid by-product can be eliminated from further reaction. To overcome this limitation, commercial transamination approaches have almost invariably used aspartate as the amino donor, which upon transamination to the relatively unstable oxaloacetate will spontaneously decarboxylate to yield pyruvate. Because pyruvate is frequently not a substrate for these transaminases, this enables the reaction to proceed to a yield often exceeding 90%. The decarboxylation of oxaloacetate is more rapid at elevated temperature and has prompted the development of transaminase bioconversions carried out in excess of 50°C. Aminotransferases from thermophiles have been explored in this respect as well as chemical and enzymatic steps to accelerate oxaloacetate decarboxylation and thereby optimize product yield and substrate conversion.

Role of Phenylalanine Dehydrogenase

A similar process using the phenyl pyruvate precursor has also been developed using the enzyme phenylalanine dehydrogenase. Unlike the aminotransferases, phenylalanine dehydrogenase uses ammonia as a substrate in the reductive amination of phenyl pyruvate, thereby realizing a cost advantage. In addition, the reaction equilibrium lies heavily in favor of phenylalanine synthesis. These advantages in the reaction mechanism are diminished by the dependence of the dehydrogenase on an NADH cofactor that in turn requires enzymatic regeneration. This has been addressed through the use of a coupled enzyme system in which formate dehydrogenase is used along with phenylalanine dehydrogenase.

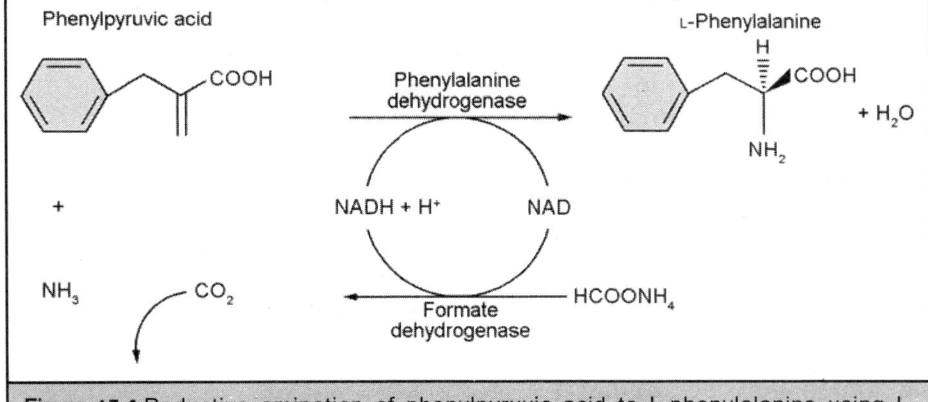

Figure 17.4 Reductive amination of phenylpyruvic acid to L-phenylalanine using L-phenylalanine dehydrogenase with cofactor regeneration by coupled formate dehydrogenase.

The reaction proceeds quantitatively because of the evolution of CO_2 generated in the oxidation of formate. Using partially purified enzymes in membrane reactors, this reaction has achieved phenylalanine titers and substrate conversions comparable to the aminotransferase based systems. Despite the high efficiencies obtained in this bioconversion, commercial production of L-phenylalanine using this method has not been economically competitive with fermentative processes, mainly because of the costs of phenylpyruvic acid preparation. Nevertheless, in a similar process operated by Degussa using leucine dehydrogenase, this approach has been successfully applied to the synthesis of compounds of higher intrinsic value such as the unnatural amino acid tertiary leucine.

STEREOSELECTIVE RESOLUTIONARY SYNTHESIS

In contrast to the various fermentation and bioconversion approaches that have used enzyme stereoselectivity to synthesize L-phenylalanine as a single isomer, additional methods have been developed that rely on the same stereoselectivity to resolve chemically synthesized amino acid DL-racemic mixtures. One such general approach that has been successfully commercialized in L-phenylalanine production relies on cleavage of the L-isomer of a DL-α-amino acid amide mixture by a highly enantiospecific amidase enzyme. The general procedure operated by DSM uses an amidase containing strain of *Pseudomonas putida* to specifically hydrolyze the L-isomer of the DL-amino acid amide mixture.

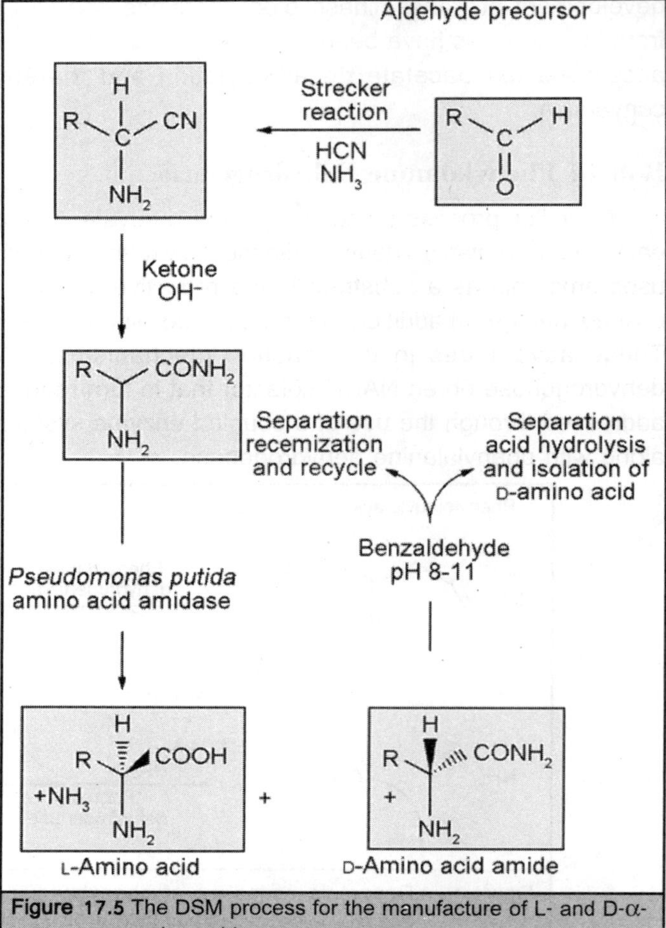

Figure 17.5 The DSM process for the manufacture of L- and D-α-amino acid.

The racemic amino acid mixture is prepared inexpensively from an aldehyde precursor via the Strecker reaction, with alkaline hydrolysis of the resulting aminonitrile. Treatment with benzaldehyde after selective hydrolysis of the L-amino acid amide yields an insoluble Schiff base that forms between benzaldehyde and the residual D-amino acid amide. In this way, the residual D-amino acid amide may be readily separated from the resulting L-amino acid and can then be chemically hydrolyzed and purified or racemized and recycled. The advantages of this process include general applicability to the preparation of a range of amino acids and the opportunity to recover the D-amino acid if desired from the reaction in addition to the L-amino acid. Conversely, in the preparation of only the L-amino acid by this approach, the need to recycle the unreacted substrate is considered a drawback of the process, necessitating additional complexity and cost. This aspect has been considered in a further development of this process by DSM in which *P. putida* strains possessing D- or L-amino acid amidase as well as amino acid racemase activity have been identified. The in situ coupling of either amidase with the racemase activity avoids the recycling requirement of the process and facilitates direct conversion of the DL-amino acid amide mix into the desired isomer.

Enzymatic Hydrolysis

An earlier strategy involving stereoselective enzyme hydrolysis of chemically prepared racemic precursors has been operated successfully for many years by Tanabe Seiyaku. In this process, a

DEAE-Sephadex immobilized L-specific aminoacylase selectively hydrolyzes *N*-acetyl-L-amino acids from an *N*-acetyl-DL-amino acid mixture. After hydrolysis, the L-amino acid is isolated by selective crystallization and the remaining *N*-acetyl-D-amino acid is racemized either chemically or enzymatically and recycled. The preferred aminoacylase for this reaction was obtained from the mold *Aspergillus oryzae* and is readily isolated and immobilized in the bioreactor. The enzyme has broad substrate specificity and has been used by Tanabe in the production ofL-methionine and L-valine and addition to L-phenylalanine. More recently, this process has been further developed by Degussa/Rexim to prepare bulk quantities of L-phenylalanine and a large number of L-amino acids. In addition, Tanabe has reported on the further development of the aminoacylase process through the coupling of acetamidocinnamate aminohydrolase and aminotransferase activity. In this process, L-phenylalanine is produced in two steps from acetamidocinnamic acid through a phenylpyruvic acid intermediate using immobilized cells of two independent bacterial strains.

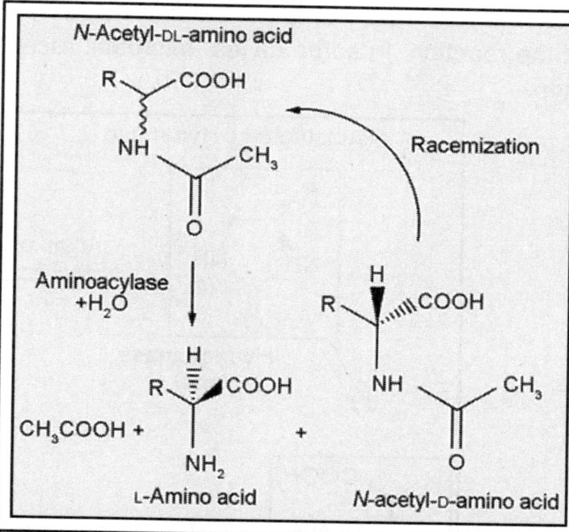

Figure 17.6 Resolution of N-acetyl-DL-amino acids using aminopeptidases yielding L-amino acid and recycling of residual N-acyl-D-amino acid.

Recycling Strategies

The recycling aspect of resolution-based amino acid syntheses such as the amidase and aminoacylase processes is again considered and largely overcome in an alternative strategy using DL-5-monosubstituted hydantoins. Inexpensive chemical routes such as the Bucherer-Bergs synthesis or the condensation of aldehydes with hydantoin are used to prepare the racemic hydantoin derivatives that are then resolved using either L- or D-enantiospecific hydantoinase and *N*-carbamoyl amino acid amidohydrolase enzymes to yield the desired amino acid isomer. In the case of phenylalanine, both L- and D-phenylalanine can be prepared from racemic DL-5-benzylhydantoin. The advantage of the hydantoinase process is that in many cases a dynamic resolution can be established whereby the unreacted isomer can be racemized in situ either chemically or in the presence of a racemase enzyme to enable the process to approach 100% yield.

Enzymes with L-and D-specific hydantoinase activity have been described in a number of microbial species, with the D-specific activity having been identified much more frequently, often attributable to dihydropyrimidinases involved in pyrimidine catabolism. Substrate specificities vary significantly between enzyme isolates and contribute to the broad applicability of this approach. Molecular cloning has been applied in at least one case to isolate a D-hydantoinase encoding gene from a thermophilic bacillus and to subsequently overproduce the enzyme. The D- and L-specific *N*-carbamoyl amino acid amidohydrolases have also been shown to be widely distributed in microorganisms. The bioconversion steps are typically carried out using whole cells in immobilized systems with racemization of the unreacted hydantoin occurring under the normal alkaline conditions of the reaction. In some cases, microbial racemases have been used to optimize the racemization step.

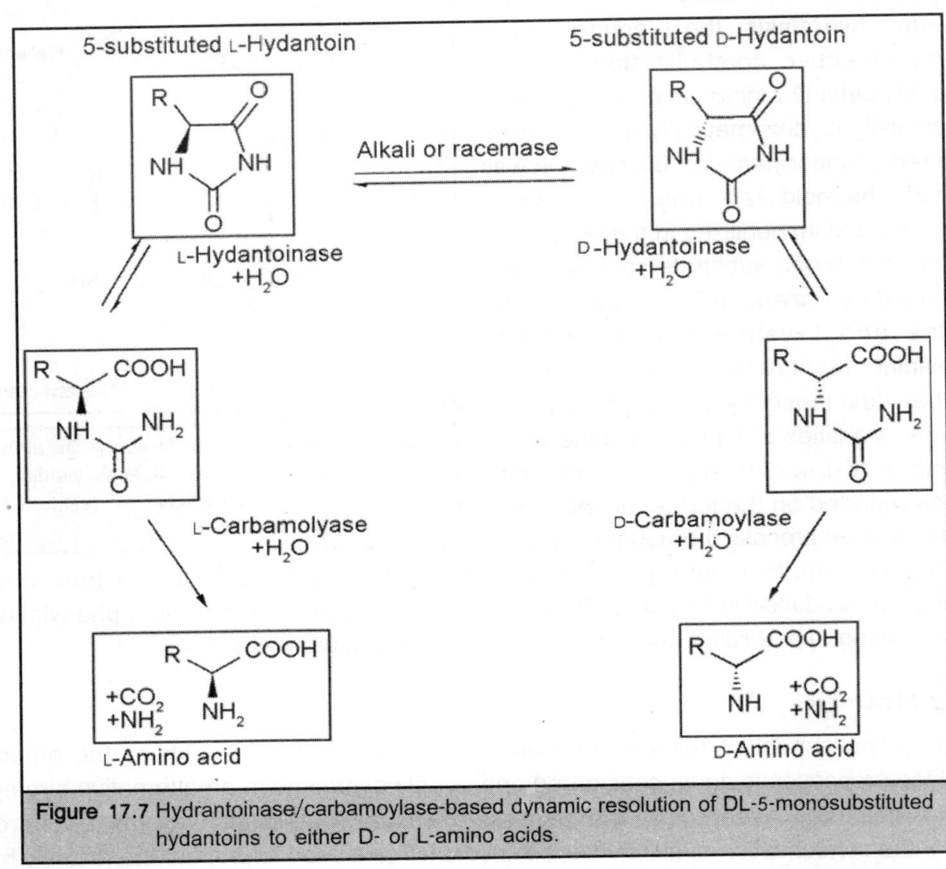

Figure 17.7 Hydrantoinase/carbamoylase-based dynamic resolution of DL-5-monosubstituted hydantoins to either D- or L-amino acids.

The hydantoinase process offers the advantage of relatively inexpensive precursors and high conversion rates through the in situ racemization of the substrate. In addition, through the broad substrate range of hydantoinase and carbamoylase enzymes, the process is broadly applicable in L- and D-amino acid biosynthesis. However, it is in D-amino acid production that large-scale commercial application of the hydantoinase process has been more significantly applied, most notably by Recordati and Kanegafuchi in the manufacture of D-phenyl glycine and D-*p*-hydroxyphenyl glycine as components of semisynthetic penicillins and cephalosporins but also by Degussa and Ajinomoto in the production of additional D- amino acids, including D-phenylalanine.

PHENYLALANINE DEHYDROGENASE

Hummel et al. screened for a PheDH-producing strain of bacteria from soil and isolated a *Brevibacterium* sp. by an enrichment culture technique. Culture conditions for PheDH formation and some reaction conditions to synthesize Phe and Tyr with the crude enzyme were optimized.

Asano et al. noted PheDH production in the gram-positive spore-forming bacteria *S. ureae* SCRC-R04, *B. sphaericus* SCRC-R79a, and *B. badius* IAM 11059 and described its purification, crystallization, and characterization; no activity was found in yeasts. Following these studies, PheDH was given a new entry number. Later, *Rhodococcus maris, Nocardia* sp., *Thermoactinomyces intermedius, Rhodococcus* sp., *Bacillus cereus,* and *Microbacterium* sp. were found to produce the enzyme. The occurrence of PheDH appears to be limited in some groups of gram-positive spore-forming bacteria and thermophilic Actinomycetes. The distribution of PheDH seems similar to that of LeuDH, although the distribution of the former is much more limited than that of the latter, which may be implicated in microbial sporulation, which shuffles carbon and nitrogen metabolism of amino acids by their reversible nature.

Because PheDH is specifically distributed in *B. sphaericus* and *B. badius,* which are characterized to be poor in sugar oxidation among the genus *Bacillus,* PheDH seems involved in the catabolism of Phe and other amino acids in the medium. Hummel et al. suggested that PheDH is responsible for the degradation of Phe, not its synthesis; however, in *S. ureae,* L-phenylalanine aminotransferase activity with oxaloacetic acid as an amino acceptor was detected during growth on a medium containing Phe at a range between 0.001 and 0.006 units/mg, while PheDH was highly induced (0.033 units/mg, measured in the oxidative deamination reaction). The formation of PheDH was growth associated, indicating that the enzyme is responsible for the catabolism of Phe. On the other hand, when the strain was grown in M9 medium supplemented with 0.1% yeast extract, the PheDH activity was detected only in the exponential phase at a level as high as 0.059 units/mg, which is much higher than L-phenylalanine aminotransferase activity (less than 0.001 units/mg). Because the equilibrium of PheDH favors the synthesis of the amino acid, the enzyme probably also functions in the anabolism of Phe, depending on the environment.

Properties of Phenylalanine Dehydrogenase

The molecular weight of the PheDH subunits are in the range of 36,000 to 42,000. The higher structures vary; PheDHs from *Sporosarcina, Bacillus,* and *Microbacterium* are octameric enzymes, whereas PheDH from *Rhodococcus* sp. is tetrameric, that from *Rhodococcus maris* is dimeric, and that from *Nocardia* is monomeric. The molecular weight of PheDH ranges from 42,000 (the monomeric *Nocardia* enzyme) to 331,480 (the octameric *B. sphaericus* enzyme). The optimum pH for the reversible oxidative deamination and reductive amination varies by about 1 pH unit; the former reaction varies in the range of pH 10.4 to 11.3, whereas the latter is in the range of pH 9 to 10.3. The velocity toward the reductive amination to synthesize Phe generally is several times faster than that for the reverse reaction at optimum pH; the V_{max} values for the oxidative deamination reaction of Phe catalyzed by *Sporosarcina* and *B. sphaericus* PheDH are both 114, whereas those for the reductive amination reaction for phenylpyruvate are 598 and 416, respectively. The K_m values toward the preferred substrates of L- amino acids and keto acids are in the range of 0.1 to 1 mM, though much higher K_m values of around 100 mM were observed toward ammonia.

Specificity

In the oxidative deamination reaction, Phe and L-norleucine were active as substrates for the *Sporosarcina* enzyme. The from enzymes *B. badius* and *S. ureae* show narrower substrate specificity than that from *B. sphaericus,* which has almost equal affinity toward Tyr and its keto analog, thus *B. sphaericus* PheDH may be called *tyrosine dehydrogenase*. On the other hand, the enzymes showed wider substrate specificities in the reductive amination reaction than observed in the oxidative deamination reaction.

Table 17.2 Substrate specificity of PheDH from *S. ureae* SCRC-R04 and *B. sphaericus* SCRC-R79a

Amino acid	*S. ureae SCRC-R04* Relative activity (%)	K_m (mM)	V_{max} (units/mg)	*B. sphaericus SCRC-R79a* Relative activity (%)	K_m (mM)	V_{max} (units/mg)
Phe	100	0.096	114	100	0.22	114
Tyr	5.4			72	0.50	112
Trp	5.0			1.2		
His	0.1			ND		
Met	4.1			3.0		
Ethionine	7.0			3.1		
Val	3.1			1.4		
Leu	2.3			1.3		
Ile	0.7			0.45		
L-*allo*-Isoleucine	4.3			0.3		
L-α-Amino-*n*-butyric acid	1.6			NT		
L-Norvaline	6.3			1.3		
L-Norleucine	15	0.25	20.0	3.9		

Table 17.3 Substrate specificity of PheDH from *S. ureae* SCRC-R04 and *B. sphaericus* SCR-R79a in the reductive amination reaction

Keto acid	*S. ureae SCRC-R04* Relative Activity (%)	K_m (mM)	V_{max} (units/mg)	*B. sphaericus SCRC-R79a* Relative activity (%)	K_m (mM)	V_{max} (units/mg)
Phenylpyruvate	100	0.16	598	100	0.40	416
p-Hydroxyphenylpyruvate (5 mM)	24			136	0.34	1240
Indole-3-pyruvate (2 mM)	0.73			0.39		
Imidazole pyruvate	0.04			ND		
α-Keto-γ-methylthiobutyrate	27			11		
α-Ketoisovalerate	2.1			5.5		
α-Ketoisocaproate	13			7.8		
DL-α-Keto-β-methylvalerate	3.2			2.9		
α-Ketovalerate	8.8			6.2		
α-Ketocaproate	32	2.44	379	ND		

The substrate specificity of PheDH has been extensively characterized with the *B. sphaericus* enzyme by using chemically synthesized substrates and substrate analogs. The relative rates of the reductive amination of α-keto acids and their analogs catalyzed by PheDH from *S. ureae* and *B. sphaericus* are shown in Table 17.3. The substrate specificity of *B. sphaericus* was examined in detail with various synthetic compounds as shown in Table 17.4. The enzyme was found to accommodate α-keto acids with large substituents, including compounds substituted at the β-position of pyruvic acid with a longer or bulkier group, although the relative velocity of the reductive amination reaction was low. Phenylpyruvate analogs substituted at the phenyl ring were relatively good substrates, whereas the substitution at the β-position of phenylpyruvate with a bulkier group, such as a hexyl group, greatly lowered the reaction velocity.

Table 17.4 Substrate specificity of PheDH from *B. sphaericus* SCRC-R79a

Substrate (10 mM)	*Relative activity (%)*
4-Hydroxyphenylpyruvate	136
Phenylpyruvate	100
4-Vinylphenylpyruvate	52
4-Fluorophenylpyruvate	39
α-Keto-γ-methylthiobutyrate	11
α-Keto-β-DL-phenylvalerate	8.8
α-Ketoisocaproate	7.8
α-Ketobutyrate	6.3
α-Ketovalerate	6.2
α-Ketoisovalerate	5.5
α-Keto-β-DL-(4-fluorophenyl)butyrate	5.0
α-Keto-β-DL-phenylbutyrate	4.8
α-Keto-4-phenylbutyrate	3.4
α-Keto-β-DL-methylvalerate	2.9
α-Keto-ε-phenylvalerate	2.0
α-Keto-β-DL-(3-methylphenyl)butyrate	1.0
α-Ketononanoate	0.73
α-Keto-β-(2-naphthalene)propionate	0.46
Indolepyruvate (2 mM)	0.39
α-Keto-β-DL-phenyl-γ-methylvalerate	0.10

On the other hand, pyruvate, α-ketoglutarate, α-hydroxypyruvate, benzoylformate, α-keto esters such as ethyl phenylpyruvate and ethyl α-keto-γ-phenylbutyrate, β-keto esters such as ethyl β-keto-γ-phenylbutyrate, β-keto acids such as β-keto-γ-phenylbutyrate, and α-ketoalcohols such as α-keto-β-phenylpropanol were inactive as substrates. Thus, it was shown that the enzyme acts solely on α-keto acids, and a free carboxylic acid moiety is required for a compound to be recognized as a substrate.

It was also revealed that the enzyme does not differentiate the configuration of the substituent at the β-position of pyruvic acid; for example, a diastereomeric mixture of L-α-amino-β-DL-methyl-β-phenylpropionate was synthesized from α-keto-β-DL-methyl-β-phenylpropionate. Gln, Asp, methylamine, dimethylamine, trimethylamine and ethylamine (each at 200 mM) did not replace the ammonium ion in the reductive amination reaction. Thus, PheDH from *B. sphaericus* showed wide substrate specificity toward substituted pyruvic acids. The finding that the enzyme utilized α-keto-γ-phenylbutyrate and α-keto-ε-phenylvalerate, but not β-keto-γ-phenylbutyrate, shows it has a definite requirement for a distance between the carbonyl carbon and the carboxyl group of the substrates.

For the *Sporosarcina* enzyme, phenylpyruvate, α-ketocaproate, α-keto-γ-methylthiobutyrate, *p*-hydroxyphenylpyruvate, and α-ketoisocaproate were active as substrates. NAD^+ was active as a cofactor for both enzymes, whereas $NADP^+$ was not. The mean value of the apparent equilibrium constant (K_{eq} = [phenylpyruvate][NADH][NH_4^+][H^+]/[Phe]-[NAD^+][H_2O]) was determined using the *Sporosarcina* enzyme to be 1.4×10^{-15} at the pH range of 7.0 to 11.5 and 25°C, assuming the conventional concentration of water ([H_2O] = 1M). For the *B. sphaericus* enzyme, it was calculated to be 2.0×10^{-15} at pH 8.40 to 10.38.

The substrate specificities of PheDH from *B. badius* and *R. maris* are similar to that of *S. ureae,* which shows more limited activities than the *B. sphaericus* enzyme, although *R. maris* PheDH has a slightly higher affinity toward Tyr than *S. ureae* PheDH. It is notable that the PheDH of *Nocardia* sp. is 0.54 and 2.4 times more active toward indolepyruvate and α-ketoisocaproate, respectively, than toward phenylpyruvate, and *T. intermedius* PheDH is inactive toward Tyr and its keto analog.

STRUCTURE OF PHENYLALANINE DEHYDROGENASE

The *pdh* genes have been cloned and sequenced, and over-production of the *Bacillus* enzymes has been achieved. *Escherichia coli* JM 109/pBPDH1-DBL expresses about 120-fold higher activity of PheDH (7,200 units/L) than the wild type *B. sphaericus*. The nucleotide sequences for the *pdh* genes from *B. sphaericus, B. badius, S. ureae, T. intermedius,* and *Rhodococcus* sp. have been reported. Computer assessment of the data showed that the deduced primary structure of *B. badius* PheDH is similar to PheDH of *B. sphaericus* (67.9% identical over 377 amino acids) and *T. intermedius* (56.6% identical over 346 amino acids), LeuDH from *B. stearothermophilus* (49.9% identical over 353 amino acids), $NADP^+$-dependent valine dehydrogenase from *Streptomyces coelicolor* (46.6% identical over 352 amino acids), PheDH from *Rhodococcus* sp. (34.7% identical over 334 amino acids), and GluDH from *Clostridium symbiosum* (25.9% identical over 185 amino acids).

Similarities between these proteins are seen in the catalytic domains of amino acid dehydrogenases, G-G-(G or A)-K, and the glycine-rich nucleotide-binding domain G-X-G-X-X-(G or A), connecting β-strand with α-helix in the region of adenine dinucleotide phosphate (ADP)-binding *bab* folds, which is strongly conserved among NAD(P)$^+$-dependent dehydrogenases and FAD-containing oxidoreductases.

Molecular Reactions

Speculation concerning the reaction mechanism of PheDH is based on the findings with respect to the substrate specificity already described, and on other experiments with GluDH and LeuDH. The formation of 2-iminoglutarate, catalyzed by GluDH, has been proved. The *pro-S* hydrogen at position 4 of the reduced pyridine ring of NADH is incorporated into the product Phe in the reductive amination reaction of PheDH (B-stereospecific). Studies on the steady-state kinetics of the PheDH reaction revealed that the reaction proceeds sequentially; that is, after all three substrates (phenylpyruvate, ammonia, and NADH) bind to the enzyme, the product Phe is released. In a homology search of the deduced primary structure of PheDH, a sequence similar to the active site of GluDH, which also resembles LeuDH, has been found. The reaction catalyzed by PheDH would be analogous to that of GluDH from *Clostridium symbiosum* discussed on the basis of X-ray crystallographic studies by Rice et al. and to that of *B. stearothermophilus* LeuDH estimated by the kinetic studies done with wild-type and mutant enzymes by Tanizawa et al.

The first step would be initiated by the deprotonation of the α-amino group of Phe by an acidic residue (^{126}Asp in *B. sphaericus* PheDH, which corresponds to ^{165}Asp in GluDH) acting as a general base. Subsequently, a hydride transfer from the hydrogen attached to the α-carbon to the *Si* face of NAD^+ (a hydride transfer between the *pro-S* hydrogen of NADH and the *Re* face of the imino acid) occurs. Next, the iminophenylpyruvate produced is hydrated by the action of a general base (^{90}Lys in PheDH, which corresponds to ^{125}Lys in GluDH and ^{80}Lys in LeuDH) in the consensus sequence (G-G-X-K, X = G or A or S).

The iminophenyl-pyruvate breaks into phenylpyruvate, ammonia, and a proton. The orientation of phenylpyruvate is fixed, with the phenyl group accommodated in the hydrophobic pocket and the carboxyl group as an anchor, as evidenced by the finding that the enzyme does not catalyze the reductive amination of neutral α-keto-β-phenylpropanol and the α-keto esters. Tanizawa et al. generated LeuDH mutants of *B. stearothermophilus* in which ^{80}Lys was changed to Ala, Arg, or Gln and found that the mutants showed reduced activities, but their K_m values for substrate and coenzyme did not change significantly. They analyzed the pH dependence of the reaction and observed a solvent isotope effect, suggesting that ^{80}Lys participates in catalysis as a general base and assists a nucleophilic attack of water to the substrate α-carbon atom in the oxidative deamination reaction. PheDH of *T. intermedius* acts preferentially on Phe and Tyr, whereas LeuDH of *B. stearothermophilus* acts almost exclusively on Leu and some other branched-chain L-amino acids. The two enzymes share a similar sequence (47%). The inherent hexapeptide segment (^{124}FVHAAR) in the substrate-binding domain of a mutant PheDH of *T. intermedius* was replaced by the corresponding part of LeuDH (MDIIYQ).

The catalytic efficiencies (K_{cat}/K_m) of the mutant enzyme with aliphatic amino acids and aliphatic keto acids as substrates were 0.5 to 2% of those of the wild-type enzyme. However, the efficiencies for Phe and phenylpyruvate were greatly reduced to 0.008 and 0.035% of those of the wild-type enzyme, respectively. Soda et al. constructed a chimeric enzyme consisting of an N-terminal domain of PheDH containing the substrate-binding region and a C-terminal domain of LeuDH containing the NAD^+-binding region. The chimeric enzyme showed 6% of that of the parental PheDH activity on Phe and had broader substrate specificity in the oxidative deamination. The substrate specificity of the chimeric enzyme in the reductive amination was an admixture of PheDH and LeuDH. The chimeric enzyme has a Gly corresponding to position 124 of PheDH and a Val corresponding to position 307. Engel et al. showed the same effect by changing only two sites. [124]Gly and [307]Leu of PheDH from *B. sphaericus* were altered by site-specific mutagenesis to the corresponding residues in LeuDH, Ala and Val, respectively. These two residues have previously been implicated (by molecular modeling based on the X-ray crystallographic analysis of *C. symbiosum* GluDH), as important residues in determining the degree of substrate discrimination. Of the single mutants L307V and G124A, and the double mutant G124, L307V displayed lower activities toward Phe and enhanced activity toward almost all aliphatic amino acid substrates tested compared to the wild-type; thus an amino acid dehydrogenase that shows closer resemblance to LeuDH was made from PheDH by changing only two residues.

SIGNIFICANCE OF ENZYME

Application of thermostable AlaDH and LeuDH is reviewed by Ohsima and Soda. A review on regeneration of NADH for organic synthesis was published by Chenault and Whitesides. By carefully choosing the differences in the substrate specificities of PheDHs, *B. sphaericus* enzyme was applied to the synthesis of various L-amino acids. PheDH from *B. badius* was shown be narrower in substrate specificity and, therefore, was applicable to the microdetermination of Phe and phenylpyruvate in blood.

Biosynthesis of L-Amino Acids

The production of Phe as a starting material for the artificial sweetener aspartame has been a target of industrial development. Several enzymatic processes of Phe synthesis have been reported: L-specific hydrolysis of benzyl-hydantoin, amination of *trans*-cinnamic acid, transamination from an amino donor to phenylpyruvate, and two-step conversion starting from acetamidocinnamic acid *via* phenylpyruvate. Because the α-keto acids are generally more expensive than the L-amino acids, it has been impractical to use the α-keto acids as starting materials for the synthesis of L-amino acids. With the development of an efficient method of phenylpyruvic acid synthesis by double carbonylation of benzylchloride in the presence of a cobalt catalyst, a new enzymatic method for the synthesis of Phe from phenyl-pyruvate was sought. Previously, the enzymatic synthesis of Phe from phenyl-pyruvate was accomplished with the transamination reaction. However, this process requires an L-amino acid, such as Asp, as an amino donor, and after the reaction, a by-product such as oxaloacatate or its degradation product, pyruvic acid, are formed. With PheDH, ammonia can be used as a ammonium source. The discovery of PheDH made it possible to devise a reaction scheme for the synthesis of Phe from phenylpyruvate with NADH as a reducing agent.

The enzyme from *B. sphaericus* was chosen to study the application of the enzyme to the synthesis of various L-amino acids because it is very stable and shows broader substrate specificity than other PheDHs, acting on Tyr as well as Phe. Using PheDH from *B. sphaericus* and formate dehydrogenase (FDH) from *Candida*, with a catalytic amount of NAD^+, Phe was synthesized in a good yield (120 g/L). In the study on substrate specificity of the enzyme in the reductive amination of α-keto acids, various optically pure L-amino acids were quantitatively synthesized using PheDH and FDH. The products from α-keto-β-DL-methylbutyrate and α-keto-β-DL-methyl-β-phenylpyruvate were identified as diastereomeric mixtures of Ile and L-*allo*-isoleucine, and L-α-amino-β-DL-methyl-β-phenylpropionic acid, respectively. L-α-Amino-γ-phenylbutyric acid (L-homophenylalanine) and other unnatural L-amino acids could be efficiently synthesized. L-Homophenylalanine is a building block of some of the angiotensin-converting-enzyme inhibitors. The solubility of these Phe homologs, such as Tyr, L-α-amino-γ-phenylbutyric acid, L-α-amino-ε-phenylvaleric acid, and so on, is so low that they are easily separated in crystalline forms from the reaction mixture by filtration. The filtered enzyme solution can be used for further repeated synthesis.

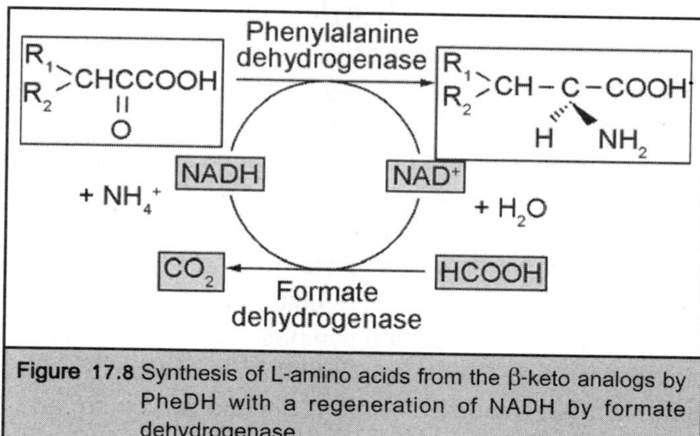

Figure 17.8 Synthesis of L-amino acids from the β-keto analogs by PheDH with a regeneration of NADH by formate dehydrogenase.

Table 17.5 Synthesis of L-amino acids from α-keto acids by using *B. sphaericus* PheDH and FDH.

Substrate	*Product*	*Yield (%)*
Phenylpyruvate	L-Phe	>99
4-Hydroxyphenylpyruvate	L-Tyr	>99
4-Fluorophenylpyruvate	L-4-Fluorophenylalanine	>99
α-Keto-γ-phenylbutyrate	L-α-Amino-γ-phenylbutyric acid	99
α-Keto-ε-phenylvalerate	L-α-Amino-ε-phenylvaleric acid	98
α-Keto-β-methyl-β-phenylpropionate	L-α-Amino-β-DL-methyl-β-phenylpropionate	98
α-Ketononanoate	L-α-Aminononanoic acid	99

L-β-Chloroalanine is a useful intermediate for the synthesis of several L-amino acids. Conditions for the synthesis of optically pure L-β-chloroalanine from β-chloropyruvate using AlaDH, LeuDH, and PheDHs from *B. badius, S. ureae,* and *T. intermedius* with a regeneration of NADH by FDH were investigated. The enzymatic reaction was carried out at neutral pH because of the chemical instability of β-chloropyruvate under alkaline conditions. AlaDH from *B. stearothermophilus* IFO 12550 showed the highest activity for the production of L-β-chloroalanine at pH 7.5. L-β-Chloroalanine was produced with high chemical (>90%) and optical yields (100% enantiomeric excess) and at a high concentration (43 g/L).

Immobilized *Nocardiaopaca* cells were used to synthesize Phe from phenylpyruvate and ammonia under hydrogen pressure, although no enzymological study was done. Thermostable L-amino acid dehydrogenases such as PheDH and formate dehydrogenase were simultaneously expressed in *E. coli,* which was utilized for the synthesis of L-amino acids from their keto analogs.

Regular Synthesis of Phenylalanine

Membranes are efficiently used in the enzyme-catalyzed syntheses of optically active compounds on laboratory and industrial scales. This technique can be applied for the repeated use of the enzyme in the synthesis, especially if smaller amounts of enzyme are available. Small amounts of PheDH from *B. sphaericus* and FDH were placed in a dialysis tube to check their durability and efficiency in the continuous synthesis of Phe. The *B. sphaericus* PheDH proved to be very stable in the synthesis, with high efficiency. PheDH maintained its activity for up to 28 changes. During the 34-day operation, 10.74 g (6.51×10^{-2} mol) of Phe was synthesized, using 45 μg (5 units, 1.35×10^{-10} mol) of PheDH, ca. 7.5 mg (15 units) of FDH, and ca. 35 mg of $NAD^+ \cdot Na$ (5×10^{-5} mol). Thus, 1 mol of PheDH catalyzed the synthesis of 4.8×10^8 mol of Phe, which is 2.4×10^5 times the weight of the enzyme. Continuous synthesis of Phe by the use of membrane reactors has been demonstrated with PheDHs from *Brevibacterium* sp. and *Rhodococcus* sp. To avoid the multistep purification procedure, the use of enzymes in whole cells is preferable. Acetone-dried cells of *B. sphaericus* and *C. boidinii* were also effective for Phe synthesis, providing a simple microbial method of synthesis.

In a typical course of the reductive amination reaction of phenylpyruvate using the mixture of acetone-dried cells as catalysts, the concentration of Phe reached 61.5 mg/mL with a yield of more than 99%. Tyr and Phe labeled with a positron-emitting radionuclide would be useful for positron emission tomography (PET) studies of tissues and neoplasms, such as measuring protein synthesis in tumor and brain tissue. ^{13}N-Labeled Phe, Tyr, and L-dopa were synthesized for PET. PheDH catalyzes the reductive ^{13}N amination of either phenylpyruvate or *p*-hydroxyphenylpyruvate to form [^{13}N]Phe or [^{13}N]Tyr, respectively, with NADH as a reducing agent. After short incubation of ^{13}N ammonia and either phenylpyruvate or *p*-hydroxyphenylpyruvate with the enzyme immobilized on the CNBr-activated Sepharose column, 83% of the ^{13}N ammonia was converted to [^{13}N]Phe, and 38% of the ^{13}N was converted to [^{13}N]Tyr, respectively. The labeled amino acids were purified by passage of the solution through an ion exchange column. Yields of each labeled amino acid were >30 mCi. L-Dopa was labeled by the exchange reaction between L-dopa and ^{13}N ammonia in the presence of NAD^+ and PheDH; 9% of the label was transferred to L-dopa. In a separate experiment involving the exchange reaction between Phe and ^{13}N ammonia, transfer of label from ^{13}N ammonia to Phe was 50%. Thus, utilizing *B. sphaericus* PheDH is useful for preparing [^{13}N]Tyr and [^{13}N]Phe.

Quantitative Determination

Quantitative determination of Phe in the plasma is important in diagnosing PKU. Several groups have been investigating the use of PheDH for the microdetermination of Phe and phenylpyruvate in blood samples. *B. badius* PheDH was found to be structurally closely related to that from *B. sphaericus*, although the former has a narrower substrate specificity suitable for the microdetermination, whereas the latter is more stable and shows a wider specificity suitable for the synthesis of various L-amino acids. Utilizing *B. sphaericus* PheDH, enzymatic cycling assay for the determination of Phe and phenylpyruvate in deproteinized tissue extracts was devised by Cooper et al. Assay 1 couples glutamine transaminase K with PheDH. Assay 2 combines PheDH, L-amino acid oxidase, and catalase. In both assays, Tyr and some other amino acids (or their α-

keto acid analogs) can replace Phe (or phenylpyruvate) to a small extent. Thus, when measuring Phe, a correction must be made for the nonspecificity of the reaction.

By removing Phe on a cation exchange column, it was possible to measure phenylpyruvate in tissue extracts. Concentrations of phenylpyruvate (μmol/kg) in normal rat liver, kidney, and brain were 2.1 ± 1.1 (n = 8), 1.8 ± 0.4 (n = 4), and 3.3 ± 0.6 (n = 4), respectively. A spectrophotometric recycling assay for the quantitation of Phe (and phenylpyruvate) involves the coupling of PheDH with rat kidney cytosolic glutamine transaminase K. The latter enzyme possesses high affinity for phenylpyruvate. Recycling results in a 50-fold increase in sensitivity over that of a conventional spectrophotometric end-point analysis procedure. The spectrophotometric recycling procedure has now been adapted to the measurement of Phe in microliter quantities of human blood. The procedure is 10 times more sensitive than the end-point assay for the spectrophotometric measurement of Phe in human blood. The findings suggest that the recycling procedure adapted for fluorometry will be even more sensitive.

Enzymatic Microplate Method

Naruse et al. have introduced a sensitive enzymatic microplate method using PheDH from *B. badius* for the mass screening of PKU among neonates as an alternative method for the Guthrie test. The principle of this system involves two steps. First, Phe is oxidized by PheDH to phenylpyruvate, reducing NAD^+ to NADH, which then serves as the electron donor in a colorimetric reaction as follows. In the presence of NADH and the electron carrier l-methoxy-5-methylphenazium methylsulfate (l-MPMS), Co^{3+} is reduced to Co^{2+}, which produces a colored chelate with the metal indicator 2-(5-bromo-2-pyridyazo)-5-(*N*-propyl-*n*-sulfopropyl-amino)phenol (5-BrPAPS). This assay system detected Phe in filter paper blood in the concentration range of 0.02-1.50 mM. This method uses a simple TCA extraction method and a PheDH reaction with high substrate specificity that produces very sensitive colorimetric quantitation.

The absorbance at 585 nm of this chelate can be measured by a colorimetric micro-plate reader. This highly specific and sensitive method has been used to screen neonates for PKU in Japan since 1992. Since the recent discovery of PheDH in 1984, several microbial producers of the enzyme have been isolated and their enzymological properties and primary structures have been studied. Indeed, PheDH has been used to detect PKU in the blood of neonates in Japan since 1992. However, PheDH has many further applications to be studied. If the structure and function of the amino acid dehydrogenases are thoroughly understood, it might be possible to create a new amino acid dehydrogenase with a wider scope of application.

SUMMARY

Numerous commercial processes have been developed for the large-scale commercial production of L-phenylalanine since the early 1980s, fueled by the enormous increase in L-phenylalanine demand for the dipeptide sweetener aspartame. The most successful processes use microbial enzymes at some stage, either in an enantioselective hydrolysis or biotransformation step using chemically derived precursors or in whole cell fermentations using microorganisms engineered through classical and recombinant strain improvement methodology. In general, the most economically competitive processes have resulted from large-scale fermentation approaches. This is mainly because of the incremental savings achieved in such processes through increased understanding of the physiology and molecular genetic regulation of L-phenylalanine biosynthesis in microbes and the low raw material costs attained through media and process development. In these processes, the concentration of L-phenylalanine can often exceed 50 g/L in crude fermentation broth with the overall incorporation of carbon into phenylalanine from a source such as glucose, equaling or exceeding the theoretical maximum calculated at around 25 to 30%. With this level of efficiency, it is unlikely that most major manufacturers of L-phenylalanine will continue to invest significantly in the development of fermentation processes as the return on investment is diminished.

Several elegant biotransformation approaches such as the PAL and dehydrogenase processes have been optimized through molecular genetic strategies but have been impacted by the drawbacks of greater precursor cost and in some cases limited biocatalyst stability. As a result, none of these approaches are currently used to produce L-phenylalanine. Nevertheless, a number of manufacturers, including DSM, Degussa/Rexim, and Tanabe, successfully produce L-phenylalanine through chemoenzymatic routes that are generally more versatile and are often part of more broadly applicable processes in natural and unnatural amino acid manufacture. This is apparent in the manufacture of D-phenylalanine where, in contrast to the preponderance of fermentative routes to L-phenylalanine, the synthesis is currently carried out using the hydantoinase and amidase resolution strategies mentioned earlier. Ultimately, through a greater understanding of microbial pathway engineering and the increasing availability of novel enzymatic activities, it is quite possible that the favorable economics of large-scale fermentation may also become more applicable to D-phenylalanine biosynthesis.

18

BIOLOGICAL POLYMERS

Plastics have been produced by the chemical industry since the early 1930s. They constitute a very important group of materials, because their high molecular weight and low reactivity make them especially suited for applications for which durable, inert materials are required. However, these very same properties have recently been recognized as major drawbacks of plastic materials. Whenever plastic is disposed of, it remains at the site of disposal, and its resistance to biological breakdown causes it to accumulate in the environment. The recognition of this undesired property has led to a search for new plastic materials that disappear after their disposal, the so-called biodegradable plastics. Several biological polymers that could fulfill this requirement are currently under investigation. Most of these biopolymers belong to the class of polysaccharides or polyhydroxyalkanoates (PHAs). The first example of PHAs to be discovered was polyhydroxybutyrate (PHB). In 1926, Lemoigne isolated and characterized PHB from *Bacillus megaterium*.

Since this first report, PHB accumulation has been found in many microorganisms: representatives of Gram-negative and Gram-positive species (i.e., autotrophs, heterotrophs, phototrophs, aerobes, anaerobes) and archaebacteria. Bacteria synthesize and accumulate PHAs as carbon and energy storage materials or as a sink for redundant reducing power under conditions of limiting nutrients in the presence of excess carbon source. When the supply of the limiting nutrient is restored, the PHA can be degraded by intracellular depolymerases and subsequently metabolized as a carbon and energy source. In 1976, Imperial Chemical Industries recognized the potential applicability of PHB to replace some of the oil-derived commodity plastics. Although the bacterially produced PHB was relatively expensive compared to petrochemical plastics, it was expected that the price of crude oils would rise to high levels, making the production of PHB economically feasible. Because no rise in oil prices occurred, PHB came to be seen as a biodegradable and biocompatible plastic material.

Mainly because of their high price, PHB and PHB/HV (a copolymer consisting of hydroxybutyrate and hydroxyvalerate) known under the trade name Biopol have only recently found their way to the market. In 1990, the German company Wella released a new shampoo packaged in bottles made of Biopol. Early in 1996, Monsanto bought the Biopol business from ZENECA BioProducts, a daughter company of ICI. Monsanto is now in the process of generating transgenic plants (i.e., soybean and rapeseed) for future agricultural production of PHB and PHB/HV, which is expected to result in a significantly lower price of the final product. The discovery of a polyester that consists mainly of hydroxyoctanoate monomers by de Smet et al. in 1983 was the first example of a new group of polymers that can contain a wide variety of different monomers.

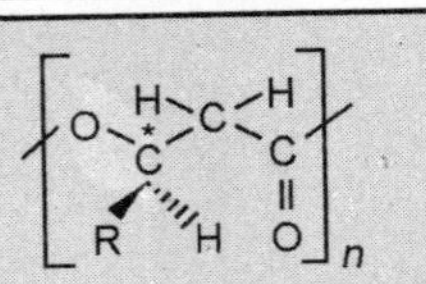

R = CH_3–C_2H_5 Short-chain-length PHAs
R = C_3H_7–$C_{14}H_{29}$ Medium-chain-length PHAs

Figure 18.1 General structure of polyhydroxyalkanoates (PHAs).

Thus far, more than 90 different monomers have been found in the polymers. Among these are 3-hydroxy acids of 6 to 14 carbon atoms with saturated, unsaturated, straight, or branched aliphatic side chains. Furthermore, monomers with various different functional groups in the side chain, such as halogen atoms, hydroxy-, epoxy-, cyano-, carboxyl-, and esterified carboxyl groups, have been introduced into medium-chain-length (MCL) PHAs. PHAs can be classified into two groups depending on the number of carbon atoms in the monomer units: short-chain-length (SCL) PHAs, which consist of 3–5 carbon atoms, and MCL PHAs, which consist of 6-14 carbon atoms. The monomer units in these microbial polyesters are all in the *R*-configuration because of the stereospecificity of the biosynthetic enzymes. The molecular weights of PHAs range from 2×10^5 to 3×10^6, depending on the specific PHA, the microorganism from which the PHA was obtained, and the growth conditions. In this review, the term *PHA* refers to polyhydroxyalkanoates generally. PHB (or PHV) refers to specific SCL PHAs, and MCL PHA refers to all PHAs that contain monomers with six or more carbon atoms.

POLYHYDROXYALKANOATE

PHB, the best-known of the SCL PHAs, not only appears in microorganisms as storage material but is also ubiquitous in nature. A fascinating development in recent years has been the discovery of the very wide distribution of PHB as a low molecular weight oligomer (120-200 monomers) in microorganisms, plants, and animals, including humans. In many cases this form of PHB is found as a PHB-calcium polyphosphate complex in membranes that seems to function as an ion channel through cell membranes. In *Escherichia coli,* the complex has been found to occur in large amounts in the membrane of cells made competent for genetic transformations. In addition, PHB of similar molecular weight has been found in eukaryotes, for example, in various plant tissues, in organs of animals (such as beef heart, chicken liver, porcine kidney), and in human blood plasma. In this review, we focus on the high molecular weight polymer that has considerable potential as an industrial biodegradable plastic.

Polyhydroxybutyrate

The biochemistry of PHB biosynthesis is very well studied. In most bacteria, such as *Alcaligenes eutrophus,* PHB is synthesized in a three-step reaction starting with acetyl-CoA when cultivated on carbohydrates, pyruvate, or acetate. Two acetyl-CoA molecules are coupled to form acetoacetyl-CoA in a condensation reaction catalyzed by 3-ketothiolase. The product is subsequently stereoselectively reduced to (*R*)-3-hydroxybutyryl-CoA in a reaction catalyzed by NADPH-dependent acetoacetyl-CoA reductase. Finally, PHB is synthesized by polymerization of *(R)*-3-hydroxybutyryl-CoA molecules by the PHB synthase.

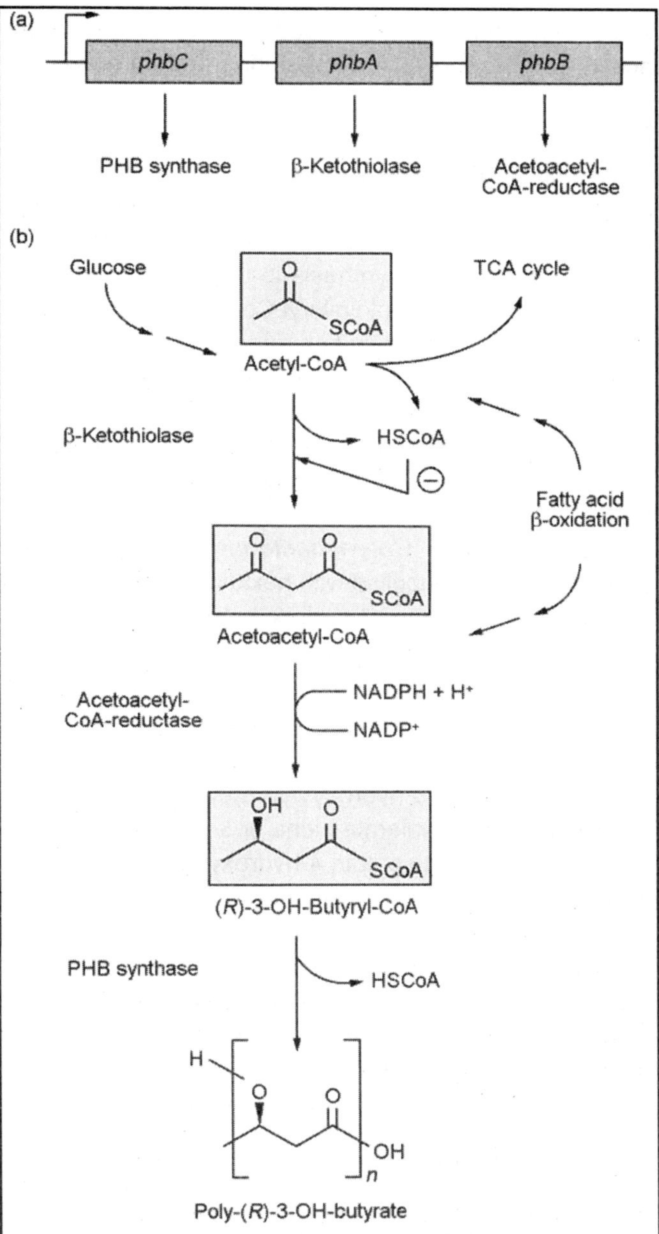

Figure 18.2 Polyhydroxybutyrate (PHB) synthetic genes and PHB synthesis pathway of *Alcaligenes eutrophus*.

A few exceptions to this general pathway are known, such as *Rhodospirillum rubrum*, where (*S*)-3-hydroxybutyryl-CoA is generated and converted to *(R)*-3-hydroxybutyryl-CoA by two stereospecific enoyl-CoA hydratases. As pointed out in the introduction, PHB biosynthesis and accumulation occur in most bacteria if a carbon source is provided in excess and if another nutrient, such as nitrogen, sulfur, phosphate, iron, magnesium, potassium, or oxygen, is limiting.

PHB synthesis is regulated at the enzymatic level, and the availability of suitable substrates for PHB synthesis routes results from an imbalanced supply of nutrients to the cell via physiological regulation pathways than have been studied in *Azotobacter beijerinckii* and *Alcaligenes eutrophus*. It was found that the intracellular concentration of acetyl-CoA and free coenzyme A play a central role in the regulation of polymer synthesis. Under balanced growth conditions acetyl-CoA is oxidized via the tricarboxylic acid (TCA) cycle. In this reaction sequence, NADH is generated and is used for biosynthetic purposes. When growth ceases the NADH concentration increases, which reduces the activity of the TCA cycle enzymes citrate synthase and isocitrate dehydrogenase. As a result acetyl-CoA cannot be oxidized via the TCA cycle and enters the PHB synthetic pathway. The 3-

ketothiolase enzyme of this pathway is inhibited by free CoA, which is generated by oxidation of acetyl-CoA via the TCA cycle during normal growth.

Synthesis of Copolymer

A copolymer of PHB and PHV can be synthesized by *Alcaligenes eutrophus* and many other microorganisms from either a mixed substrate of glucose and propionic acid or a direct precursor of 3-hydroxy-valerate (e.g., valeric acid). If propionate is fed, essentially the same biochemical pathway as for PHB synthesis is used, but propionyl-CoA and acetyl-CoA are condensed by the ketothiolase to give 3-ketovaleryl-CoA, which leads to the incorporation of 3-hydroxyvalerate monomers in the polymer. 3-Hydroxy-butyrate arises in the normal way. Alkanoic acids of odd-number carbon chain length can also serve as carbon source. In this case, the 3-hydroxy-valerate in the polymer arises directly from the β-oxidation of these fatty acids. Recently, it was shown that low dissolved oxygen concentrations enhance 3-hydroxy-valerate incorporation into the polymer produced by *A. eutrophus*. Several microorganisms have been isolated that are capable of accumulating PHB/HV from substrates apparently unrelated to, and not immediate precursors of, 3-hydroxyvalerate. These organisms include *Corynebacterium, Nocardia,* and *Rhodococcus* strains and mutants of *A. eutrophus*. When supplied with hexanoic acid as substrate, *Rhodococcus* sp. was able to accumulate a terpolymer containing 3-hydroxybutyrate, 3-hydroxyvalerate, and 3-hydroxyhexanoate monomers.

In addition, *Rhodococcus* sp. NCIMB40126 and *Chromobacterium violaceum* are capable of producing a homopolymer consisting of 3-hydroxyvalerate when grown on valeric acid as sole carbon source. Copolymers of 3-hydroxybutyrate and 4-hydroxybutyrate can be synthesized by *A. eutrophus* from 4-hydroxybutyric acid, 1-4-butanediol, butyrolactone, and 4-chlorobutyric acid. Terpolymers of 3-hydroxybutyrate, 3-hydroxyvalerate, and 5-hydroxyvalerate can be produced by *A. eutrophus* fed with either 5-chlorovalerate alone or 5-chlorovalerate or 5-hydroxyvalerate in combination with valeric acid. Polymers that contain 4-hydroxyvalerate and 3-hydroxypropionate have been synthesized by a variety of organisms, including *A. eutrophus*. The substrates used were 4-hydroxyvalerate, 4-valerolactone, and 3-hydroxypropionic acid. Most *Alcaligenes* strains tested and *Pseudomonas oxalaticus* accumulated a terpolymer of 3-hydroxybutyrate, 3-hydroxyvalerate, and 4-hydroxyvalerate from 4-valerolactone or 4-hydroxyvalerate as sole carbon source. *Alcaligenes eutrophus* and *Burkholderia cepacia* are able to synthesize terpolyesters that consist of 3-hydroxybutyric acid, 3-hydroxyvaleric acid, and 2- methyl-3-hydroxybutyric acid when tiglic acid is used as sole carbon source or in combination with gluconic acid.

Synthesis of SCL PHA at Molecular Level

In *A. eutrophus,* the structural genes for PHA synthesis are organized in the *phbCAB* operon, coding for PHB synthase, β-ketothiolase, and NADPH-dependent acetoacetyl-CoA reductase, respectively. The transcription start of *phbC* was identified 307 base pairs upstream from the translation initiation codon by S1 nuclease protection assay, and a sequence that exhibits a striking homology to the *E. coli* σ^{70} promoter consensus sequence was found. The PHB synthase structural genes have been cloned and sequenced from various bacteria. Comparison of the protein primary structures revealed that PHB synthases represent a group of highly homologous enzymes. The

PHB synthase of *A. eutrophus* has been studied in detail in several laboratories. It is a large enzyme with identical subunits that exhibit a relative molecular mass of 63.9 kDa and specificity for 3-hydroxyacyl-CoA of C_3 to C_5 units. Site-directed mutagenesis demonstrated that the $cysteine_{319}$ residue is required for PHB synthase activity in *A. eutrophus*.

Furthermore, evidence has been obtained that the enzyme requires posttranslational modification by phosphopantetheine to provide a second thiol group per PHB synthase subunit, which could

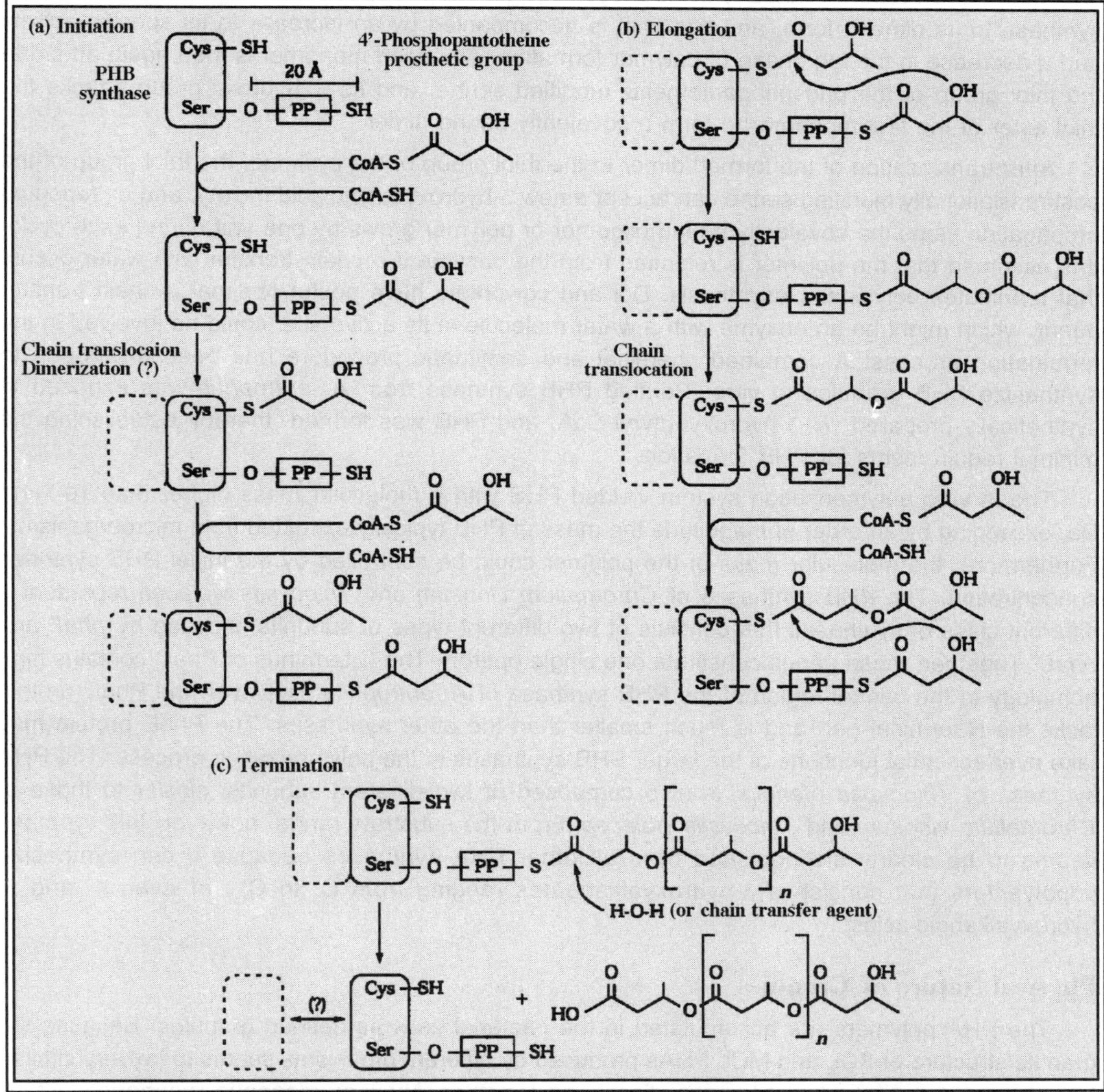

Figure 18.3 The hypothetical model of the catalytic PHB synthase mechanism is based on analogy to the de novo fatty acid synthesis and on information available about PHB synthases.

support the mechanistic model for the polymerization process that depends on the presence of two thiol groups. A putative model for the polymerization process based on de novo fatty acid synthesis illustrates only one possible reaction mechanism, and other models are also possible. In an initiation or priming process, the 3-hydroxybutyric acid moiety from 3-hydroxybutyryl-CoA is transferred to the thiol group of the phosphopantotheine-modified serine, coupled with the release of coenzyme A. This 3-hydroxybutyryl moiety is then translocated to the thiol of the cysteine. It has been shown recently that acylation of $cysteine_{319}$ causes a shift of the monomeric form of the synthase to its dimeric form, and this shift is accompanied by an increase in its specific activity and a decrease in the lag-phase of polymer formation. A second monomer is then again added to the thiol group of the phosphopantotheine-modified serine, and its 3′-hydroxy group attacks the thiol ester of the first monomer to form a covalently bound dimer.

After translocation of the formed dimer to the thiol group of the cysteine, the thiol group of the posttranslationally modified serine can accept a new 3-hydroxybutyric acid moiety, and by repeated propagation steps the covalently bound oligomer or polymer grows by one unit during each cycle. It is assumed that the polymer is released from the enzyme if a chain transfer with water occurs that terminates polyester biosynthesis. Doi and coworkers have postulated that a chain transfer agent, which might be an enzyme with a water molecule in its active site, could be involved in the termination process. A combined chemical and enzymatic procedure has been developed to synthesize PHB granules in vitro. Purified PHB synthase from *A. eutrophus* was exposed to synthetically prepared *(R)*-3-hydroxybutyryl-CoA, and PHB was formed, thereby establishing the minimal requirements for PHB formation.

The in vitro polymerization system yielded PHB with a molecular mass higher than 10×10^6 Da, exceeding by an order of magnitude the mass of PHB typically extracted from microorganisms. Furthermore, the molecular mass of the polymer could be controlled by the initial PHB synthase concentration. The PHB synthases of *Chromatium vinosum* and *Thiocystis violacea* represent a different class of synthases that consists of two different types of subunits encoded by *phaE* and *phaC*. Together, these genes constitute one single operon. The C-terminus of PhaC contains high homology to the central region of the PHB synthase of *A. eutrophus*. However, the PhaC protein lacks the N-terminal part and is much smaller than the other synthases. The PhaE protein may take over essential functions of the larger PHB synthases in the polymerization process. The PHA synthase of *Thiocapsa pfennigii* is also composed of two different subunits, similar to those of *Chromatium vinosum* and *Thiocystis violacea*. From the substrate range, however, this synthase seems to be clearly distinguished from all other PHA synthases because it can synthesize copolyesters that consist of 3-hydroxyalkanoates ranging from C_3 to C_{14}, or even 4- and 5-hydroxyalkanoic acids.

Physical Nature of Granules

The PHA polymers are accumulated in the bacterial cells as defined granules. Because the granule structure of SCL and MCL PHAs produced by different organisms seems to be very similar, both types of polymer granules are discussed together. Nuclear magnetic resonance (NMR) spectroscopy of various bacteria has clearly demonstrated that the polyester in the cells occurs in a metastable amorphous state and that it is stored as a mobile elastomer. The bouyant density of PHB granules is about 1.18–1.24 g cm^{-3}, in contrast to MCL PHA granules, which have a bouyant density of approximately 1.05 g cm^{-3}. The isolated granules consist of polyester, proteins, and lipids. The composition was first determined for PHB granules of *Bacillus megaterium*, which consist of 97.7% polyester, 1.87% proteins, and 0.46% lipids or phospholipids. Several different models of granule structure have been proposed. First indications that the native PHA granules in *Pseudomonas oleovorans* might be surrounded by a phospholipid monolayer in which the polymerase enzymes are embedded have been found by Witholt and coworkers. In contrast to this model, Stuart et al. presented a model in which the hydrophobic PHA core is surrounded by two protein layers with a phospholipid layer in between. In this model, the polymerase is supposed to be embedded only in the inner layer.

Recently, determination of the thickness of the boundary layer surrounding the polymer inclusions of different bacteria confirmed that a lipid monolayer is more likely than two protein layers with a lipid layer between them. In addition to the polymerases, other proteins such as depolymerases, granule-associated proteins (GAPs, also called phasins), and proteins of unknown function seem to be embedded in the phospholipid layer. Phasins are in general low molecular weight proteins and are assumed to form a close protein layer at the surface of the granules, providing the interface between the hydrophilic cytoplasm and the hydrophobic core of the PHA inclusion. In *A. eutrophus*, the first evidence was found by mutant studies for the structural importance of a 24-kDa GAP (called GA24). Mutants lacking the GA24 protein synthesized only one large PHB granule per cell in contrast to 10-30 granules as found in the wild type. The amino acid sequence of the GA24 protein revealed two closely related stretches consisting exclusively of nonhydrophilic amino acids at the C-terminal region, which are presumably involved in the binding of GA24 to the granules. Similar granule-forming proteins were identified and isolated from various other strains, such as *Acinetobacter* sp., *Rhodococcus ruber, Chromatium vinosum, Methylobacterium extorquens*, and *Pseudomonas oleovorans*.

PHB and PHB/HV in Industries

Many bacteria have been screened to produce PHB or PHB/HV. However, only a few have been used for biotechnological production of these polyesters on a large scale. Here, we focus on the two examples that have found application in an industrial process. The suitability of a bacterium for PHA production depends on many different factors such as stability and safety of the organism, growth and accumulation rates, achievable cell densities and PHA contents, extractability of the polymer, molecular weights of accumulated PHA, range of utilizable carbon sources, costs of the carbon source and the other components of the medium, and occurrence of by-products. Recently, the theoretical yield (the yield based on the reaction stoichiometry) of PHB has been estimated for several substrates. Regeneration of nicotinamide nucleotides, which are used as cofactors for PHA synthesis, has been taken into account in this analysis. In the production plant of ZENECA BioProducts at Billingham, the PHB homopolymer and the copolyester PHB/HV have been produced at scales up to 200,000 L. The process utilized a mutant of *A. eutrophus*, and the fermentation was carried out in a two-step fedbatch process. During the first step, the cells were grown in a mineral salt medium with glucose as sole carbon and energy source and with a calculated amount of phosphate based on the known requirements of the organism to allow production of a given amount of biomass.

As the culture grew, phosphate was depleted from the medium, and during the second step when phosphate was limiting, the culture started to store polymer. Glucose was fed to the culture, and the fermentation was continued until the required polymer content was reached. Each phase lasted for approximately 48 h, and final biomass dry weights of 100 g/L were achieved. The copolymer of PHB/HV was made by substituting a mixed feed of glucose and propionic acid in the polymer accumulation phase. The 3-hydroxyvalerate content of the copolymer was controlled by adjusting the ratio of the two substrates in the feed. A different industrial process has been developed at the Biotechnologische Forschungsgesellschaft in Linz (Austria) for the production of PHB homopolymer. In this process strain *Alcaligenes latus* DSM1 124 was employed, which can accumulate PHB during balanced cell growth up to 80% of the cellular dry matter. Therefore, a one-step fed-batch fermentation process was carried out using a mineral salts medium with sucrose as sole carbon source. A biomass of 60 g/L was achieved. In 1993, however, the company stopped the production of PHB/HV.

Separation Methods

The cells that contain PHB/HV must be separated from the broth by conventional procedures such as centrifugation, filtration, or flocculation–centrifugation. After the biomass is harvested, cells must be disrupted to recover the polymer. Several methods have been developed for the recovery of PHAs. The method most often used involves extraction of the polymer from the biomass with solvents (e.g., chloroform, methylene chloride, propylene carbonate, dichloroethane). Large amounts of solvent are required in this process. Several other methods have been developed, such as sodium hypochloride treatments for the differential digestion of non-PHA cellular materials. Although this method is effective, it causes severe degradation of PHB, resulting in a reduction in the molecular weight. ZENECA has developed a nonsolvent-based recovery process as an alternative to solvent extraction for the commercial production of PHB and PHB/HV by *A. eutrophus*. In this process the cells were first exposed to 80°C and subsequently treated with a cocktail of various hydrolytic enzymes consisting of lysozyme, phospholipase, lecithinase, the proteinase alcalase, and others. Most of the cellular components were hydrolyzed by these enzymes, whereas the polymer remained intact. After washing, flocculation, and drying, the polymer was recovered as a white powder that was melted, extruded, and converted into chips, the form in which the polymer was supplied to the fabricators. In contrast, the Biotechnologische Forschungsgesellschaft has applied a solvent-based process for the recovery of PHB from *A. latus*. The cells were harvested by centrifugation, and subsequently the PHB was extracted with methylene chloride from the suspended cells and precipitated out of the solvent by the addition of water. After drying of the polymer, a polyester with 99% purity was obtained. The process also included recovery of the solvent.

Physical Characters

Comprehensive overviews on the physical properties of PHB and its copolymers, or on the material properties of the manufactured objects from those polymers, have been published previously and are summarized next. The PHB homopolymer is a completely stereoregular polyester, with all asymmetric carbon atoms in the (*R*)-configuration, and it is therefore highly crystalline. The high crystallinity (typically 55–80%) makes it relatively stiff and brittle. The glass transition temperature (T_g of PHB lies between 5 and 9°C, and the melting point (T_m) lies between 173 and 180°C. PHB polymer decomposes at approximately 200°C, which is close to its melting temperature. In its solid, crystalline state, the polymer is a right-handed helix, a condition that is retained in chloroform solution. The mechanical properties of PHB including Young's modulus (3.5 GPa) and the tensile strength (40 MPa) are similar to those of polypropylene. However, the elongation to break for PHB is about 3%, which is significantly lower than that of polypropylene (400%).

Table 18.1 Comparison of polymer properties of polyhydroxybutyrate and its copolymers with polypropylene

Physical properties	*P(3HB)*	*P(3HB/3HV) 90:10*	*P(3HB/3HV) 80:20*	*P(4HB)*	*P(3HB/4HB) 90:10*	*P(3HB/4HB) 10:90*	*PP*
Meltingpoint (°C)	179	150	135	53	159	50	170
Tensile strength (MPa)	40	25	20	104	24	65	34.5
Young's modulus (GPa)	3.5	1.2	0.8	149	—	100	1.7
Elongation tobreak (%)	3.0	20	100	1000	242	1080	400

The flexibility of the material is greatly improved when 3-hydroxyvalerate units are incorporated into the polymer, resulting in a decrease of Young's modulus below 0.7 and a decrease of tensile strength below 30 MPa. The elongation to break also increases as the comonomer fraction increases. The melt temperature is greatly depressed, down to 130°C, dependent on the 3-hydroxyvalerate content, but the degradation temperature is little affected. This property generates a relatively wide window of conditions that allow thermal processing as a melt without degradation of the material. PHB/HV copolymers with different 3-hydroxyvalerate content, ranging from 0 to 30%, are commercially available as Biopol resins. Copolymers of 3-hydroxybutyrate and 4-hydroxybutyrate monomers cannot form isomorphic crystals because of the extent of the structural differences between the monomer units. The glass transition temperature decreases from 5 to –50°C, and the melting temperature decreases from 180 to 54°C as the mol% of 4-hydroxybutyrate increases from 0 to 100%. Young's modulus, tensile strength, and % elongation at break (for 94 mol% 4-hydroxybutyrate) are reported as 55 MPa, 39 MPa, and 500%, respectively. The T_g and T_m of copolymers consisting of 3-hydroxybutyrate and 3-hydroxypropionate decrease with an increase in the 3-hydroxypropionate (from 0 to 26 mol%) from 4 to –8.1°C and 180 to 85°C, respectively. The degree of crystallinity also decreases with an increase of 3-hydroxypropionate.

Genetic Recombination

Successful biosynthesis of PHB in *E. coli* was independently achieved by two different groups. Polymer formation was only observed when all three PHB biosynthesis genes of *A. eutrophus* were expressed in *E. coli*; expression of *phbC* alone did not result in PHB production. These studies were extended by the production of the copolymer PHB/HV in recombinant *E. coli* (*fadR atoC*) mutants containing the PHB biosynthetic genes and using glucose and propionic acid as carbon source. The *E. coli* (*fadR atoC*) mutants produce the enzymes required for the catabolism of SCL fatty acid constitutively. 3-Hydroxyvalerate contributed up to 39 mol% to the accumulated polyester. Synthesis of PHB by recombinant *E. coli* does not require limitation of a specific nutrient, but is dependent on the amount of acetyl-CoA available and is significantly enhanced by the addition of various complex nitrogen sources, amino acids, or oleic acid. Application of recombinant *E. coli* in fed-batch culture resulted in cell densities of 101 g L^{-1}, containing approximately 80% PHB per cell dry weight after 39 h of cultivation.

Table 18.2 Enzymes involved in the pathways leading to MCL PHA formation and their respective genes in *Escherichia coli*

Enzyme	*Gene in E. coli*
1. Acyl-CoA synthetase	*fadD*
2. Acyl-CoA dehydrogenase	*fadE*
3. Enoyl-CoAhydratase or crotonase	*fadB*
4. (*S*)-3-OH-Acyl-CoA dehydrogenase (NAD dependent)	*fadB*
5. 3-Ketoacylthiolase	*fadA*
6. 3-ketoacyl synthase I; II; III	*fabB, F, H*
7. Ketoacyl-ACP reductase	*fabG*
8. 3-OH-Acyl-ACP dehydratase	*
8a. 3 -OH-Decanoyl-ACP dehydratase	*fabA*
9. Enoyl-ACP reductase	*
10. Fatty acid thioesterase I; II	*tesA, B*
11. Ketoacyl-CoAreductase (NADPH dependent)	*
12. (*R*)-3-OH-Acyl-CoA dehydrogenase (NAD dependent)	*
13. 3-OH-Acyl-CoAepimerase	*fadB*
14. Enoyl-CoA hydratase II	*
15. Enoyl-CoA isomerase	*fadB*
16. (*R*)-3-OH-Acyl (ACP $\rightarrow$ CoA) transferase	*
17. Several acyl transferases or acylases	*
18. PHApolymerase	None

Under specific conditions, such as pH 6.0, *E. coli* produced PHB with an extremely high molecular mass of 20 MDa. Furthermore, it was shown that the amount of PHB synthase in recombinant *E. coli* plays a key role in controlling the molecular weight and the polydispersity of the polymer. With an increased amount of PHB biosynthetic enzymes, a decrease in polyester molecular weight was observed. For the first time, PHB biosynthesis was successfully conferred to *Arabidopsis thaliana*, a small oil seed plant, by introducing the *phbB* and *phbC* gene from *Alcaligenes eutrophus*. When the enzymes necessary to catalyze the synthesis of PHB from acetyl-

CoA were produced in the cytoplasm, only small amounts of PHB were synthesized. Plants were viable but growth was very poor, possibly because of a reduction of acetyl-CoA needed for growth. Targeting of all three required proteins to the plastids of *Arabidopsis thaliana* resulted in normal growth, active enzymes, and PHB accumulation up to 14% of total cell dry weight. However, the molecular weight distribution of the plant-produced PHB was much broader than that of typical bacterial PHB. The *phbC* gene of *Alcaligenes eutrophus* has also been introduced into *Saccharomyces cerevisiae*, which enabled this yeast to accumulate PHB to up to 0.5% of cell dry weight. A novel pathway for PHB synthesis has been engineered in insect cells (*Spodoptera frugiperda*). In these insect cells, a heterologous dehydrase-deficient fatty acid synthase from rats is produced in addition to the PHB synthase of *A. eutrophus*. The multifunctional fatty acid synthase effectively replaces the β-ketothiolase and acetoacetyl-CoA reductase activities of the natural prokaryotic PHB pathway and, therefore, provides the precursor 3-hydroxybutyryl-CoA for conversion to PHB. This strategy suggests a new method for generating structurally diverse polyhydroxyalkanoates by metabolic engineering.

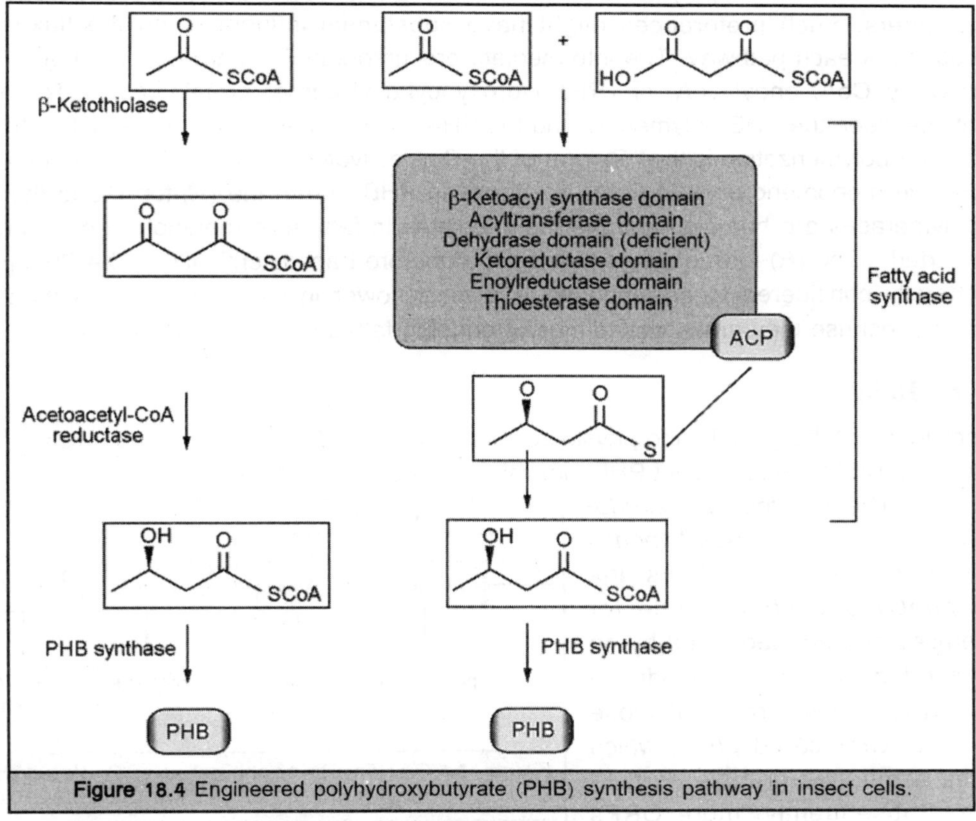

Figure 18.4 Engineered polyhydroxybutyrate (PHB) synthesis pathway in insect cells.

SYNTHESIS OF MCL PHA

Although many species of bacteria are reported to synthesize PHB and PHV, including some *Pseudomonas* strains, the ability to store polymers that contain monomers of 3-hydroxy acids of carbon chain length C_6 and longer, here referred to as MCL PHAs, is unique to the group of pseudomonads. Three main pathways are involved in the synthesis of PHA precursors in *Pseudomonas putida*. These were revealed by labeling studies, NMR spectroscopy, and gas chromatography-mass spectroscopy (GC-MS) analyses of PHA from cells that were grown on simple carbon substrates such as glucose and/or fatty acids. *P. putida* derives PHA from precursors made via the following routes. (i) Chain elongation, in which acyl-CoA is extended with acetyl-CoA, is comparable with the major route for PHB monomer synthesis, where the acyl-CoA is one of the two acetyl-CoA molecules. It has been found that about 3-10% of the monomeric units of PHA are one C_2 unit longer than the fatty acid used as substrate. However, this route is less important for PHA synthesis than it is for PHB synthesis. (ii) Fatty acid degradation by β-oxidation is the main pathway when fatty acids are used as substrate. (iii) De novo fatty acid biosynthesis is the main route during growth on simple carbon compounds.

The enzymes involved in the pathways do not have the same activity for different chain length substrates. Thus, the PHA polymerase seems to have a preference for the incorporation of C_8 to C_{10} monomers. Such preferences might have substantial influences on the fluxes and the contributions of each pathway. The intermediate precursors of PHA are either ketoacyl-CoA, (*S*)-3-hydroxyacyl-CoA, enoyl-CoA, or (*R*)-3-hydroxyacyl-acyl carrier protein (ACP). Regarding the similarity between the PHB polymerase and the PHA polymerases, it seems likely that the ultimate substrate for polymerization is the (*R*)-form of the CoA-activated 3-hydroxy fatty acid intermediates. Because the monomeric units in PHA, like those in PHB, are in the (*R*)-form, and the PHB and PHA polymerases are homologous, the intermediates in fatty acid metabolism must presumably be converted to the (*R*)-form of 3-hydroxyacyl-CoA before transesterification. The PHA precursors are generally considered to be available in a much lower in situ concentration than the PHB precursors because they are intermediates of ongoing fatty acid metabolic cycles.

Catalytic Role

Knowledge of the genetic and physiological regulation of PHA biosynthesis is relatively limited compared to what is known about PHB biosynthesis because PHA synthesis has only been studied since 1983. The *pha* loci of *Pseudomonas oleovorans* and *Pseudomonas aeruginosa* consist of four ORFs: the ORFs *phaC1* and *phaC2* code for polymerases and are separated by an ORF called *phaZ*, which encodes a depolymerase; downstream of these genes is an ORF called *phaD*, which encodes a putative granule surface protein. Subsequently, more ORFs have been identified downstream and in opposite orientation to the *pha* locus of *P. putida* and *P. oleovorans,* which presumably encode granule-associated proteins. The sequence of this region has not yet been published. At least two promoters are involved in the transcription of the *pha* genes in *P. aeruginosa*. Two promoters are upstream of the *phaC1* gene and resemble the

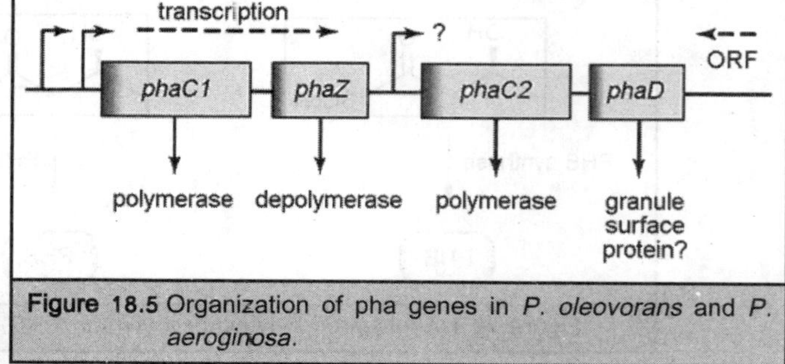

Figure 18.5 Organization of pha genes in *P. oleovorans* and *P. aeroginosa*.

consensus sequences for σ^{54}-dependent and σ^{70}-dependent promoters. It is not clear whether a third promoter upstream of the *phaC2* gene in *P. aeruginosa* is active. Northern blot experiments indicated the presence of two mRNAs, one carrying *phaC1* and *phaZ*, the other *phaC2* and *phaD*. However, a transcriptional start site accompanying the putative promoter upstream of *phaC2* could not be experimentally shown, suggesting that this promoter may not be functional.

The two PHA polymerases show identities of 54-59% at the protein sequence level when compared within each species. When corresponding PHA polymerases from different species are compared, they show 69-80% identity at the amino acid level. The identities between these PHA polymerases and the *A. eutrophus* PHB polymerase are nearly 40%. According to the gene sequences, the *Pseudomonas* PHA polymerases have molecular weights of 62-64 kDa. The enzymes show a small difference in substrate specificity. The *phaZ* gene product, a PHA depolymerase with a predicted molecular weight of 31.5 kDa, shows homology with a range of hydrolytic enzymes. It contains the common lipase box Gly-X-Ser-X-Gly, which is also found in extracellular PHA depolymerases.

Role of *Pseudomonas*

Pseudomonas strains such as *P. oleovorans, P. putida, P. aeroginosa,* and *P. resinovorans* are the major strains used for MCL PHA production. Batch, fed-batch, and continuous cultivation techniques have been extensively investigated at the laboratory scale, but MCL PHA production up to a 200-L scale has also been carried out. Efficient PHA production requires rapid formation of biomass and product. In batch and fed-batch fermentations, similar to the PHB production process, cells are first grown to a desired concentration without nutrient limitation, after which an essential nutrient is limited to trigger efficient PHA synthesis. During this nutrient limitation stage the residual cell concentration (defined as the cell concentration minus the PHA concentration) remains almost constant, and the cell concentration increases only because of the intracellular accumulation of PHA. *P. oleovorans* has been used to produce MCL PHAs in two-liquid-phase cultivations. The organism was cultured in media consisting of an aqueous phase containing all minerals necessary for growth and a second liquid phase containing *n*-alkanes as sole carbon and energy source.

Two-liquid-phase, single-stage chemostat, and fedbatch cultivations of *P. oleovorans* using *n*-octane as the carbon source resulted in highest volumetric productivities of 0.58 g L^{-1} h^{-1} and 0.25 g L^{-1} h^{-1}, respectively. In a two-liquid-phase, two-stage continuous cultivation a volumetric productivity of 1.06 g L^{-1} h^{-1} with a final cellular PHA content of 66% per total cell dry weight has recently been reported. The monomer composition of the synthesized polymer is mainly dependent on the strain used and the carbon source applied. The cultivation conditions have only a minor influence on the composition of the polymer. However, it was shown that temperature and concentration of the substrate has an influence on the molecular weight of the polymer. Recently, it was reported that *Pseudomonas putida* U grown on octanoic acid as the sole carbon source accumulated a homopolymer of poly-3-hydroxyoctanoate. Different *Pseudomonas* strains have been used to produce MCL PHAs from renewable resources, such as fatty acids and carbohydrates. In fedbatch cultivations, *P. putida* KT2442 was cultivated to high cell densities (92 g/L) with oleic acid as a substrate. With this process, an overall volumetric productivity of 1.6 g L^{-1} h^{-1} was achieved with a final cellular PHA content of 40% per total cell dry weight. Furthermore, MCL PHAs could be produced from glucose, fructose, and glycerol using *P. putida* KT2442. In addition to the major constituent 3-hydroxydecanoate, six other monomers were found to be present: 3-hydroxyhexanoate, 3-hydroxyoctanoate, 3-hydroxydodecanoate, 3-hydroxydodecenoate, 3-hydroxytetradecanoate, and 3-hydroxytetradecenoate. The degree of unsaturation of PHA was slightly influenced by the cultivation temperature.

Table 18.3 Composition of MCL polyhdroxyalkanoates produced by *Pseudomonas oleovorans* grown on different alkonoates as sole carbon source

	Composition (mol% of total 3-hydroxy fatty acids)						
Carbon source	*C6*	*C7*	*C8*	*C9*	*C10*	*C11*	*C12*
Hexanoate	95		5				
Heptanoate		100					
Octanoate	8		91		1		
Nonanoate		35		65			
Decanoate	8		75		17		
Undecanoate		28		59		13	
Dodecanoate	6		57		32		5
Tridecanoate		32		48	5	14	
Tetradecanoate	7		59		30		4
Pentadecanoate		32		47	8	13	
Hexadecanoate	8		50		30		12

Various functional groups can be introduced into the polymer by applying different substrates in the growth medium. Olefin groups were introduced by adding *n*-alkenes or alkenoates to the medium. Furthermore, different halogen atoms were incorporated in the polymer when halogenated alkanoates were used as substrates.

Table 18.4 Influence of cultivation temperature on MCL polyhydroxyalkanoate composition in *Pseudomonas putida*

			Composition (mol% of total PHA)						
Substrate	*T (°C)*	*% PHA*	*C6*	*C8*	*C10*	*C12:1*	*C12*	*C14:1*	*C14*
Glucose	15	20.5	0.4	8.4	64.8	16.7	6.0	4.0	<0.1
	30	17.0	<0.1	6.9	74.3	8.8	7.7	1.6	<0.1
Decanoate	15	48.5	5.9	50.9	43.2	—	—	—	—
	30	38.1	5.3	52.3	42.4	—	—	—	—

A homopolymer of 99% purity consisting of 3-hydroxy-5-phenylpentanoate is synthesized by *P. oleovorans* when grown on 5-phenylvalerate. Polymers containing different mol% of phenyl groups, or other aromatic compounds such as phenoxy groups or cyanophenoxy groups, were produced by adding phenyl alkanoates, phenoxy alkanoates, or cyanophenoxy alkanoates, respectively, as a cocarbon source. Recently, it was shown that MCL PHAs can be produced by specific recombinant *E. coli* strains that are deficient in fatty acid β-oxidation and in addition contain the PHA polymerase-encoding gene of *P. aeroginosa* or *P. oleovorans*. The recombinant strains were cultured on LB medium or mineral salt medium supplemented with yeast extract and alkanoates as additional carbon source. Up to 21% PHA per cell dry weight could be achieved by these recombinant *E. coli* strains.

Separation

Recovery procedures for MCL PHA resemble those originally developed for the production of PHB. A number of solvent extraction processes have been assessed to separate MCL PHAs from biomass. These usually involve the use of a chlorinated solvent such as chloroform or methylene chloride. An alternative, nonsolvent-based extraction process was developed as well to make the overall production process more attractive. The biomass is separated from the medium by centrifugation, and treated with a protease cocktail and a detergent to solubilize all cell components. Removal of the solubilized cell material and concentration of the resulting PHA suspension is achieved by cross-flow microfiltration. The submicrometer MCL PHA granules display a density close to that of water, as a result of which a MCL PHA suspension does not settle; in fact, it forms a highly stable polymer latex.

The overall purity of the latex amounts to 95%. MCL PHAs produced for applications that require higher purity, like those in biomedicine, must be further purified by extraction of PHA from the latex using methylene chloride and subsequent precipitation in a methanol–water mixture. The applicability of MCL PHA as a latex product has been evaluated by Marchessault et al. Atypical latex product would be a coating. Unlike that reported for PHB and its derivatives, additional treatments (e.g., heat, solvent vapor) to enhance coalescence into a film are not necessary, because MCL PHA does not readily crystallize. Heat, UV, or electron beam sources have been applied to cross-link polymer chains and fix the film. Mcl-PHA coatings are transparent with good clarity and exhibit excellent barrier properties to water.

Physical Characters

Owing to their enzymatic synthesis, as is the case for PHB, MCL PHAs have an exceptional stereochemical regularity and therefore are fully isotactic. This allows MCL PHAs to achieve some crystallinity, which can be as high as 25%. The molecular weight is of the order of 2×10^5 g/mol. The glass transition temperature is usually well below room temperature, ranging from –43°C to –25°C for MCL PHAs. Melting temperatures depend on both the thermal history and the nature of the pendant chain. Values reported vary between 39°C and 61°C with MCL PHA, based on *n*-octane or *n*-octanoate (PHO). The glass transition temperature of PHO is –35°C. Depending on the thermal history and the test conditions, the tensile modulus varies between 2.5 and 9 MPa, the tensile strength between 6 and 10 MPa, and the maximum elongation between 300 and 450%. These data classify PHO as a thermoplastic elastomer.

Table 18.5 Physical characteristics of saturated and unsaturated MCL polyhydroxyalkanoates isolated from *pseudomonas* strains

Strain	*Substrate*	*Mw*	*Mn*	*Mw/Mn*	T_g(°C)	T_m(°C)
P. oleovorans	*n*-Octane	178,000	99,000	1.8	–36.5	58.5
P. oleovorans	*n*-Octene	242,000	101,000	2.4	–36.6	—
P. putida KT2442	Oleic acid	140,000	70,000	1.9	ND	ND
Pseudomonas sp. NCIMB40135	Glucose	143,000	61,900	2.32	<– 60	54.8

Because the low melting temperature and crystallization rate hamper the applicability of MCL PHAs as thermoplasts, cross-linking was anticipated to overcome both limitations. Cross-linking was achieved by conventional techniques, namely electron-beam irradiation, peroxide cross-linking, or vulcanization. In all three studies, the presence of olefin groups was shown to enhance chemical cross-linking.

Sufficient cross-linking appeared to prevent polymer crystallization, so a true rubber was produced. Properties were therefore constant over a wide temperature range from the glass transition temperature (T_g, –30°C) up to the decomposition temperature (–180°C). The rubber modulus could be controlled up to 10 MPa by varying the cross-link density. Similar to PHO, maximum elongations amounted to more than 300% for the irradiated material, but other cross-linking techniques produced lower values. Peroxide cross-linking even resulted in a loss of material integrity. The tensile strength of the cross-linked material never exceeded 1 MPa and tear resistance was poor.

BIOLOGICAL BREAKDOWN

PHAs can be degraded either intracellularly by intracellular depolymerases of the accumulating strain, to supply accumulated carbon and energy in the absence of a suitable exogenous sources, or extracellularly by extracellular depolymerases for consumption of oligomers or monomers by fungi, yeast, and bacteria. Relatively little is known about intracellular degradation of PHAs. It was shown that intracellular PHA depolymerases are unable to hydrolyze extracellular PHAs, and extracellular depolymerases cannot hydrolyze intracellular granules, apparently as the result of differences in the physical structures of the completely amorphous intracellular granules and the partially crystallized extracellular PHA. Under sterile or aseptic conditions, PHA is degraded by a hydrolytic mechanism, especially at high pH and high temperature. This type of degradation is important for medical applications, such as the use of PHA in drug release carriers or surgical sutures. Because PHAs are so abundant, a wide range of bacteria and fungi have developed the capability to utilize PHAs. It has been shown that PHA degradation occurs in a large variety of complex ecosystems, including oxic and anoxic environments such as soil, seawater, or sludge.

Microbiological parameters of the ecosystem, like population density, microbial diversity, microbial activity, and spatial distribution of microorganisms, can influence the degradation of the polymers. Furthermore, physicochemical parameters of the ecosystem, such as temperature, pH, water content, oxygen content, redox potential, and nutrient supply, can all affect biodegradation. The polymer itself influences biodegradability as well. The most important factors are (i) the stereoregularity, (ii) the crystallinity, (iii) the molecular mass, and (iv) the monomeric composition of the polymer. Finally, material processing can also affect the extent of biodegradability. The effect of different environments on the degradation rate of PHB and PHB/HV has been studied by different groups. It was shown that the polymer (1-mm molding) was completely degraded after 6 weeks in anaerobic sewage, after 75 weeks in soil, and after 350 weeks in seawater. Concerning functional MCL PHAs, it could be shown that even after cross-linking of unsaturated side chains the polymer remained biodegradable. However, it should be noted that at higher cross-link densities the rate of degradation was reduced. Eleven groups of PHA-degrading bacteria can be distinguished by their substrate (PHA) specificities. In addition, 95 genera of fungi capable of degrading PHAs have been identified. The synthesis of PHA depolymerases is highly regulated. Most PHA-degrading bacteria repress depolymerase synthesis in the presence of a soluble carbon source that permits high growth rates. However, after exhaustion of the nutrients, the synthesis of depolymerase is derepressed.

The extracellular PHA depolymerases of many bacteria have been purified and characterized. Depending on the depolymerase, the hydrolysis products are monomers [e.g., *Comamonas* sp.] or monomers and dimers [e.g., *Pseudomonas fluorescens*] or a mixture of oligomers (monomers to trimers) as in the case of *Alcaligenes faecalis* and *Pseudomonas lemoignei* depolymerases. In a second step, the oligomers are hydrolyzed to monomers by oligomer hydrolases. All PHA depolymerases analyzed so far are specific for polymers that consist of monomers in the (*R*)-configuration. Chemically synthesized poly-(*S*)-hydroxybutyrate is not degraded by PHB depolymerases. Nine bacterial PHA depolymerase genes that encode depolymerases with specificities for SCL PHAs have been cloned and analyzed. The enzymes contain a signal peptide (which is cleaved off during passage across the plasma membrane), a large catalytic domain at the N-terminus, a C-terminal substrate-binding domain, and a linking region in between. Three strictly conserved amino acids, namely serine, aspartate, and histidine, constitute the active center of the catalytic domain. The serine is part of the lipase-box motif Gly-X-Ser-X-Gly, which has been found in all known serine hydrolases such as lipases, esterases, and serine proteases.

The oxygen atom of the serine side chain is the nucleophile that attacks the ester bond and is supported by the imidazole ring of the histidine. The positive charge of the latter is stabilized by the carboxyl group of the aspartate. Only one gene that encodes an extracellular MCL PHA depolymerase has been cloned and analyzed so far. In contrast to most SCL PHA depolymerases, the MCL PHA depolymerase of *P. fluorescens* GK13 consists of two identical subunits, and except for small regions adjacent to the lipase-box motif there are no significant homologies to SCL PHA depolymerases.

BENEFICIAL CHARACTERISTICS

PHAs display an unique combination of features. First, in contrast to synthetic polymers, the natural PHAs have the fundamental advantage of being renewable resources not dependent on the supply of petroleum. Second, as described in the previous section, they are genuinely biodegradable, that is, they can be completely metabolized into harmless, naturally occurring molecules. Although complete degradation can easily be achieved, the shelf life of the PHA products is virtually unlimited. Colonization of the hydrophobic surface is an arduous task for the degrading organisms, and thus a high biological activity is a prerequisite for biodegradation. Cheaper biodegradables, such as the starch-based plastics, are available however. Therefore, the most important feature of these polyesters may well be their hydrophobicity, which allows them to surpass carbohydrate-based biodegradable competitors in moisture resistance.

Applications of PHB

As mentioned previously, PHB and its derivates are all highly crystalline materials and therefore possess a high modulus, making them suitable for polypropylene-type applications including injection molding, extrusion blow molding, and fiber-spray gun molding. The polymers can be manufactured into many different materials with a wide range of applications. The first commercial application of PHB or PHB/HV, sold under the trade name Biopol, was in packaging for personal hygiene products, for example, as biodegradable shampoo bottles. Later applications included packing containers, bottles, tubes, wrappings, bags, tubes, and disposable razors. In these cases the rather costly Biopol was used for its environmentally friendly image to increase product sales. In addition, applications that require biodegradability for functional reasons were developed. Among those, garbage bags, which should ideally be composted together with their contents, fishing nets, which can cause severe damage once they are lost in the marine environment, and agricultural delivery devices for nutrients, fertilizers, herbicides, and insecticides are important examples.

Applications of PHA

No applications for MCL PHAs have been reported yet. Complementing the intrinsically rigid and brittle properties of PHB and its derivates, MCL PHAs are expected to support applications that require flexibility and elastomeric properties. MCL PHAs will have very limited applicability as thermoplasts. Amorphous MCL PHAs, cross-linked and non-cross-linked, are more likely to penetrate the plastics market. The capability of pseudomonads to incorporate many different monomer units into MCL PHAs suggests various possibilities to produce tailor-made polymers. Incorporation of unsaturated pendant groups to enhance cross-linking is one example. A recent study reported that nitrophenoxy groups have been introduced, aiming to create chiral polymers for nonlinear optical applications. Specific functionalities for surface modification could be introduced as well. In the long run, it should be feasible to custom design new polymers and then program the bacteria to produce them.

Blending Techniques

Polymer blending offers interesting possibilities of preparing inexpensive (biodegradable) materials with useful mechanical properties. In this context, PHAs could form an important source of biodegradable additives. Several blending studies have been performed on PHB or PHB/HV copolymers with other dissimilar compounds. For example, the miscibility and compatibility of these polyesters with chlorinated polymers, polyamides, polycarbonates, cellulose derivatives, and other functional polymers are well documented. At present, applications of PHAs as commodities are primarily limited by their high cost, and thus attention is being focused on composite products with plastics constituting only a minor part of the product, such as paper coatings like the plastic film moisture barrier in food or drink cartons and in sanitary napkins. In these cases, total product biodegradability can be attained for little extra cost.

Pharmaceutical Utility

A notable feature of PHB is that it is biocompatible, producing a mild foreign-body response to any implant. Degradation of PHB in vivo is significantly faster than the in vitro hydrolysis in neutral buffer at body temperature, indicating that enzymes present in vivo catalyze the degradation process. The only degradation product is (*R*)-3-hydroxybutyric acid, which is a normal mammalian metabolite and is resorbed without metabolic or histological disturbance. Like PHB, MCL PHA might be biocompatible and useful in biomedical applications. The rigid PHB might be useful in bone fracture fixation, whereas the flexible elastomeric MCL PHAs might find uses in vascular grafts, surgical swabs, wound dressings, breast prostheses, and other implants.

Hydrolysis of PHAs yields chiral (*R*)-3-hydroxy fatty acids, which could be a feedstock for stereospecific synthesis and the associated pharmaceutical industry. Concerning the enormous number of monomers that have been incorporated into various polymers, many different synthon precursors could be provided by hydrolyzing the polymers.

MARKET VALUES

The market price of Biopol when produced by ZENECA BioProducts was about \$16 kg^{-1}. Current advances in fermentation and purification technology, as well as the application of metabolic engineering techniques for expanding the spectrum of utilizable substrates and for improving polymer concentration in the cells, are likely to lower the price of PHB to \$4 kg^{-1}. A preliminary economic study for the production of mcl-PHAs showed that the price for octanoate-based PHAs would be of the order of \$20 kg^{1} for an annual production of 1,000 tons. Optimizing the polymer content of the cells, using cheaper carbon sources as substrates, reducing the chemicals used during downstream processing, and reducing energy consumption is expected to lower the MCL PHA price to a minimum of \$5 kg^{-1}. It can be assumed that production of PHAs by using native or recombinant bacteria is in the long term only required for development purposes and for the production of specific functionalized PHAs. It is more likely that efforts in generating transgenic plants for PHA production purposes will succeed in ultimately reducing the price of the polymer to that of other large-scale agromaterials such as starch or conventional commodity plastics, or less than \$1 kg^{-1}. Therefore, an approach to insert the PHB biosynthesis genes into crop plants (e.g., soybean and rapeseed), which have the agronomic properties to provide high yields of PHB per hectare, has been undertaken by CEREGEN/Monsanto and is currently under way.

CONCLUDING REMARKS

The development of PHAs as a commercial product remains limited to PHB and its copolymer, PHB/HV. As yet, no applications in the fabrication of articles made of MCL PHAs have been reported. There are probably at least two reasons for this. There has been little material available for development work, and the lower polymer concentration in the cells and the different recovery process means that the material cost of MCL PHA will remain even higher than for PHB for the foreseeable future. Future large-scale production of PHB (and MCL PHA) in transgenic plants should ultimately provide a cheaper and sustainable technology for the production of these materials. This advance is essential if PHAs are to be used for commodity purposes and must compete on the plastic market.

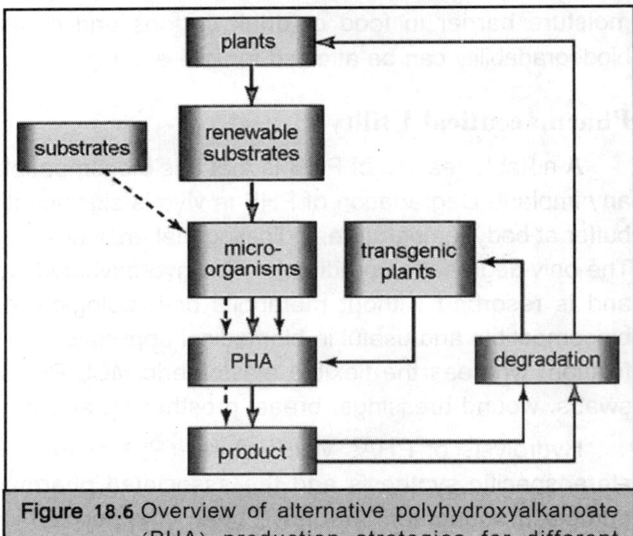

Figure 18.6 Overview of alternative polyhydroxyalkanoate (PHA) production strategies for different applicaions.

Improvements in the competitiveness of PHAs on the market could also derive from environmental legislation and from decreasing availability of oil and concomitant rising prices of synthetic plastics. The development of polymers containing exotic monomers into commercial products will depend on the identification of niche applications that will support the high cost of production arising from both the cost of the monomers used and the small scale of the operation to make these specialty materials. Such applications are likely to include nonlinear optical devices or medical implants of the isotactic polymers, and the use of the stereospecific monomers as chiral synthons for the production of fine chemicals. Furthermore, capsules made of PHA for controlled release of bioactive agents, such as drugs or fertilizers, could find broad application in medicine and agriculture.

19

GRAM-POSITIVE BACILLUS

Members of the genus *Bacillus* are gram-positive, endospore-forming bacteria that include both mesophiles and extremophiles. They are widely distributed in nature but are most commonly found in soil and associated water sources. Their extraordinary metabolic diversity has been exploited by old and new biotechnologies—from the production of Natto, a traditional Japanese food obtained by fermenting soybeans with *Bacillus subtilis* var. *natto,* to engineered industrial enzymes used in the food and detergents industries.

SYSTEMATIC POSITION

The genus *Bacillus* comprises gram-positive, endospore-forming, aerobic or facultatively anaerobic bacteria. Members of the genus show extraordinary metabolic diversity and lifestyles and include thermophiles, psychrophiles, alkalophiles, and acidophiles. They are metabolically chemoorganotrophs, being dependent on organic compounds as sources of carbon and energy. *B. subtilis,* the most extensively characterized species, is a prototroph capable of growing at mesophilic temperatures on chemically defined salts media with glucose or other simple carbon sources. In contrast, some insect pathogenic species are nutritionally fastidious and require highly specialized growth media. *Bacillus* species are among the most widely distributed microorganisms, found commonly in soil and associated environments, including plants and water sources such as rivers, estuaries, and coastal waters.

Most species are harmless to humans and animals, and only a few pathogens have been identified. The latter include *B. anthracis,* the causative agent of anthrax, *B. cereus,* which causes food poisoning, and several insect pathogens. The primary reservoir for most *Bacillus* species is the soil, where they secrete a variety of carbohydrate- and protein-degrading enzymes that allow them to grow at the expense of plant material and other nutrients. These macromolecular hydrolases are the source of much of the commercial interest in these species. Their ability to form spores means that they can often be detected in an environment long after periods of metabolic activity.

The major soil types are members of the *B. subtilis* and *B. sphaericus* groups. However, representatives of the more fastidious species in the *B. polymyxa* group tend to accumulate in soils with a high organic content. Some species form a close relationship with plant roots, particularly representatives of the *B. polymyxa* group. Several species (e.g., *B. azotofixans, B. macerans,* and *B. polymyxa*) fix nitrogen under anaerobic conditions.

Certain strains of *B. subtilis* are categorized as plant growth–promoting rhizobacteria (PGPR). In return for stimulating plant growth by increasing the availability of inorganic nutrients, or secreting plant growth hormones or compounds (often peptide antibiotics) that inhibit the growth of plant pathogens, PGPR receive nutrients from the plant in the form of plant exudates. The metabolic diversity of the genus is matched by significant genetic diversity, and the %GC content of their genomic DNA varies from about 33 to 67 mol%. This variation is much larger than would be expected for a well-defined genus, and the 60 to 70 currently recognized species of *Bacillus* are likely to be reassigned to an increased number of more clearly defined genera. This process has already started with the reassignment of *B. acidocaldarius* and some other acidophilic thermophiles to the genus *Alicyclobacillus*. Numerical taxonomic and 16*S* ribosomal (r)RNA sequence analyses show good but not perfect congruence. These studies have resulted in *Bacillus* species being assigned to five groups.

The evolutionary distance between these groups is large enough to affect the outcome of inter species gene transfer and cloning experiments, particularly when there are large discrepancies between the %GC contents of the species involved. The *B. subtilis* group includes many of the better-known bacilli. These include facultative anaerobes such as *B. licheniformis,* which grow fermentatively in the absence of exogenous electron acceptors, and aerobes such as *B. subtilis* that grow weakly in the absence of oxygen except in the presence of nitrate, which they can use as an alternative electron acceptor. Members of this group produce acid from several sugars, and several (e.g., *B. lentus, B. firmus*) are halotolerant. The group includes toxigenic species such as *B. anthracis, B. cereus,* and *B. thuringiensis*. The *B. sphaericus* group is strictly oxidative, using acetate or amino acids rather than sugars as carbon and energy sources. Members of the *B. sphaericus* group are virtually unique among bacilli in having the *meso*-diaminopimelic acid that is usually present in cell walls replaced by lysine or ornithine.

The *B. polymyxa* group (renamed *Paenibacillus*) is made up of facultative anaerobes that use sugars and polysaccharides. *B. macerans* grows better in the absence of oxygen with an acidogenic type of fermentation that changes to an ethanolic fermentation at acidic pHs. *B. polymyxa* exhibits a butanediol type of fermentation. Members of the *B. brevis* group (renamed *Brevi bacillus*) are taxonomically heterologous. They are strict aerobes that generally do not produce acid from sugars. Some species, notably *B. azotoformans,* can grow anaerobically by using nitrate as an electron acceptor.

The thermophilic bacilli, which grow optimally at temperatures above 50°C, represent a heterologous group with at least three lines of descent that range from anaerobes to strict aerobes. Some are able to grow fermentatively on lactic acid (*B. coagulans*) whereas others are chemolithotrophic autotrophs. The group includes *B. acidocaldarius* and the other acidophilic thermophiles that have been reassigned to the genus *Alicyclobacillus*.

Long Storage

Although most strains of *Bacillus* survive well on agar plates, either at room temperature or at 4°C, it is recommended that they are subcultured on a weekly basis. Because *Bacillus* species are able to sporulate, viable cells may even be recovered from severely dehydrated agar plates, particularly from minimal agar plates that encourage sporulation. Long-term stocks of *Bacillus* may be preserved as glycerol or lyophilized cultures, in the form of spores or vegetative cells. Strains that sporulate well may be preserved as spore suspensions in sterile water. Spore suspensions are generally stable for many years. Most *Bacillus* strains can be transported on freshly inoculated nutrient agar slopes or as spore suspensions spotted on sterilized filter paper disks encased in sterilized aluminium foil.

Rearing in Laboratory

The *Bacillus* Genetic Stock Center has an extensive collection of mutant strains, bacteriophages, and plasmids. Although mainly devoted to *B. subtilis,* the collection also includes strains of *B. cereus, B. licheniformis, B. megaterium, B.pumilus, B. stearothermophilus, B. thuringiensis,* and a few others. Strains of *Bacillus* are also available from a variety of international culture collections and a comprehensive list has been published. The publication of the genome of *B. subtilis* 168 has provided the opportunity to investigate the function of open reading frames (ORFs) with no known function (unknown reading frames (URFs). A collection of isogenic mutants has been established in which the URFs of *B. subtilis* 168 have been systematically inactivated with a purpose-designed integrational vector, pMUT in. These mutants also permit the expression of the target genes to be monitored and any gene(s) downstream in the same transcriptional unit to be induced.

MANIPULATING TECHNIQUE

A wide range of genetic techniques has been developed for *B. subtilis,* making it among the most amenable bacteria to genetic manipulation. These techniques include classical in vitro and in vivo mutagenic methods and extremely efficient in vitro genome manipulation methods that are in advance of those available for other bacteria, including *Escherichia coli. Bacillus* vectors have been designed to allow recombinant or mutant genes to be maintained autonomously in the host or integrated into the chromosome by double-crossover or single-crossover recombinations. In the case of a single-crossover recombination, tandem copies of the gene, one wild type and one mutated, can be engineered into the chromosome, if required, each under the control of independently controlled promoters. The techniques for the manipulation of other *Bacillus* species are limited by their transformability.

Change in Nucleotides

A number of DNA repair systems have been detected and studied in *B. subtilis,* including error-prone SOS-like translesion DNA synthesis but not photoreactivation. Consequently, classical in vivo mutagenic methods generate mutants efficiently. Limitations in whole-cell mutagenic techniques, including lack of specificity and the generation of multiple mutations that may influence the cell's phenotype in an unpredictable manner, can be overcome by use of an alternative approach in which cloned copies of the target DNA are subjected to in vitro mutagenesis and subsequently reintegrated into the chromosome. This approach is aided by the publication of the genome sequence of *B. subtilis* strain.

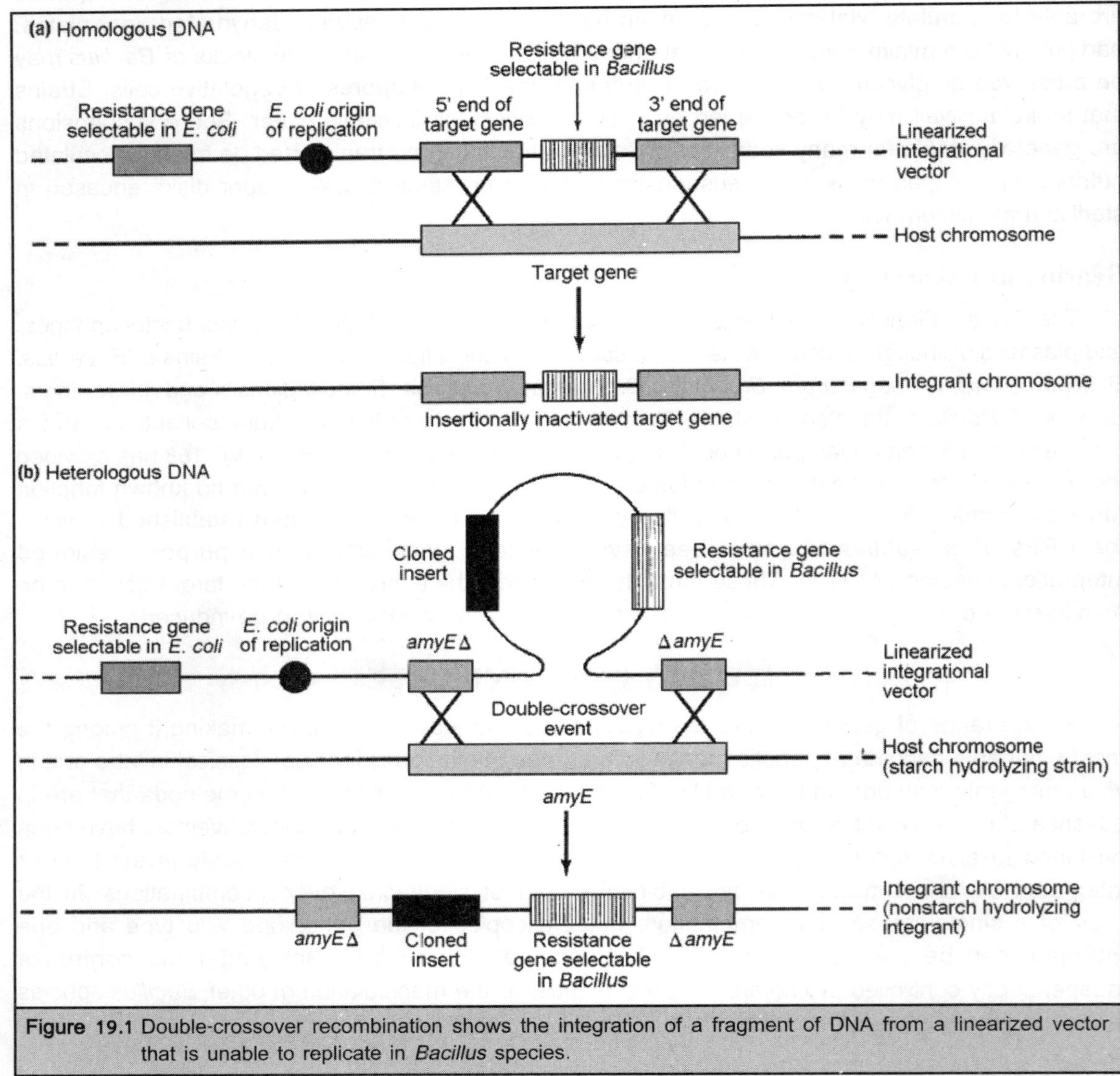

Figure 19.1 Double-crossover recombination shows the integration of a fragment of DNA from a linearized vector that is unable to replicate in *Bacillus* species.

Specific changes are currently introduced into DNA sequences by PCR techniques, with more extensive changes being engineered by splicing PCR methodologies (e.g., splicing by overlap extension). The modified DNA can then be introduced into the host bacterium either on replicating or integration plasmids. The ability to generate insertion mutants is one of the great technical strengths of *B. subtilis*. The techniques rely on the ease with which *Bacillus* DNA can be cloned into plasmids that replicate in *E. coli* but not in *B. subtilis* and the high frequency with which homologous recombination occurs within *B. subtilis*. The identification and analysis of *B. subtilis* genes involved in specific functions or responding to particular stimuli has been facilitated by the use of transposons and transpositional mutagenesis. Although the value of transposons as a tool for the analysis of *B. subtilis* has diminished with the publication of the genome sequence, nevertheless the technology remains important for the analysis of other *Bacillus* species. Because of the lack of suitable native *Bacillus* transposons, Tn*917* from *Enterococcus faecalis* and Tn*10* from *E. coli* have been used.

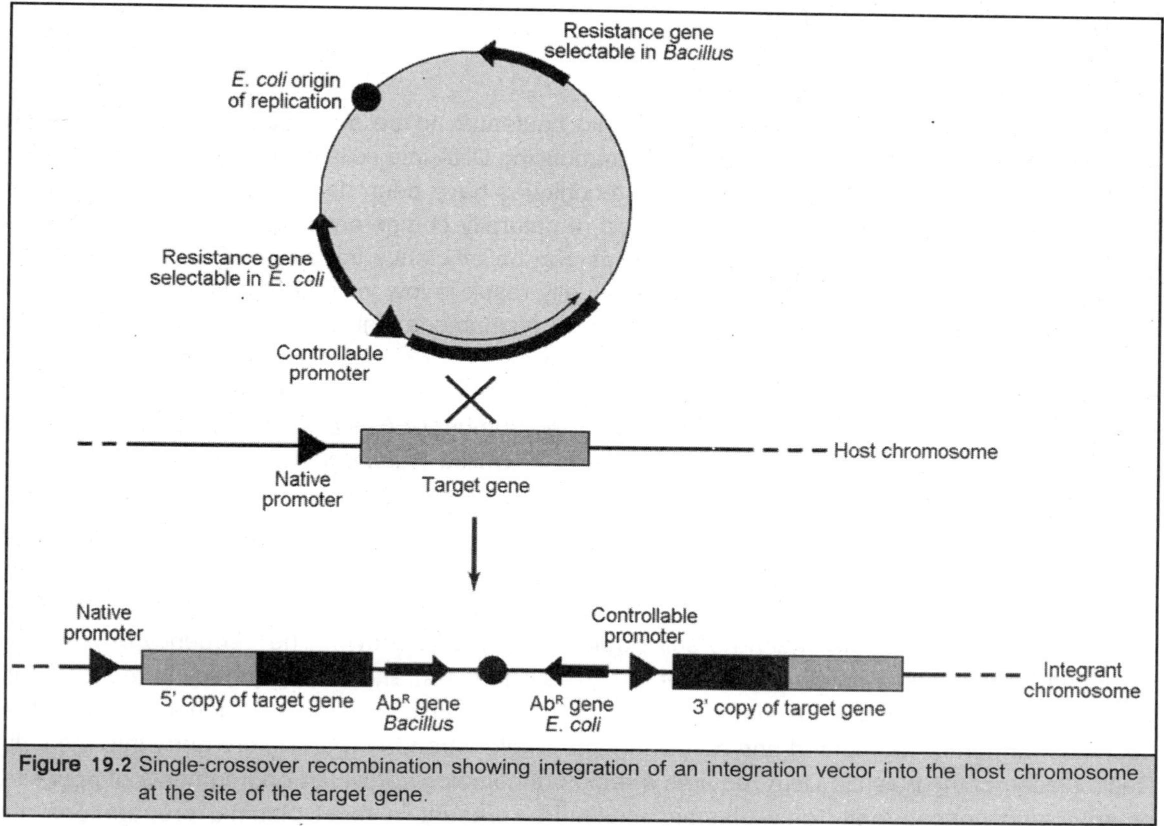

Figure 19.2 Single-crossover recombination showing integration of an integration vector into the host chromosome at the site of the target gene.

In addition to their value for mutagenesis, transposons have been used to clone DNA adjacent to the site of integration, to generate transcriptional fusions to reporter genes, to control the expression of adjacent genes, and as phenotypic tags for mapping and cloning studies. Transposon Tn*917* is a 5.3-kb Tn*3*-like transposon conferring inducible erythromycin resistance (MLS-type). This transposon has a number of properties that make it a valuable genetic tool, including (i) the ability to insert relatively randomly into bacterial chromosomes, (ii) a relatively high transposition frequency, (iii) the ability to accommodate at least 8 kb of DNA without markedly affecting transposition frequency or insertional stability, and (iv) a host range that includes gram-positive and gram-negative bacteria. In addition to *B. subtilis*, Tn*917* has been used in *B. amyloliquefaciens*, *B. licheniformis* and *B. megaterium*. Insertion of Tn*917* is not entirely random, and a limited preference for specific loci has been observed in the case of *B. subtilis* and *B. licheniformis*.

An alternative to Tn*917* is an engineered mini-derivative of Tn*10*. This transposon consists of a *Staphylococcus aureus*–derived *cat* (chloramphenicol resistance) gene flanked by 307-bp fragments derived from IS*10*. The gene encoding the transposase is incorporated into a delivery vector rather than the mini-transposon itself. Tn*10*-derived transposons have two potential advantages over Tn*917*-based transposons. They appear to transpose at a significantly higher frequency and insert randomly without bias to specific targets. A series of special-purpose delivery systems have been developed for gene inactivation, the recovery of adjacent chromosomal sequences, and transcription fusions.

Transformation

Bacillus subtilis was the first nonpathogenic bacterium to be successfully transformed with extracted DNA. The availability of methods for introducing DNA into cells is essential for recombinant DNA technology, and several transformation techniques have been developed for *B. subtilis*. The most frequently used method is transformation of naturally competent cells, although protoplasts of *B. subtilis* and several other *Bacillus* species can be efficiently transformed. Current methods for electro-transformation of *Bacillus* species usually result in low yields of transformants. Natural competence is one of several post-exponential phase phenomena that are characteristic of *B. subtilis* strain 168. In glucose-minimal media, maximal competence develops shortly after the transition from exponential to stationary phase, and only about 10% of the cells are able to take up DNA. Natural competence has been observed with a limited number of *B. subtilis* strains, and few natural competence systems are known for *Bacillus* species. Unlike *Haemophilus influenzae,* competent *B. subtilis* cells do not discriminate between native and foreign DNA.

In recombination-proficient strains, homologous chromosomal DNA is integrated in the host chromosome at the site of homology. If saturating amounts of DNA (>1 μg/mL of culture) are used, the cotransfer of unlinked genetic markers occurs. This phenomenon, called congression, can be used to introduce genes for which no selectable phenotype is known. The frequency of plasmid-mediated transformation is usually between 0.001% and 0.1%, and this frequency is reduced still further with ligation mixtures. A major reason for these low efficiencies is that donor plasmid DNA becomes randomly fragmented and converted into single- stranded DNA before entry into the cell. Recirculation of the plasmid DNA requires a region of homology provided on fragments of opposite polarity, and consequently only plasmid multimers, present in most plasmid preparations, or monomers containing internal repeats are effective in transformation. *Bacillus* protoplasts can incorporate DNA if treated with chaotropic agents such as polyethylene glycol. The fragility of the wall-less protoplasts means that transformation needs to be carried out in an isosmotic medium; the cell walls are subsequently regenerated on complex regeneration media.

Protoplast transformation was originally developed for *B. subtilis* by Chang and Cohen but has subsequently been modified in many laboratories. A major advantage of protoplast transformation is that plasmid DNA remains double stranded throughout the procedure and consequently is less likely to undergo rearrangements. Despite its advantages, protoplast transformation methods are laborious and difficult to reproduce; transformation frequencies up to 10% per protoplast are claimed but in practice are usually a great deal lower. In addition, the complexity of the regeneration media precludes the selection for prototrophic markers. Protoplast transformation protocols have been developed for several other *Bacillus* species including *B. amyloliquefaciens, B. licheniformis, B. stearothermophilus, B. anthracis,* and *B. firmus.*

Inducing DNA Cloning

The genome of *B. subtilis* strain 168 is 4.2 Mbp in length, has a G+C content of 43.5%, and encodes slightly more than 4,100 protein-coding genes. The genome shows evidence of 10 prophages or remnants of prophages, which may account for its susceptibility to lysis at the end of exponential growth in the absence of an energized membrane. Nearly half the protein-coding genes have paralogues in the genome, and several expanded gene families were detected, particularly for two-component signal transducers (34 genes) and ATP-binding transporters (77 genes). Although analysis of a complete microbial genome sequence reveals many new ORFs, it is not possible to ascribe functions to all the detected ORFs on the basis of currently available sequence databases and bioinformatics tools. Even in well-characterized species such as *E. coli* and *B. subtilis,* URFs represent about 40% of the detected ORFs. Given the intensive in vitro studies on bacteria, it is likely that the function of a significant proportion of the URFs will relate to growth and survival in natural habitats. The inability to ascribe function to more than a third of the ORFs in *B. subtilis* has required the development of new strategies for analyzing gene function. In the case of *B. subtilis,* a highly coordinated systematic functional analysis program involving groups with a wide range of technical expertise has been established.

Approximately 2,000 unknown genes have been systematically disrupted with an integrational vector to generate a collection of isogenic mutants strains. The mutant construction strategy includes the introduction of a *lacZ* transcriptional reporter for monitoring the expression of the target gene and a controllable promoter that facilitates the expression of any genes downstream in the same operon. The availability of the *B. subtilis* genome sequence has revolutionized the methods of analysis and manipulation of this bacterium, and consequently traditional methods of genetic linkage mapping involving generalized transducing for long-range mapping and transformation for fine structure mapping have largely been rendered redundant. This is not the case for species such as *B. megaterium* for which generalized transducing phages such as MP13 and protoplast transformation remain important analytical tools. Increasingly, integration systems are used for the analysis and manipulation of the *B. subtilis* genome.

Transformation of cells with homologous DNA fragments is very efficient, and if heterologous sequences are flanked by sequences homologous to chromosomal DNA they can be inserted efficiently at the sites of homology via a double-crossover recombination event. In the case of circular molecules, a single site of homology is all that is required and integration is via a single-crossover recombination event. Single or Campbell-type integration events generally use *E. coli*–based plasmid vectors that are not able to replicate in gram-positive bacteria but that include an antibiotic resistance gene that can be selected in *B. subtilis.* A fragment of *B. subtilis* DNA (≥0.4 kb) is cloned via *E. coli* into the integrational vector, which is then transformed into *B. subtilis.* Selection for an antibiotic-resistant gene results in transformants in which the vector has integrated into the host chromosome via a single-crossover recombination between homologous regions on the vector and bacterial chromosomes. Integration results in a duplication of the cloned fragment at the flanking ends of the integrated vector.

These tandem repeats represent an amplifiable unit; increasing the amounts of antibiotic used for selection can result in the amplification of the intervening sequences up to 70 copies. The integrational vector system can be used to generate different mutational outcomes. If the cloned fragment carries an entire operon, including promoter and transcription terminator, two fully functional copies of the gene are present in the integrant. If 5′- or 3′-end fragments of an operon are used, only one functional copy of the target gene is generated. If the 5′-end fragment includes a ribosome-binding site and start codon, and is placed downstream of a controllable promoter, then the expression of the single functional copy of the target gene is under the control of this promoter. Finally, if regions internal to the target gene are used, no functional copies are formed after integration. With double-crossover integrations, chromosomal DNA is replaced by either mutationally altered homologous DNA or heterologous DNA. Only a single copy of the target gene is introduced. The incoming fragment, which must encode a selectable phenotype, is flanked by DNA that is homologous to sequences at the chromosomal target site. The *amyE* gene, encoding a nonessential α-amylase, is often used as a target of double-crossover integrations because its disruption provides a positive selection phenotype on starch plates.

Manipulation

The development of recombinant DNA techniques for *B. subtilis* is now well advanced, and previously encountered problems of vector stability have largely been overcome.

Plasmid vectors

B. subtilis 168, the strain that has been sequenced and that is used for most genetic studies, does not contain endogenous plasmids, whereas most plasmids present in other *Bacillus* strains are cryptic. Consequently, plasmid vectors designed for use in *B. subtilis* have been taken from other gram-positive bacteria, notably *Enterococcus faecalis, Lactococcus lactis,* and *Staphylococcus aureus*. Staphylococcal plasmids such as pUB110 and pE194, which formed the basis of the first generation of *Bacillus* vectors, are still in common use for *B. subtilis*. The kanamycin resistance plasmid pUB110 is the most widely used replicon but suffers from problems of structural and segregational instability that result from its single-stranded (rolling circle) mode of replication. The second generation of *Bacillus* vectors involved the construction of chimeras between gram-positive and gram-negative replicons. These shuttle or bifunctional vectors enabled initial cloning experiments to be carried out in *E. coli,* exploiting this organism's highly efficient transformation system. In recent years, stable single-stranded replicating plasmids and theta-replicating plasmids have been identified and are being developed as a new generation of *Bacillus* vectors.

The former includes derivatives of plasmids such as pWVO1 derived from *Lactococcus lactis* and certain endogenous *Bacillus* plasmids, the latter plasmids based on pAMβ1, originally isolated from *Enterococcus faecalis*. An alternative to using autonomously replicating plasmid vectors is to use plasmid integration vectors. Such integrants are usually stable; reversal of the process, by recombination in the duplicated sequences, occurs at a frequency of about 10^{-4} to 10^{-5} per cell generation. Various special-purpose plasmid vectors have also been developed, including expression and secretion vectors. Although the expression of many genes has been studied in *B. subtilis*, relatively few promoter systems have been developed for controlled, high-level expression. The majority of extracellular enzymes synthesized for industrial purposes are expressed from native promoters induced at the end of exponential growth and for extended periods in the stationary phase. During this time they can be subjected to a variety of control pathways, including transition-state regulators and catabolite repression. Industry has developed promoter–expression systems (including growth regimes) that can direct the synthesis of extracellular proteins to concentrations of about 20 g/L, although for commercial reasons the details of these systems are not generally available.

P_{spac} promoter

The P_{spac} promoter was constructed by fusing the 5′ sequences of a promoter from SPO1, a lytic *B. subtilis* bacteriophage, and the 3′ sequences of the *lac* promoter, including the entire *lac* operator. The point of fusion is within the -10 region of the hybrid promoter. The controllability of P_{spac} is dependent on the presence of the *lac* repressor, expressed in *B. subtilis* by placing it downstream of a promoter and ribosome-binding site (RBS) derived from a *B. licheniformis* penicillinase gene. The P_{spac} promoter can be induced 50 fold with the lactose analog IPTG.

XylR-controlled promoters

The genes for the utilization of xylose polymers are controlled in *B. subtilis* at the level of transcription through the activity of a repressor, the product of the *xylR* gene. The xylose-inducible promoter–operator elements have been used without modification to control expression from chromosomal and plasmid locations. A copy of the *xylR* gene is usually included on high-copy-number expression vectors to maintain a balance between the number of repressor molecules and operator sites. Although genes in the xylose regulon are subject to catabolite repression, the catabolite-responsive element is located well downstream of the promoter and operator, and is therefore not included in the expression vectors. In contrast to IPTG, xylose is a cheap and readily available substrate and can be used for the induction of large-scale fermentations.

SacR-controlled promoters

The inducible expression of *sacB*, the gene encoding extracellular levansucrase by sucrose, involves a number of regulatory mechanisms, not all of which are fully understood. The *sacB* gene is controlled positively by sucrose, the *sacY* anti-terminator, and the products of *degQ* and *sacU* and negatively by *sacX*, a putative PTS enzyme $II^{sucrose}$. The *sacR* regulatory region located upstream of the *sacB* promoter contains a target sequence for DegQ and SacU, which together enhance transcription. Downstream of this promoter, but before the *sacB* RBS, a termination-like stem-loop structure is present, which is acted on by the SacY anti-terminator to mediate sucrose-induced expression. Expression vectors based on the *sacB* promoter have the advantage that this promoter can be activated during exponential growth when the expression of most *Bacillus* extracellular proteases is repressed and is not subject to catabolite repression. The level of induction is also geared to the level of sucrose in the medium; the promoter is only weakly induced in the presence of 1 mM sucrose and is fully induced at 30 mM.

Phage vectors

The development of phage vectors for *B. subtilis* has lagged well behind that of *E. coli,* and no phage vectors with the range and versatility of those of lambda exist for this organism. Libraries have been generated in derivatives of *Bacillus* phages φ105 and SPβ, although the former has been used more extensively. Attempts to develop a cosmid system for *B. subtilis* have failed because of the lack of an efficient in vitro packaging system for any of the candidate phages. Bacteriophages φ105 and PBSX (a defective phage of *B. subtilis* strain 168) have both been developed for the production of proteins in *B. subtilis.*

In both cases the phages have been modified to make them temperature inducible by the introduction of mutations in the genes encoding the immunity repressor proteins. Target genes are introduced downstream of a strong prophage promoter. In the case of φ105MU333, temperature induction can lead to the production of the target gene product to 0.5 mg/mL culture supernatant. Bacteriophage-based expression systems have the advantage of being relatively stable because they are maintained as a single copy during growth. Copy number increases at the time of induction due to the replication of the phage. The promoter is tightly controlled but is induced cheaply and in a manner favored in industrial fermentations by increasing temperature. However, vector development is still required to maximize the full potential of this system for high-level protein production.

SECRETORY BEHAVIOUR OF BACILLUS

Bacillus subtilis and its close relatives secrete proteins directly into the growth medium efficiently and to high concentrations. The productivity of *Bacillus* species for the production of extracellular proteins has been improved by a variety of techniques including random and directed mutagenesis, recombinant techniques, and growth regimes. Yields of tens of gram per liter are expected from industrial strains secreting native *Bacillus* proteins. The secretion process is highly specific, involving functional sequences on the secretory protein and recognition between the secretory proteins and the host's protein secretion machinery (secretory preprotein translocase). For convenience, the process in *B. subtilis* can be divided into three stages: intracellular processes, involving the targeting of secretory proteins to the cytoplasmic membrane–secretory translocase; translocation across the cytoplasmic membrane; and post-translocational stages, involving the processing of the secreted protein, its folding into the native conformation, and passage through the cell wall into the growth medium.

A major limiting step in the production of native secretory proteins is the level of transcription of their genes. The transcription of genes encoding native secretory proteins is tightly controlled and involves a number of complex, often interacting, regulatory networks, including transcriptional activators (e.g., DegQ, DegR, DegU, and SenS) and repressors (e.g., AbrB, Hpr, and Sin). For example, the nucleotide sequence upstream of *apr,* encoding the major alkaline protease of *B. subtilis,* includes binding sites for Hpr, Sin, SacU/Q, and AbrB. Most *Bacillus* secretory proteins are synthesized as precursors with amino-terminal extensions known as signal peptides, or pre-peptides. The signal peptide is required to target the precursor to the cytoplasmic membrane and for initiating the interaction with the secretory translocase. During or shortly after translocation across the membrane, the signal peptide is removed through the action of signal peptidases.

Bacillus signal peptides vary in length from 18 to 35 amino acid residues; their average length of about 30 residues is 5 to 7 residues longer than their counterparts in gram-negative bacteria and eukaryotes. The C-terminal end of the signal peptide contains the signal peptidase recognition sequence. Lipoproteins, which remain attached at the *trans* side of the membrane after translocation, are cleaved by a prolipoprotein-specific signal peptidase that has a different recognition sequence. Many exported *Bacillus* proteins are synthesised as pre-pro-proteins. The pro-peptide, which is located between the signal peptide (pre-) and the mature protein, can vary in length from about 8 (e.g.,*B. subtilis* α-amylase) to more than 200 amino acids (various neutral proteases). Pro-peptides have been implicated in the folding of the mature protein into its native conformation following translocation. Cleavage of the pro-peptide occurs after translocation, autocatalytically in the case of proteases or via extracellular proteases in the case of other secretory proteins. Long (8–20 residues) hydrophobic domains, flanked by positively charged residues, are usually absent from secretory proteins because such domains act as “stop transfer” or “membrane anchor” sequences. Similarly, few positively charged residues are found in the amino terminus of the mature protein as these reduce export efficiency. Surprisingly little is known about early events in the *B. subtilis* secretory pathway. Secretion-specific molecular chaperones, required to maintain secretory precursor proteins in an unfolded conformation, to prevent their aggregation, and to target them to the secretory machinery in the membrane, have not been detected in *B. subtilis*. This could indicate that secretion is cotranslational, and indeed elements of the signal recognition particle–like (SRP-like) pathway have been found in this bacterium.

The secretory translocase complex appears to be similar to that of *E. coli*. The SecAEY complex is primarily responsible for the ATP-dependent transport of proteins across the membrane, and homologs of other components of the *E. coli* translocase, such as SecD, F, and G, also appear to be present, in the former two cases as a SecDF fusion. A unique feature of *B. subtilis* strain 168 is the presence of five type I signal peptidases and a single type II lipoprotein signal peptidase; some industrial strains contain plasmids specifying additional type I signal peptidases. The release of the translocated mature protein from the membrane is usually accompanied by its folding to the native conformation. For many *B. subtilis* secretory proteins, this requires folding factors such as Fe^{3+} and Ca^{2+} and/or the extra cytoplasmic lipoprotein PrsA. Efficient folding of the mature protein on the *trans* side of the membrane appears to be important for avoiding or reducing electrostatic interactions with negatively charged components of the cell wall and degradation by extracellular proteases. Although there are limits to the amounts of native proteins secreted by bacilli, such proteins are usually capable of being produced at commercially significant levels. In contrast, the extracellular production of heterologous proteins, especially from eukaryotic sources, is frequently inefficient. The reasons for this are not fully understood but are likely to include that the inherent incompatibilities with the secretory apparatus and their susceptibility to the many extracellular proteases, cell associated and released, produced by these bacteria.

ENZYMATIC CONTROL

Most of the extracellular hydrolase enzymes produced by *Bacillus* species are induced during the transition from vegetative growth to stationary phase, where they appear to function to increase the range of substrates that may be used to sustain growth. The expression of the genes encoding some of these enzymes appears to be linked to, but not necessary for, sporulation itself (e.g., alkaline and neutral proteases), while others (e.g., α-amylases and alkaline phosphatases) are linked to the exhaustion of specific substrates.

Production of Resistant Spores

Sporulation is a developmental process in which bacilli develop into heat-, chemical-, solvent-, radiation-, and desiccation-resistant spores. Their transition from dormancy to vegetative cells involves germination and outgrowth. Under laboratory conditions at 37°C, sporulation takes approximately 8 h. The process has been divided into a series of stages that are defined primarily on the basis of their morphology. These stages, labeled 0, II, III, ... through VII (stage I is no longer recognized), broadly correlate with the time (in hours) after the induction of sporulation. Sporulation is induced in response to nutrient deprivation and is only initiated after the failure of transition phase processes (e.g., motility and chemotaxis, production of macromolecular hydrolases, development of competence) to restore vegetative growth. Sporulation is induced by activated SpoOA (SpoOA–P) via a phosphorelay that facilitates multifactorial signal processing. The first morphologically distinct stage is stage II, which is defined by the formation of an asymmetric cell division event that divides the cell into two compartments, the larger sporangium (or mother cell) and the prespore, or forespore. During stage II, the prespore becomes completely engulfed by the sporangium so that, at stage III, the two compartments are separated by a double membrane system. During stage IV, peptidoglycan is laid down between the two membrane layers to form the spore cortex and the germ cell wall. During this stage the prespore becomes refractile to light.

In stages V and VI, spore coat proteins are laid down on the surface of the inner and outer membrane of the prespore. As these mature the spore becomes increasingly resistant to chemicals and heat. In the final stage, stage VII, the by-now-dehydrated spore is released from the sporangium by cell lysis. The fully mature spore is metabolically inactive and capable of surviving as such for certainly tens, and possibly hundreds, of years. Sporulation is a tightly controlled process requiring differential expression of genes on the chromosomes in each of the two compartments, as well as developmental check-points to ensure synchrony between their patterns of development. This is achieved through the differential activation of sporulation-specific sigma (σ) factors at or about stages II to III. σ^F is active in the prespore, and σ^E is active in the sporangium. These sigma factors are, in turn, responsible for the induction of sporangium-specific (σ^K) and prespore-specific (σ^G) sigma factors. Synchrony between the two compartments is maintained by signal communication processes. Much less is known about germination and outgrowth. In the laboratory at 37°C, the time between induction of germination and the first cell division event is about 90 min. Germination is triggered by a variety of chemical and physical stimuli; spores of *B. subtilis* germinate in response to L-alanine or to a mixture of asparagine, glucose, fructose, and KCl. Germination is accompanied by a series of rapid chemical and physiological changes including the degradation of the spore cortex, disruption of the spore coat, rehydration, and swelling. These changes are accompanied by a loss of heat and chemical resistance. It takes some 30 min before de novo RNA synthesis takes place. During outgrowth the cell begins to increase in mass and to replicate its DNA, forming an elongated cylindrical cell that is only subsequently reduced to a normal-sized cell by a delayed cell division.

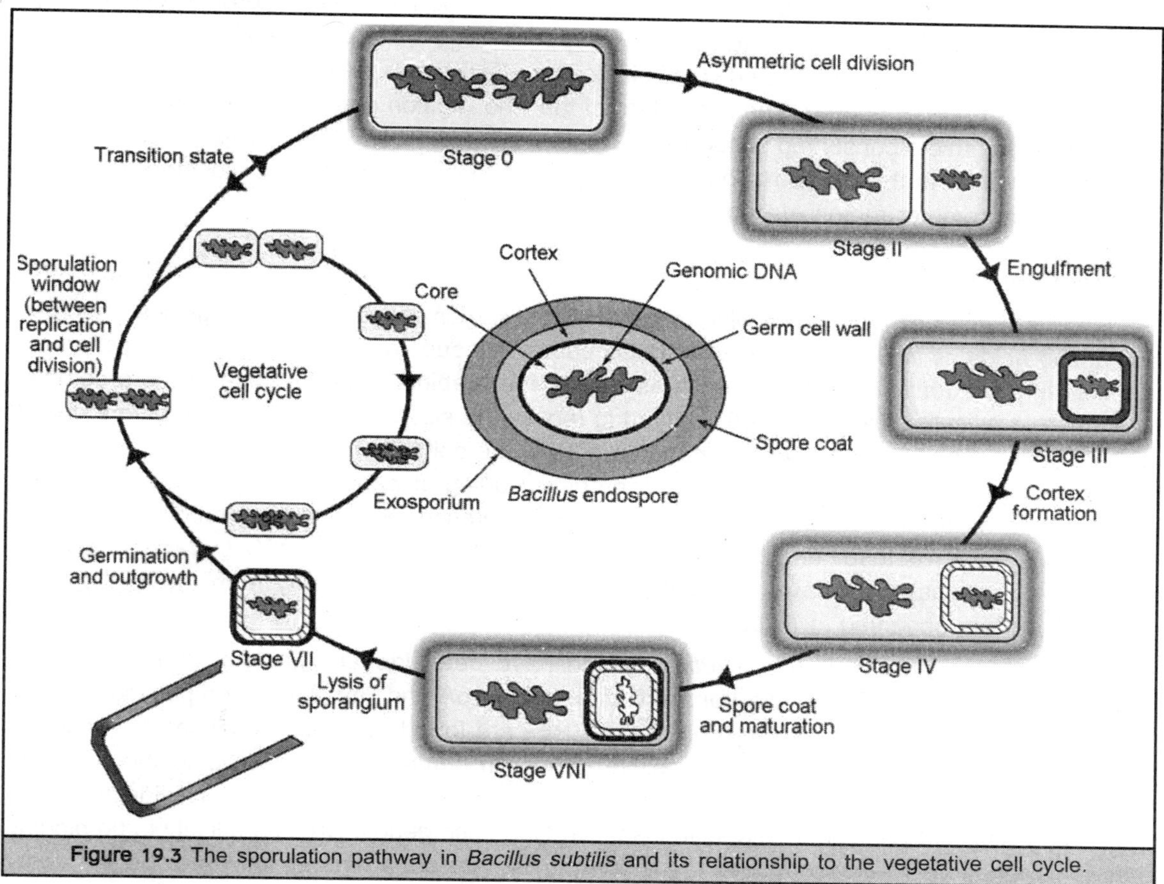

Figure 19.3 The sporulation pathway in *Bacillus subtilis* and its relationship to the vegetative cell cycle.

Growth Limiting Substrate

Phosphorus is a major growth-limiting substrate in natural environments such as soil; consequently, under phosphate limitation, *B. subtilis* induces a programmed series of responses to reduce the cell's requirement for phosphorus, increase its affinity for phosphate uptake, and produce extracellular hydrolytic enzymes to recover inorganic phosphate from organic sources. *B. subtilis* responds to phosphate stress by the coordinated induction of approximately 40 genes. Prominent among these are genes of the Pho regulon, whose products include alkaline phosphatases (APase), alkaline phosphodiesterase (APDase), a high-affinity phosphate transporter, and enzymes required for teichuronic acid synthesis. The Pho regulon of*B. subtilis* is controlled by two *trans*-acting regulators, PhoP and PhoR, that together form the components of a two-component, environment-sensing, signal transduction system. PhoR is a membrane-spanning sensor protein with histidine protein kinase (HPKase) activity that is activated at low phosphate concentrations to PhoR–P. This in turn acts as a substrate for the activation of PhoP, a response regulator required for the induction of the Pho regulon.

B. subtilis genes activated by PhoP–P, which include the bicistronic operon encoding PhoP and PhoR, constitute the Pho regulon. The response regulators of at least two other signal transduction systems, SpoOA–P and ResD, indirectly affect the expression of the Pho regulon by repressing the *phoP/R* operon. The promoters of the Pho regulon genes/operons that have been studied to date have a conserved sequence, TTAACA, located around position –22. This sequence may be equivalent to the Pho box observed in *E. coli,* which also tends to be located toward the 5′ end of the promoter region of Pho regulon genes/operons.

Carbon Compounds

Bacillus species can grow on a wide range of simple carbon sources, including glucose, sucrose, and amino acids as well as a range of complex substrates such as proteins and polysaccharides, particularly plant-derived substrates such as starch and arabinans. In the case of more complex substrates, extracellular enzymes are required to reduce the size of these substrates before uptake and utilization. Carbon substrates are transported into the cell either in an unmodified form or via a PTS system. In *B. subtilis* many genes/operons encoding enzymes involved in the catabolism of substrates as diverse as starch and histidine are repressed by readily metabolizable carbon sources such as fructose, glucose, glycerol, or mannitol—so-called PTS sugars. Catabolite repression functions at the level of transcription, but differs substantially from the cyclic AMP-mediated system observed in *E. coli.*

Three components of the *B. subtilis* system have been identified: a *cis*-acting catabolite-responsive element (CRE) and two *trans*-acting factors, CcpA and Hpr. The 14-bp palindomic CRE sequence is essential for catabolite repression. CREs are found at diverse locations in relation to the transcriptional units of catabolite-sensitive genes, in some cases overlapping the promoter (*amyE*), in others being located within the structural gene itself (*xylA*). The latter positioning makes it extremely difficult to engineer catabolite insensitivity in such genes. CREs are likely to be binding sites for a catabolite repressor protein. Although it has not been directly shown to be the case, CcpA, which is related to *E. coli* LacI-type repressors, is the most likely candidate for this protein. Mutants in the *ccpA* gene abolish catabolite repression, including glucose repression of sporulation, without affecting substrate-specific induction of the individual genes/ operons concerned. They also severely affect growth on substrates such as glucose, fructose, glucitol, and glycerol that are metabolized via glycolysis or the TCA cycle, but not on substrates such as xylose which are metabolized via the pentose phosphate pathway. In the former case, growth is restored if TCA cycle intermediates are added to the growth medium. Interestingly, a CRE appears to be responsible for glucose-mediated activation of the acetate kinase gene (*ackA*) and in genes encoding enzymes of central metabolism.

HPr is a component of the phosphotransferase system (PTS) which, when activated by phosphorylation to HPr(Ser-P), binds to CcpA. The key signaling component of the catabolite repression pathway appears to be the Embden–Meyerhof pathway (EMP) intermediate fructose-1,6-diphosphate (FDP). High concentrations of this compound activate the ATP-dependent HPr serine kinase and possibly stimulates an interaction between CcpA and HPr(Ser-P). This complex may then bind to the CREs of catabolite-responsive genes/operons.

Nitrogen Compounds

Bacillus spp. can use ammonium and a number of other nitrogen-containing compounds (notably certain amino acids) as sole sources of nitrogen. Ammonium is assimilated via glutamine synthetase (GS), glutamine: 2-oxoglutarate amidotransferase (GOGAT), and glutamate dehydrogenase (GDH). Most members of the genus are able to use all three enzymes, while a smaller group, including many (but possibly not all) strains of *B. subtilis,* use GS and GOGAT, while nitrogen-fixing *Bacillus* species use GDH. In *B. subtilis,* the relative levels of GS and GOGAT are dependent on the available sources of nitrogen. The levels of GS synthesis are dependent on the need for glutamine synthesis; they are lowest in the presence of glutamine, a nitrogen source favored by this bacterium, and highest in the presence of non-favored nitrogen sources such as ammonium. No global regulatory network analogous to that of Ntr of *E. coli* has been detected in *B. subtilis,* although GS synthesis is autoregulated via the GlnR repressor and has also been implicated as a key signal for the nitrogen status of the cell modulating the activity of a variety of other metabolic pathways. GOGAT levels are low in the presence of glutamate and amino acids (e.g., glutamine, arginine, and histidine) that can be catabolized to glutamate, and high in the presence of ammonium and amino acids that are not broken down to glutamate and other sources of nitrogen such as nitrate. In addition to its use as a nitrogen source, nitrate (and nitrite) can be used by *B. subtilis* in place of oxygen as an electron acceptor, allowing growth under anaerobic conditions. Nitrate respiration requires the products of the *narGHJI* operon.

PRODUCTS

Bacillus species are an important source of industrial enzymes, fine biochemicals, antibiotics, and insecticides. Additionally, the ability of *B. subtilis* and close relatives to secrete grams per liter quantities of protein directly into the growth medium, their ease of growth, and their well-proven safety have also made them prime candidates for the production of heterologous proteins. Many commercially important enzymes have traditionally been produced in their native hosts. However, when organisms with novel enzymic activities are newly isolated, the costs and uncertain outcome associated with process optimization are often prohibitive. This even applies to well-established strains of *B. amyloliquefaciens* and *B. licheniformis* that have been used extensively in commercial enzyme fermentations. Both these organisms are generally refactory to genetic engineering, and relatively little information is available concerning their biochemistry and physiology. Instead, companies are increasingly seeking to develop well-characterized strains such as *B. subtilis* strain 168 as universal hosts for commercial applications.

Biological Catalysts

The world annual sales for industrial enzymes was recently valued at $1 billion, with strong growth in the paper, textiles, and waste treatment markets. Three-quarters of the market is for enzymes involved in the hydrolysis of natural polymers, including proteolytic enzymes used in the detergents, dairy, and leather industries and carbohydrases used in the baking, brewing, distilling, starch, and textile industries. About two-thirds of these enzymes are produced by fermentation from *Bacillus* species. Two *Bacillus* enzymes dominate the industrial enzymes markets alkaline (serine) protease and α-amylase. Although many species of *Bacillus* secrete enzymes of these types, their catalytic properties vary from one producer strain to the next.

Moreover, commercially significant enzymes may be modified to improve their performance or stability in particular industrial processes. Alkaline proteases are the single largest enzyme market and are used extensively as additives to detergents. They are relatively nonspecific endoproteases that convert substrate proteins into small, readily soluble fragments. The two main proprietary products, subtilisin Novo (BPN') from *B. amyloliquefaciens* and subtilisin Carlsberg from *B. licheniformis*, have been the subject of extensive development programs. Their three-dimensional structures have been solved, and functionality testing has resulted in the identification of the catalytic triad, an oxyanion binding site, and substrate-binding determinants. Subtilisins have been engineered to improve their characteristics in washing detergents by increasing their stability in the presence of oxidizing chemicals, at temperatures above 60°C and at pH values up to 12. They have pH optima of about 11.0 and half-lives of approximately 50 min at 40°C. They are used at 0.4-1% in formulations that includes anionic and nonanionic surfactants, oxidants, soap, and optical brighteners. Although these were originally used simply in the form of a powder, hypersensitivity reactions among producers and users meant that the enzymes had to be added to formulations in a microencapsulated form. This process not only reduces dust formation but also protects the enzymes from other components of the formulation during storage.

α-Amylases are used extensively in the starch industry where, because of the physical properties of starch, they need to be used at high temperatures and good thermostability is an important property of these enzymes. Industry has sought to obtain thermostable amylases by screening for new sources and by improving the thermostability of existing enzymes by protein engineering. For example, the α-amylases from *B. amyloliquefaciens*, *B. stearothermophilus*, and *B. licheniformis* are closely related but have very different thermostabilities of 2, 50, and 270 min, respectively, at 90°C and pH 6.5. Extra salt bridges caused by specific lysine residues in the more-thermostable *B. licheniformis* amylase are responsible for its increased stability, and the equivalent residues in the less- stable enzymes have been the target for protein engineering. Additionally, the extensive homology between the genes encoding these amylases has also been used for in vivo recombination experiments in which hybrid enzymes combining beneficial characteristics of these enzymes have been screened.

The three-dimensional structure of the *B. licheniformis* α-amylase has recently been published, and the three amino acid residues that constitute the calcium-binding site (i.e., asparagine-104, aspartate-200, and histidine-235), have been identified. The α-amylase from *B. acidocaldarius* strain A2 shows greater thermostability under acid conditions. Additionally, *B. subtilis* strain KP1064 produces an enzyme that is active against starch and pullulan. The industrial production of α-amylase was developed in 1917 using a strain subsequently reclassified as *B. amyloliquefaciens*. This α-amylase could be used to a maximum temperature of about 85°C. Because starch needs to be cooked to 100°C to rupture the starch granules, the introduction of a thermostable α-amylase from *B. licheniformis* that can operate at about 105°C was a distinct improvement.

Table 19.1 Commercially significant products produced by Bacillus species

Product	*Producer organism*	*Properties/use*
Glucose isomerase	*B. coagulans*	Conversion of glucose to fructose in manu-facture of low calorie sweeteners
α-Amylase	*B. subtilis*	Limited hydrolysis of starch
	B. amyloliquefaciens	Liquefication of starch
	B. stearothermophilus	Thermophilic liquefaction of starch
	B. licheniformis	Thermostable liquefaction of starch
β-Amylase	*B. polymyxa*	Formation of low-dextrose-equivalent syrups from starch
β-Glucanase	*B. amyloliquefaciens*	Hydrolysis of barley β-glucans in brewing and animal feeds
	B. circulans	
	B. subtilis	
Pullulanase	*Bacillus* species	1,6-Bond starch debranching enzyme
Alkaline protease	*B. alkalophilus*	Detergents and leather industries
	B. licheniformis	
Neutral protease	*B. thermoproteolyticus*	Temperature-stable enzyme
Thermolysin	*B. thermoproteolyticus*	Condensation of L-aspartic acid and D,L-phenyl-alanine to produce aspartame
β-1,4-Galactosidase	*B. stearothermophilus*	Dairy industry
	B. subtilis	
Chymosin (rennin/pepsin)	*B. mesenterius*	Rennet substitutes
	B. cereus	
	B. polymyxa	
	B. megaterium	
	B. licheniformis	
Glucose dehydrogenase	*B. megaterium*	Glucose assay
Glycerol kinase	*B. stearothermophilus*	Glycerol/trigycerides assay
Phospholipase C	*B. cereus*	Phosphatidylcholine assay
Restriction endonucleases	*B. aneurinolyticus*	*Ban*I
	B. amyloliquefaciens	*Bam*HI
	B. subtilis	*Bsu*
	B. caldolyticus	*Bcl*
	B. globiii	*Bgl*
	B. stearothermophilus	*Bst*
	B. sphaericus	*Bsp*
	B. brevis	*Bbv*
δ-Endotoxin	*B. thuringiensis*	Biocontrol agent
Peptide antibiotics	*B. licheniformis*	Bacitracin
	B. brevis	Edeines, gramicidin, tyrocidine
	B. subtilis	Fengycin, subtilin, surfactin
	B. polymyxa	Polymyxins

B. coagulans is an important source of glucose isomerases, required for the conversion of glucose (0.75 times as sweet as sucrose) to fructose (twice as sweet as sucrose) in the production of high-fructose corn syrup. In contrast to the proteases and amylases already discussed, glucose isomerase is an intracellular enzyme used either in the form of immobilized cells or extracted and immobilized on a solid matrix.

Primary Metabolites

Bacillus species are used for the production of a number of primary metabolites for the food and health care industries. Extensive knowledge of their biochemistry and regulation has allowed growth and rational mutant isolation strategies to be developed to the extent that this organism has been considered for the production of a wide range of metabolites at commercially viable levels. *B. subtilis* has been used for the production of the nucleotides xanthanylic acid (XMP), inosinic acid (IMP), and guanylic acid (GMP), which are of commercial importance as flavor enhancers. The primary fermentation products are the nucleosides hypoxanthine, inosine, and guanosine because their greater membrane permeability allows them to accumulate to higher concentrations in the culture medium. These are then converted to their respective nucleotides by chemical phosphorylation. Attempts have been made to develop strains of *Bacillus* for the production of amino acids such as tryptophan, histidine, and phenylalanine and of vitamins such as biotin, folic acid, and riboflavin. The use of analogs has led to the isolation of resistance mutants overproducing a number of amino acids, nucleosides, and vitamins. Knowledge of *B. subtilis* is such that rational approaches are now possible and attempts have been made to engineer this bacterium for the production of folic acid. Promising strains have been obtained, although their production levels cannot yet compete with those of *Brevibacterium, Corynebacterium,* or *Serratia* species.

Special Metabolites

Under conditions of nutritional stress *Bacillus* species produce a variety of special metabolites that enhance their survival in natural environments. Prominent among these are peptide antibiotics, generally short peptides (2-20 amino acid residues) that are synthesized by large multienzyme complexes (peptide synthetases) rather than ribosomes. Individual amino acid residues are often extensively modified and may be linked to each other by peptide bonds or through esters or lactones. Gramicidin-S is a cyclic decapeptide from *B. brevis* with antibacterial and surfactant properties. Bacitracin is a branched cyclic dodecapeptide produced by *B. licheniformis* and *B. subtilis* that is used as a topical antibiotic. Its antibacterial activity is directed against bacterial cell wall synthesis where it inhibits the recycling of lipid-P carriers.

Surfactin, produced by most strains of *B. subtilis,* has both antibacterial activity and powerful surfactant properties. Polymyxins are membrane-active branched cyclic acylpeptides produced by *B. polymyxa.* A minority of the peptide antibiotics produced in *Bacillus* species are synthesized on ribosomes and subsequently modified extensively by posttranslational processing. The resulting products are usually larger, but only marginally so, than those produced by the peptide synthetases. The best studied are the lantibiotics, so-called because they contain the unusual amino acids lanthionine and methyl-lanthionine. They include subtilin, a 32-residue lantibiotic produced by *B. subtilis* that shows antibacterial and antitumor activity.

Heterologous Proteins

The efficient large-scale production of heterologous enzymes and proteins by *B. subtilis* has been the focus of considerable research effort. The ability of this bacterium to secrete proteins directly into the culture medium in quantities of grams per liter offers considerable process advantages, in terms of product yield, structural integrity, and downstream processing costs, over production systems in which proteins are accumulated in the cytoplasm or periplasm, often as insoluble aggregates. Nevertheless, attempts to use *B. subtilis* for the production of heterologous proteins has met with only limited success, and there are no reports of the use of this host in an industrial fermentation. Although extracellular proteins from close relatives can be produced in this bacterium at high concentrations, the yield of proteins from unrelated species remains disappointingly low. This is likely to be due, at least in part, to the production of at least seven extracellular proteases, which have been shown to cause extensive degradation of heterologous proteins, and to incompatibilities with the *Bacillus* protein secretion pathway.

The isolation of strains defective in the identified proteases has helped in some but by no means all cases. Other species of *Bacillus* may prove useful for development as hosts for the production of heterologous proteins, particularly those naturally producing low levels of extracellular proteases. Of interest are species such as *B. brevis* that produce a crystalline surface (S) layer at the outer surface of a thin cell wall. S-layer proteins have been observed at concentrations up to 35 g/L in the culture medium and, consequently, attempts have been made to develop the expression and signal sequences of their genes into components of a new generation of secretion vectors. However, it may ultimately be necessary to develop strains that are currently used in industrial processes as production hosts by improving our understanding of their genetics and physiology. *B. subtilis* has also proved useful for the intracellular production of outer membrane proteins of gram-negative pathogens. These proteins, which have potential for use for immunodiagnostics and as vaccines, are produced in *B. subtilis* in preference to their native hosts to avoid contamination with endotoxins with which they often form tight associations.

Insecticides

The use of chemical insecticides, with a world market worth $5 billion, is increasingly seen as problematical because of the development of resistance in target insect populations, lack of specificity against the insect pest, persistence in the environment, and toxicity to man and other animals. Alternative biocontrol methods have been developed involving the use of viruses, fungi, and bacteria. *B. thuringiensis* is the most successful insecticidal biocontrol agent and accounts for about 90% of the worldwide market for bioinsecticides. Although *B. thuringiensis* was originally isolated in 1901 from an infection of a Japanese silkworm farm and has been used as an insecticidal agent since 1920, it still only represents about 1-2% of the total insecticide market. Strains of*B. thuringiensis* have been identified that are pathogenic for each of the main groups of insect pests, namely Lepidotera (moths, butterflies), Diptera (flies and mosquitoes), and Coleoptera (beetles).

More recently strains of*B. thuringiensis* have been isolated that are active against noninsect pests, notably nematodes, mites, and protozoa. The toxicity of *B. thuringiensis* results from the production of proteinaceous δ-endotoxins during sporulation. The toxins, which may represent as much as 30% of the cell's weight, form crystals within the mother cell. The toxins are synthesized from plasmid-encoded *cry* genes, and a single strain may produce several toxins encoded by more than one plasmid. They are synthesized initially in the form of an inactive pro-toxin (70-140 kDa) that is considerably larger than the active form (60 kDa). The δ-endotoxins, which have the properties of chemical agents rather than infectious agents, are approximately 300 times more potent than pyrethroids and 8,000 times more potent than organophosphates. *B. thuringiensis* toxins therefore combine high toxicity and specificity for their target pests with little or no toxicity for nontarget insects (e.g., pollinators, predators) and other animals.

A large number of *cry* genes have been cloned, and the regions responsible for the pro- and mature forms, host specificity, and toxicity have been mapped. As different toxins exhibit a range of host specificities and levels of toxicity, chimeric toxins that combine the specificities and toxic regions of different natural toxins are currently being developed. *B. thuringiensis* strains have been used extensively for the control of crop and forest insect pests and of insect disease vectors. However, as an alternative means of application, genes encoding the δ-endotoxin have been used directly to transform a variety of plant crops, including tomatoes, tobacco, potatoes, and cotton, in which field trials have confirmed their activity against target pests.

PROTECTION

Bacillus species have been used extensively in the fermentation industry for the production of a range of products including enzymes, fine biochemicals, antibiotics, and insecticides. In addition, a strain of *B. subtilis* is used for the fermentation of soybeans to Natto, an extensively consumed ($\sim 10^8$ kg/annum) traditional Japanese food product. In a review of the safety of *B. subtilis* and *B. amyloliquefaciens* (these strains were considered to be synonymous up to 1973), no reports were found of infection or toxicity relating to *B. amyloliquefaciens*. In the case of *B. subtilis,* the situation was complicated by the fact that before about 1970 diagnostic laboratories did not distinguish between *B. cereus* and *B. subtilis*. Virtually all reports of putative *B. subtilis* infections since that time either were associated with drug abuse or occurred in patients whose immune system had been compromised by treatment with immunosuppressants or chemotherapeutic agents. This low incidence of pathogenicity and the widespread use of their products in the food, beverage, and detergents industries have resulted in GRAS (generally regarded as safe) status being granted to these species by the U.S. Food and Drug Administration (FDA).

EXPANDING CULTURE

The majority of *Bacillus* species will grow at mesophilic temperatures on commercially prepared nutrient media, although in some cases it is necessary to modify the pH or salt concentration. Obligate thermophilic species such as *B. stearothermophilus* that do not grow satisfactorily at 37°C are usually grown at 60°C. Moderately thermophilic species, such as *B. coagulans,* are grown between 45°C and 50°C. The more fastidious insect pathogens, *B. larvae* and *B. popilliae,* require the addition of thiamine for growth and are usually grown at 25°C to 30°C. *B. stearothermophilus* requires additional calcium and iron, and *B. pasteuri* requires 0.5-1% urea. *B. subtilis* and many other species are able to grow in simple salts media containing ammonium or amino acids as nitrogen sources and glucose or other simple sugars as carbon sources. The most commonly used general-purpose chemically defined medium is Spizizen's minimal medium. The growth rate in this medium can be improved by the replacement of ammonium, a poor nitrogen source for *B. subtilis,* with glutamine. *B. subtilis* is able to use a number of amino acids as nitrogen sources (e.g., arginine, glutamine, glutamate, asparagine, and aspartate), and the respective catabolic pathways are induced in the presence of these compounds.

Consequently, the concentration of many amino acids recommended to overcome auxotrophy are actually growth limiting. It should also be borne in mind that derivatives of the well-studied *B. subtilis* strain 168 often show tryptophan auxotrophy, and consequently complex media in which acid-hydrolyzed casein provides the main source of nitrogen need to be supplemented with this amino acid. Although many *Bacillus* species sporulate readily, special media and growth protocols are required for efficient sporulation. Sporulation is induced in response to nutrient deprivation and occurs after the end of exponential growth. In *B. subtilis,* sporulation is repressed in the presence of glucose and at low cell densities. A widely used sporulation medium was described by Schaeffer and colleagues. In some species sporulation can also be induced by decoyinine, an analog of GTP.

Growth in Batch Culture

Although many species of Bacillus grow under anaerobic conditions, they are normally grown aerobically. Some strains (e.g., *B. subtilis*) have a tendency to lyse in the absence of an energized membrane. Growth is normally carried out in conical flasks in an orbital (reciprocating) incubator at >200 rpm, although most species can be grown in batch fermentors. Ideally, the culture volume should not exceed 20% of the flask volume. The media should be prewarmed as temperature shock can also induce autolysis. Nutrients are usually supplied in excess and growth continues until a particular substrate is exhausted, the oxygen tension or pH falls to an inhibitory level, or an inhibitor accumulates to an inhibitory level. The various phases in the growth cycle have been described in detail elsewhere.

Growth in Continuous Culture

Bacillus species undergo marked metabolic, physiological, and morphological changes during batch culture. Many of these changes occur during transition to and during stationary phase and include morphologically distinct spores and physiologically distinct competent cells. To some extent these population variations can be reduced by growth in a chemostat under continuous culture conditions. This method allows the effect of individual growth parameters (e.g., pH, oxygen tension, growth rate, limiting substrate) on particular cellular characteristics to be identified. Chemostat cultures can be used to study cells under "steady-state" conditions, or during transitions from one steady state to another. In principle any liquid culture medium that supports growth in batch culture can be used for continuous culture. However, because the pH is actively controlled during growth, it is not necessary to include components required only for buffering pH. The concentration of the limiting nutrient determines the culture density, and all other essential nutrients are supplied so that they are in excess. For aerobes, the concentration of the limiting nutrient is set so that the culture density is not so high as to impose an excessive oxygen demand. For *B. subtilis* this may mean limiting the culture density to about 1-2 g dry weight per liter. Complex media can be used if a defined limitation can be achieved. Several defined media have been described using, for example, phosphate, magnesium, potassium, glucose, and sulfate as limiting substrates.

Conditions for Large Scale

Most of the products traditionally produced from *Bacillus* species, including enzymes, antibiotics, and insecticides, are synthesized during stationary phase using a traditional batch fermentor. However, insecticides require both a high growth yield and efficient sporulation because the δ-endotoxin accumulates in the sporangium. The production of α-amylase and of *B. thuringiensis* δ-endotoxin are given as examples of the factors that need to be taken into account during commercial *Bacillus* fermentations. Most species of *Bacillus* grow well on commercial formulations of nutrient media, although in some cases the medium needs to be modified by adjustment of the pH or ionic strength or by the addition of specific nutrient supplements. This means that media can, to some extent, be based on cheap, readily available substrates.

Despite the extensive expertise in the fermentation of *Bacillus* species, commercial considerations have limited the published literature on large-scale fermentations. Culture conditions and media formulations are highly geared to the particular strains and processes used and these have been selectively reviewed elsewhere. Media components have been developed empirically over a number of years to avoid catabolite repression that prevents the synthesis of a number of the commercially important products from *Bacillus* species. For example, α-amylases are produced in media containing starch in various forms including corn, wheat malt, and potato that is noncatabolite repressing. When carbohydrates are used as a carbon source, phosphate is required at a concentration of about 100 mM. Peptones and amino acids are a better source of nitrogen than ammonium for *B. subtilis,* probably because of the efficiency with which nitrogen is assimilated into amido groups. In the case of *B. apiaries,* however, ammonium is the preferred nitrogen source. Glutamine is the nitrogen source that sustains the fastest growth rate.

The use of amino acids as a nitrogen source represses the synthesis of proteases and the onset of sporulation, allowing the productive stationary phase to be extended. The concentrations of divalent metal cations Mn^{2+}, Ca^{2+}, and Zn^{2+} are important to the final yield of α-amylase, even when they have little or no effect on the growth of the bacterium. For example, Ca^{2+} is required for both the enzymatic activity and structural integrity of the liquefying α-amylases such as those from *B. licheniformis, B. amyloliquefaciens,* and *B. stearothermophilus*. We have evidence that this results, at least in part, from the rate of folding of the enzyme on the *trans* side of the cytoplasmic membrane. Other metal ions, including Fe^{2+}, Fe^{3+}, Na^{+}, and Mg^{2+}, enhance enzyme production presumably in part as a result of their effect in increasing growth yield. The growth kinetics in *Bacillus* cultures is complicated by the operation of the stringent response and sporulation, both of which are similarly induced by nutrient limitations and other negative growth factors.

Commercial processes often use fully or partially (oligosporogenous) sporulation-deficient strains. The stringent response, modulated by variations in the levels of ppGpp(p), appears to be a cause of linear growth. Substrates with a low carbon to energy ratio gives rise to carbon-limited growth and consequential "energy spilling" and the production of overflow metabolites. Glucose excess chemostat cultures of *B. stearothermophilus* accumulate acetate up to 34% of the carbon consumed. Under these conditions, energy efficiency is sacrificed to obtain the ATP turnover required to sustain the high growth rates. Similar metabolite overflows have been observed for *B. licheniformis* (48%) and *B. subtilis* (22%).

Catabolite repression is one of the major factors limiting the production of many commercially important products. Although mutants (e.g., *ccpA*) have been isolated that lack elements of the catabolite repression pathway, these often grow poorly. An alternative approach in which CRE elements are deleted or inactivated maybe frustrated by their occurrence within DNA sequences encoding the protein itself. Often, the most effective method for overcoming catabolite repression involves a combination of appropriate media formulations and growth regimes. Many commercially significant products are produced after the end of exponential growth (transition phase) and during the early stages of sporulation. It is therefore essential that, when sporulation-deficient strains are used, they still retain the ability to synthesize the required product. For example, mutants deficient in Spo0A (the sporulation-specific response regulator) are unable to induce various extracellular proteases.

Although batch cultures are traditionally used for α-amylase production, because the main production phase is during stationary phase, Emanuilova and Toda compared the production of α-amylase by *B. caldolyticus* in batch and continuous culture. Casitone and starch were required for optimal α-amylase production. The concentration of amylase was 10- to 20-fold higher in continuous rather than batch culture, although there was an inverse relationship between growth rate and amylase production. Industrial fermentations are carried out at 30-40°C and at pH 7.0 with $CaCO_3$ as the calcium source to stabilize the enzyme. The complex medium allows growth to 10^{10} cell/ml within 20 h, after which time production continues for about 6 days. The fermentations also produce proteases that are important for the utilization of protein substrates, but which have to be separated from the α-amylase during subsequent downstream processing.

α-Amylase production can be increased by cloning the structural gene onto a multicopy plasmid, and production, possibly up to the limitation of the native secretory pathway, is achievable. However, because many of the plasmids used in *Bacillus* species are based on staphylococcal plasmids (e.g., pUB110), both segregational and structural instabilities are frequently observed. Because δ-endotoxin production of *B. thuringiensis* occurs during sporulation, growth and process conditions need to be optimized to give cell densities >5 x 10^9 per milliliter and sporulation rates greater than 90%. Although significant strain-to-strain variations are observed, certain basic requirements can be defined: these include good oxygenation, a growth temperature between 26 and 30°C (higher temperatures can lead to plasmid curing and consequential loss of δ-endotoxin production), and pH 6.5-7.5. The production of amylases and proteases by *B. thuringiensis* allows a wide range of natural raw materials to be used as growth substrates. Glucose is catabolized with the production of acetate, lactate, pyruvate, and acetoin, and pH control is required for some strains to prevent the acidic conditions restricting growth.

The production of poly-β-hydroxybutyrate during early stationary phase under carbon excess can be mistaken for δ-endotoxin but disappears during sporulation. Cultures are usually grown in submerged liquid cultures as large as 200 L using fermentors agitated with paddle blades and with static baffles on the side wall of the vessel. Occasionally vessels with impellers or airlift fermentors are used. Because the industrial media contain significant amounts of insoluble particles, sterilization is accompanied by agitation. Toward the end of the fermentation, the released spores attach to and stabilize the foam, and silicon- or polypropylene-based antifoam agents are required as foam suppressers.

Antifoam agents can form aggregates that interfere with downstream processing, and consequently their use is kept to a minimum. During stationary phase, oxygen consumption is markedly reduced and, provided saturation levels do not fall below 20%, agitation can be reduced, with a concomitant reduction in foaming. At the end of fermentation, the fermentation vessel will be heavily contaminated with spores that need to be inactivated if different strains are used in the next fermentation run. *Bacillus thuringiensis* will grow on glucose minimal medium only when supplemented with sources of amino acids (e.g., peptones, casamino acids) or proteins. Because *B. thuringiensis* produces amylases, glucose can be replaced with starch as the carbon source. Growth is stimulated by yeast extract and optimal concentrations of Fe^{3+}, Co^{+}, and Zn^{+}, and sporulation is improved by metal ions such as Ca^{2+} and Mn^{2+}. Phosphate is added both as a source of phosphorus and as a buffer. Preparations of *B. thuringiensis* are applied at a rate of about 0.5 to 2kg per hectare. Spores can be applied directly as a pest control agent, but are often processed to improve their physical characteristics. For example, granular or bait formulations are applied to the plants or soil; in other cases, concentrates are diluted in water and applied by spraying. Stickers, spreaders, or even chemical insecticides can be added to improve dispersion and activity.

INDEX